DATE DUE

12 19			
DEC 27 1996			
GAYLORD			PRINTED IN U.S.A.

IMA MONOGRAPH SERIES

IMA MONOGRAPH SERIES

Water Waves: Relating Modern Theory to Advanced Engineering Applications

Matiur Rahman

Department of Applied Mathematics,
Technical University of Nova Scotia

CLARENDON PRESS • OXFORD
1995

Oxford University Press, Walton Street, Oxford OX2 6DP

Oxford   New York

Athens   Auckland   Bangkok   Bombay
Calcutta   Cape Town   Dar es Salaam   Delhi
Florence   Hong Kong   Istanbul   Karachi
Kuala Lumpur   Madras   Madrid   Melbourne
Mexico City   Nairobi   Paris   Singapore
Taipei   Tokyo   Toronto
and associated companies in
Berlin   Ibadan

Oxford is a trade mark of Oxford University Press

Published in the United States
by Oxford University Press Inc., New York

A catalogue record for this book is available from the British Library

Library of Congress Cataloging in Publication Data
Rahman, M. (Matiur), 1940–
Water waves : relating modern theory to advanced engineering
applications / Matiur Rahman.
p.   cm. — (IMA monograph series ; 3)
Includes bibliographical references and indexes.
1. Wave-motion, Theory of.   2. Water waves.   I. Title.
II. Series.
QA927.R35   1994
627'.042—dc20   94-19734
CIP
ISBN   0 19 853478 7

Typeset by Technical Typesetting Ireland
Printed in Great Britain by
Bookcraft (Bath) Ltd, Midsomer Norton, Avon

Preface

The theory of water waves has been a subject of intense scientific research since the days of Airy in 1845. It is of great practical importance for scientists and engineers from many disciplines gaining an insight into the behaviour of oceanic and tidal water waves by the development of this theory.

The purpose of this monograph is to present a self-contained introduction to the mathematical and physical aspects of the theory of water waves, and it is intended for senior undergraduate and graduate students. Each chapter concludes with practical problems in the form of exercises and is accompanied by ample references for further studies. Analytical solution techniques are demonstrated that are required to solve problems arising in this field. It is hoped that applied mathematicians and ocean engineers will therefore find this book useful.

The book consists of ten chapters. The material in it is arranged into three parts: Part I: Basic Fluid Mechanics and Solution Techniques covers Chapters 1–3; Part II: Water Waves covers Chapters 4–7; Part III: Advanced Water Waves covers Chapters 8–10. Parts I and II are elementary in nature, whereas Part III is more advanced. Chapter 1 briefly outlines the content of the book and gives an overview of the specification of fluid motion. Described in Chapter 2 are the basic equations of fluid motion from the viewpoint of general fluid dynamics. The developments of the Euler equations of motion for inviscid fluids have been described from a mathematical viewpoint since the Euler equations form the backbone of the study of water wave motion. Navier–Stokes equations only are cited for examples as equations for viscous fluid motion. The philosophy behind the source, sink, singularities, and circulation of water particles is explained and accompanied by some examples. The Schwarz–Christoffel transformations are developed and applied to show how the complicated geometry of a fluid region can be transformed to a simple tractable domain. Chapter 3 contains the classical solution techniques for partial differential equations governing the physical situation of the problems. The method of separation of variables, the method of characteristics and d'Alembert's solution of the wave equation have been developed and applied to practical problems. Chapters 4 and 5 are devoted to the development of wave terminology: Chapter 4 illustrates the wave theory of Airy, whereas Chapter 5 illustrates the nonlinear wave theory of Stokes. Chapter 6 has been devoted to the study of tidal dynamics in shallow water. Tidal oscillations in rectangular basins are described in this chapter. For the random wave case statistical parameters must be implemented. Chapter 7 is primarily devoted to wave statistics and wave energy spectra, illustrations of the probability distribution functions used to describe individual heights and wave periods, and an introduction to the wave energy density spectrum. A variety of mathematical formulations developed for predicting the wave spectrum are illustrated with some applications.

The application of the wave theory is demonstrated in Chapter 8. Wave forces on offshore structures are shown to be determined by the Morison equation, the Froude–Krylov method and the diffraction or potential theory. This chapter includes nonlinear wave loading calculations and also describes the numerical methods for the solution of many practical problems with complicated geometrical configurations. The Green function method, alternatively known as the boundary integral method, is applied for the evaluation of wave loads on arbitrary-shaped offshore structures. In Chapter 9 nonlinear long waves in shallow water are examined from a mathematical viewpoint. Developments of Cnoidal waves, solitary waves and the Korteweg–de Vries (KdV) equation are clearly demonstrated. Chapter 10 deals with the solution method of the solitary wave problem by an inverse scattering technique. Chapters 9 and 10 are intended for applied mathematics students to gain a physical insight of nonlinear shallow water waves.

Some knowledge of vector calculus, including integral theorems such as those by Green and Stokes, is assumed on the part of the student. A familiarity with Bessel functions, Legendre polynomials and hypergeometric functions is also expected.

I am extremely grateful to Sir James Lighthill, FRS for having read through the first version of the manuscript and for his favourable recommendation for publication. Thanks are also due to Professor J. R. Blake, Editor of the IMA Monograph Series, for his encouragement and interest. It is a great pleasure to express my deepest gratitude to the staff of Oxford University Press for taking up the arduous task of producing this book.

I would like to thank Dr David Prandle of Proudman Oceanographic Laboratory and Dr L. G. Jaeger, FRSE of TUNS for stimulating discussions and inspirations during preparation of the book. The book is developed from my lecture notes of a course on the theory of waves in potential flow given to graduate students at the Technical University of Nova Scotia

(TUNS). The author is grateful to many of his students for their encouragement in writing this manuscript. Thanks are also extended to the learned referees for their constructive comments on the final draft. I am grateful to Mr D. D. Bhatta for having proofread the entire manuscript and making all the necessary corrections. Mr Weibang Weng deserves my great appreciation for his faultless typing of numerous drafts of the manuscript. Thanks are also due to Mrs Rhonda L. Coulstring for drafting all the figures of the book for publication. I also thank my family for their patience in the long preparation of the manuscript. Financial support provided by the Natural Sciences and Engineering Research Council of Canada is gratefully acknowledged.

While it has been a joy to write this book over several years, the fruits of this labour will hopefully be in learning of the enjoyment and benefits realized by readers. Thus the author welcomes any suggestions for the improvement of the text.

Halifax, Nova Scotia   M. R.
August 1994

Contents

Part I

Basic fluid mechanics and solution techniques

1
Introduction

1.1 Preliminary background

The main topics of this book relate to surface waves in oceans and tidal waves in estuaries. A proper understanding of these phenomena is crucial to much of physical oceanography and is an exciting area of study. Many scientific papers have appeared in learned journals in which mathematical models have been used to correlate predicted and experimental data. Perfect correlation is the ultimate goal of the mathematical models, and although much has been achieved, there is great scope for future work.

The inherent nonlinearity of ocean waves makes their behaviour very complicated. However, a particular wave motion may be adequately represented by a linear model provided that the ratio of particle speed to phase speed is small compared with unity. This condition, which implies that the wave amplitude is small compared with the wavelength, holds for most, but not all, oceanic wave phenomena. In a linear system the modes are uncoupled and can be classified and studied independently.

In oceans, waves interact with each other and with the mean flow: they grow due to the action of external forces or through internal instability, and they decay as a result of molecular and turbulent friction and diffusion. As a result of all these processes, the behaviour of waves cannot be represented by a discrete spectrum of undamped modes of a bounded ocean, but must be described in terms of a continuous spectrum of modes.

The effect of wind on surface waves was studied by many pioneers of theoretical fluid mechanics, including Lagrange, Airy, Stokes and Rayleigh. They attempted to account for the elementary properties of surface waves in terms of perfect fluid theory. The problem of relating the rate of wave growth to the wind was first recognized by Kelvin. However, no progress on this problem was made until 1850, when Stevenson made observations of surface waves on a number of lakes, and derived an empirical relationship between the 'greatest wave height' and the fetch. Some 75 years later, Jeffreys experimentally modelled the generation of waves by wind.

The first solutions to the problem of waves striking an internal surface were obtained by Stokes in 1847. The development of new instruments, the careful carrying out of experiments, and the more detailed data analysis have since revealed a variety of dynamical behaviour which was previously unapparent and which offers a continuing challenge to theoreticism. The greatest experimental contributions were made by Long in his work on the problem of the excitation of internal waves caused by the flow over irregular beds.

In 1883, Reynolds published his celebrated account of laboratory observations on turbulent flow. Later, with the stimulus of the development of aerodynamics, Prandtl introduced the concept of a mixing length. A more fundamental approach to the dynamics of turbulence was given by Taylor in 1935 in a series of papers to the Royal Society; this is recognized as the beginning of the modern theory of turbulence. Further advances in the subject of turbulence were made by Batchelor, Kolmogorov, Kraichnan and Townsend. Some remarkable observations on the structure of atmospheric turbulence was made by Taylor in 1915, but it was not until 30 years later that suitable instrumentation was available to allow systematic investigations to be made. A good account of these authors' work can be found in the book *The Dynamics of the Upper Ocean* by O. M. Phillips (1966). For matters concerned with waves in fluids such as sound waves, shock waves, stratified fluids and a brief description of water waves, readers are referred to Lighthill's *Waves in Fluids* (1978).

1.2 Real and perfect fluids

In fluid dynamics, most theoretical investigations start from the concept of a perfect fluid where two contacting layers experience no tangential forces, i.e. shearing stresses, but act on each other with normal forces, i.e. pressure only. It means that no internal resistance exists in a perfect fluid. On the other hand, the inner layers of a real fluid experience tangential as well as normal stresses. These frictional tangential forces in a real fluid describe the existence of viscosity. The theory of the motion of perfect fluid supplies many satisfactory descriptions of a real fluid. Owing to the absence of tangential forces, there exists, in general, a difference in the relative tangential velocities of the perfect fluid and the solid wall wetted by the fluid. Hence, there is slip. The existence of tangential stresses and the condition of no slip near a solid wall constitute the essential differences between a perfect and a real fluid. The concepts of vorticity and circulation are founded on the basic analysis of the rotation of a fluid particle in real fluid motions and are discussed in Chapter 2.

It is important to note that certain fluids which are of great practical importance, such as water and air, have very small coefficients of viscosity.

In many instances, the motion of such fluids of small viscosity agrees very well with that of a perfect fluid, because in most cases the shearing stresses are very small. Hence, the effect of viscosity is neglected in the perfect fluid theory.

1.3 Specification of the motion

Fluid motion is usually described in one of two ways: (a) the Eulerian description of motion; (b) the Lagrangian description of motion.

In an Eulerian description of motion, physical quantities such as the velocity V, pressure p and density ρ are regarded as functions of the position X and time t. Thus $V = V(X, t)$ and ρ represent the velocity and density of the fluid, respectively, at prescribed points in space-time.

In a Lagrangian description of motion, the fluid elements can be identified in terms of an initial position a at some initial time t_0 and the elapsed time $t - t_0$.

Thus the current position and initial position vectors are given by $X = X(a, t - t_0)$ and $X_0 = X(a)$, respectively.

The velocity of a fluid element is the time derivative of its position $V(a, t - t_0) = (\partial/\partial t)X(a, t - t_0)$ so that $X = a + \int_{t_0}^{t} V(a, t - t_0)\,dt$.

The fluid acceleration is then given by $f(a, t - t_0) = DV/Dt = (\partial^2/\partial t^2)X(a, t - t_0)$.

The total time derivative, or the derivative 'following the motion', can be expressed in Eulerian terms as (see Chapter 2) $(D/Dt) = (\partial/\partial t) + (V \cdot \nabla)$, the sum of the time rate of change at a fixed point and a convective rate of change.

In this book the Eulerian approach is used. Many instruments, measuring fluid properties at a fixed point, provide Eulerian information directly. On the other hand, in questions of diffusion or mass transport, if the motion of fluid elements is of interest, then a Lagrangian specification of the problem may be more natural. In observation, the marking of fluid elements by dye or other traces gives Lagrangian information.

One of the most interesting and successful applications of hydrodynamical theory, with regard to shallow water waves, is to the small oscillations under gravity of a liquid having a free surface. In certain cases which are somewhat special as regards the theory, but very important from a practical viewpoint, these oscillations may combine to form progressive waves travelling over the surface with no change of form. The term 'tides', as applied to waves, has been given various meanings, but it seems most natural to confine it to gravitational oscillations possessing the characteristic features of oceanic tides produced by the action of the sun and moon.

A description of tidal oscillations, together with relevant practical applications, is given in Chapter 6.

1.4 Outline of the book

This book is intended for senior undergraduates, graduates, scientists and engineers whose main interests are surface waves in oceans and tidal waves in estuaries. The subject matter is arranged so that the topics follow in sequence, each one progressing from the previous material. Exercises at the end of each chapter are intended to give the reader experience of the principles developed in the book.

The first three chapters give the derivation of the fundamental mathematical equations. Chapter 2 outlines appropriate differential equations to describe the physical phenomena, and Chapter 3 reviews solution techniques of some simplified partial differential equations.

Chapter 4 gives the development of wave equations, including the essential boundary conditions, and describes small-amplitude wave theory. Chapter 5 deals with finite-amplitude wave theory and Chapter 6 outlines the study of tidal dynamics in shallow water.

For the random wave case, the deterministic methods described in previous chapters do not hold good. Therefore, Chapter 7 is clearly devoted to wave statistics and the wave energy spectrum. The application of wave theory is demonstrated in Chapter 8. Chapter 9 examines the nonlinear long waves in shallow water from a mathematical viewpoint. The book concludes with Chapter 10 which illustrates the inverse scattering technique to solve the solitary wave problem.

References

Airy, G. B. (1845). Tides and waves. *Encycl. Metrop.*, Art. 192, 241–396.

Batchelor, G. K. (1967). *An Introduction to Fluid Dynamics*, Cambridge University Press, Cambridge.

Jeffreys, H. (1925). On the formation of waves by winds. II. *Proc. R. Soc.*, A 110, 341–347.

Jeffreys, H. (1931). Tidal friction in shallow seas. *Philos. R. Soc.*, A 211, 239–264.

Kolmogorov, A. N. (1941). The local structure of turbulence in an incompressible viscous fluid for very large Reynolds number. *C. R. Acad. Sci., USSR*, 30, 301.

Kraichnan, R. H. (1959). The structure of isotropic turbulence at very high Reynolds number. *J. Fluid Mech.*, 5, 497–543.

Lighthill, M. J. (1978). *Waves in Fluids*, Cambridge University Press, Cambridge.

Long, R. R. (1955). Some aspects of the stratified fluids. III. Continuous density gradients. *Tellus*, 7, 342–357.

Phillips, O. M. (1966). *The Dynamics of the Upper Ocean*, Cambridge University Press, Cambridge.

Prandtl, L. (1904). *Uber Flussigkeitsbewegung bei sehr kleiner Reibung*, Vehr. III, Tuener, Leipzig.

Rayleigh, Lord (1880). On the stability, or instability, of certain fluid motions. I. *Sci. Pap.*, 1, 474–487.

Reynolds, O. (1883). An experimental investigation of the circumstances which determine whether the motion of water shall be direct or sinuous, and of the law of resistance in parallel channels. *Philos. Trans.*, 174, 935–982.

Stevenson, T. (1850). Observations on the force of waves. *Br. Assoc. (London) Rep.* See also *New Edinburgh Philos. J.*, 53, 358 (1852).

Stokes, G. G. (1847). On the theory of oscillatory waves. *Trans. Cambridge Philos. Soc.*, 8, 441–455.

Taylor, G. I. (1920). Tidal oscillations in gulfs and rectangular basins. *Proc. London Math Soc.*, 20, 148–181.

Thompson, S. (1969). Turbulence interface generated by an oscillating grid in a stably stratified fluid. Ph.D. dissertation, University of Cambridge.

Townsend, W. (1932). *Turbulence*, Cambridge University Press, Cambridge.

2
Fundamental equations of motion of fluids

2.1 Introduction

The study of fluid motion and other related processes can begin when the laws governing these processes have been expressed in mathematical form. Usually, in investigating fluid flow problems, the physical situation is described by a set of differential equations. Thus the solution of the differential equations will predict the fluid flow pattern. For a comprehensive derivation of these equations, the reader should turn to standard textbooks including Aris (1962). In this book, we shall avoid rigorous mathematical development of the equations.

2.2 Flow along a stream tube

Sir Isaac Newton conceived the notion that a fluid consists of a granulated structure of discrete particles. However, the range of validity of Newton's method was limited, as shown by a comparison of the theoretical and experimental results. Later, Lagrange and Euler developed improved methods in which the fluid was regarded as a continuous medium. It is usual to adopt the Lagrangian method where the actual paths of fluid particles are required. The Eulerian method is based on the observation of the characteristic variation of the fluid as it flows past a point previously occupied by the fluid. Thus any quantity associated with the fluid may be functionally represented in the form $f(r, t)$.

Using the Eulerian method, the state of the fluid flow along a *streamline* is considered. A *streamline* is defined as a line drawn in the fluid such that a tangent at each point of the line is in the direction of the fluid velocity at that point at any given instant. A *stream tube* is formed by drawing a set of such streamlines through all the points of a small, closed curve. More precisely, the streamlines of a steady flow are the paths along which fluid particles move. In fact, a particle on any one streamline remains always on that streamline. The streamlines associated with the velocity field $(\mathbf{V}(\mathbf{X}) = (u(x, y, z), v(x, y, z), w(x, y, z)))$ represent the doubly infinite set of solutions of the differential equations $(dx/u) = (dy/v) = (dz/w)$, where each expression represents the very short time dt during which a particle of fluid makes a change of position (dx, dy, dz). In unsteady flow, the pathlines (paths of particles of fluid) are completely different in shape from the shapes of the streamlines at any one instant. As a particle moves along a pathline it is at each instant moving tangentially to each local streamline, but the pattern of those streamlines is changing in time. These pathlines associated with the velocity field

$$(\mathbf{V}(\mathbf{X}, t) = (u(x, y, z, t), v(x, y, z, t), w(x, y, z, t)))$$

can be determined by solving a system of ordinary differential equations $(dx/dt) = u$, $(dy/dt) = v$, $(dz/dt) = w$.

In a steady flow, of course, a stream tube has unchanging shape because the motion of each particle of fluid on its boundary is directed along the streamline on which it is situated, and these lines are the bounding surface of the tube. Instantaneously, of course, a streamline pattern exists for an unsteady flow and this allows a stream tube to be defined such that the motion of each particle of the fluid on the surface of the tube is directed tangentially to the streamline on which it is situated and therefore tangentially to the tube composed of those streamlines. Thus the characteristics of this flow in one dimension will be fully defined once the pressure P, density ρ, velocity v, and the cross-sectional area A of the tube are known as functions of the axial distance of the tube. Hence four equations are needed to evaluate these four unknowns. It should be noted here that when the cross-section A is infinitesimally small, the stream tube is known as the *stream filament*.

The first equation is obtained by the condition that the fluid is neither created nor destroyed within the tube. This condition is known as the *conservation of mass*. Hence, the equation of *continuity* for a one-dimensional flow system (see Fig. 2.1) is

$$\rho A v = \text{constant}. \tag{2.1}$$

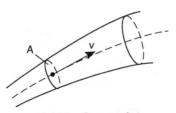

Fig. 2.1. Stream tube.

The second equation can be obtained by considering the motion of the fluid.

The motion of the fluid is described by the application of Newton's law of motion, FORCE = MASS × ACCELERATION, to a small cylindrical element of the stream tube of elementary length δs and mass $\rho A\, \delta s$. In calculating the force, the fluid acceleration must be known. To find the acceleration, the following steps are needed.

With reference to Fig. 2.2, let the velocity of the fluid at A' be v along the axis s at time t. Then $v = v(s,t)$ at time t at A'. Following the motion of the fluid, the velocity at B' after time $t + \delta t$ is $v + \delta v$ and may be written as $v + \delta v = v(s + v\, \delta t, t + \delta t)$.

Hence $\delta v = v(s + v\, \delta t, t + \delta t) - v(s,t) = v(s + v\, \delta t, t + \delta t) - v(s, t + \delta t) + v(s, t + \delta t) - v(s,t)$.

However, by using Taylor's theorem, we have

$$v(s + v\, \delta t, t + \delta t) - v(s, t + \delta t) = (\partial v/\partial s)v\, \delta t + O(\delta t^2)$$

and $v(s, t + \delta t) - v(s,t) = (\partial v/\partial t)\delta t + O(\delta t^2)$. By addition, the acceleration at B' is given by $\lim_{\delta t \to 0}(\delta v/\delta t) = (\partial v/\partial t) + v(\partial v/\partial s)$.

Therefore, $(Dv/Dt) = (\partial v/\partial t) + v(\partial v/\partial s)$ which is the axial acceleration of a fluid through a stream tube. The forces acting on this cylinder are shown in Fig. 2.3 and will be those due to the fluid pressure and any external force such as gravity. Assuming the fluid is perfect, then the pressure forces will act normal to the surfaces of the cylinder.

Let the mass of the cylinder be $\rho A\, \delta s$, the pressure at C be P and the pressure at D be $P + (\partial P/\partial s)\delta s$.

Therefore the net pressure in the increasing direction of s is $P - [P + (\partial P/\partial s)\delta s] = -(\partial P/\partial s)\delta s$. Let the net external force per unit mass be R. Thus by Newton's laws, the equation of motion will yield $\rho A\, \delta s[(\partial v/\partial t) + v(\partial v/\partial s)] = (\rho A\, \delta s)R + [-(\partial P/\partial s)\delta s]A$. Simplifying this equation yields

$$\frac{\partial v}{\partial t} + v\frac{\partial v}{\partial s} = R - \frac{1}{\rho}\frac{\partial P}{\partial s} \cdots \tag{2.2}$$

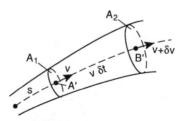

Fig. 2.2. Acceleration.

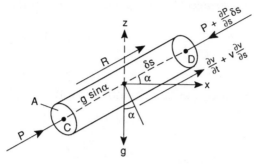

Fig. 2.3. Equation of motion.

Suppose the external force field R is conservative such that $R = -(\partial V_e/\partial s)$ where V_e is a potential function; further, assuming this external force acting on the fluid is due to gravity g, then

$$R = -g \sin \alpha = -g \frac{\partial z}{\partial s} = -\frac{\partial V_e}{\partial s}, \tag{2.3}$$

where the coordinate z is measured in the upward vertical direction and α is the angle made by the axis of the cylinder with the horizontal direction.

Integrating eqn (2.3) with respect to s yields

$$V_e = gz. \tag{2.4}$$

Define the fluid velocity v by the gradient of a scalar quantity $\phi(s,t)$ such that

$$v = \frac{\partial \phi}{\partial s}. \tag{2.5}$$

Here ϕ is called the velocity potential. Substituting relations (2.3), (2.4) and (2.5) into (2.2) and integrating the resulting equation partially with respect to s leads to

$$\frac{\partial \phi}{\partial t} + \frac{1}{2} v^2 + gz + \int \frac{dP}{\rho} = C(t), \tag{2.6}$$

where $C(t)$ is an arbitrary function of t. This equation is known as Bernoulli's equation and provides the second of the four required equations. The remaining two equations are obtained from the interpretation of the physical state of the fluid and geometry of the boundary streamlines.

For incompressible steady flow, eqn (2.6) can be written as

$$\frac{1}{2}v^2 + gz + \frac{P}{\rho} = C,$$
(2.7)

where C is an absolute constant. In hydraulics, Bernoulli's equation (2.7) may be written as

$$\frac{P}{\rho g} + \frac{v^2}{2g} + z = \text{constant},$$
(2.8)

where each term in (2.8) has the dimension of length.

Define $h_1 = (P/\rho g) = $ pressure head $=$ the height of a column of fluid density ρ, which gives rise to the pressure P, and $h_2 = (v^2/2g) = $ velocity head, where $v = \sqrt{2gh_2}$ is the velocity obtained by a body falling freely through a distance h_2, and $z = $ potential head. Therefore, the total head is $h_1 + h_2 + z = \text{constant}$.

Special cases

(a) When the fluid is at rest ($v = 0$) let the pressure be P_H, the hydrostatic pressure. Then Bernoulli's equation (2.8) becomes

$$\frac{P_H}{\rho} + gz = \text{constant}.$$
(2.9)

This is in accord with Archimedes' principle applied to the particles of fluid themselves.

(b) Subtracting eqn (2.9) from eqn (2.7) yields

$$\frac{(P - P_H)}{\rho} + \tfrac{1}{2}v^2 = \text{constant}.$$
(2.10)

The pressure $P - P_H$ is called the *gauge pressure*. From the pressure $P - P_H$ the force with which two fluid particles are pressed together may be calculated, the buoyancy effect being the same for both. But in the case of a liquid, the free surface of which is disturbed by waves, the hydrostatic pressure at a point in the liquid will vary and hence the gravitational force must be included.

2.3 Acceleration vector

As seen, the characteristics of the fluid as it flows past any fixed point in space are a function of a position vector, $\mathbf{r}$ (of the point in space), and of the time variable, t. In particular, the velocity vector, $\mathbf{V}$, at such a point will be a function of the independent variables $\mathbf{r}$ and t. In obtaining the general equations of motion, the general acceleration vector of a moving fluid particle corresponding to a fixed set of orthogonal axes is established and subsequently applied to a moving set of orthogonal axes.

Let a fluid particle in space occupy the position P at time t, the position vector of P being $\mathbf{r}$. Let the resultant velocity at P be $\mathbf{V} = \mathbf{V}(r, t)$. Suppose after an interval of time, δt, the fluid particle moves to the position P', the position vector of which is $\mathbf{r} + \mathbf{V}\delta t$. The velocity vector at P' may be written as (see Fig. 2.4) $\mathbf{V} + \delta \mathbf{V} = \mathbf{V}(\mathbf{r} + \mathbf{V}\delta t, t + \delta t)$. Hence,

$$\delta \mathbf{V} = \mathbf{V}(\mathbf{r} + \mathbf{V}\delta t, t + \delta t) - \mathbf{V}(\mathbf{r}, t)$$
$$= \mathbf{V}(\mathbf{r} + \mathbf{V}\delta t, t + \delta t) - \mathbf{V}(\mathbf{r}, t + \delta t) + \mathbf{V}(\mathbf{r}, t + \delta t) - \mathbf{V}(\mathbf{r}, t).$$

However, by Taylor's theorem

$$\mathbf{V}(\mathbf{r} + \mathbf{V}\delta t, t + \delta t) - \mathbf{V}(\mathbf{r}, t + \delta t) = (\mathbf{V}\delta t \cdot \mathrm{grad})\mathbf{V}(\mathbf{r}, t + \delta t) + O(\delta t^2),$$

and $\mathbf{V}(\mathbf{r}, t + \delta t) - \mathbf{V}(\mathbf{r}, t) = (\partial \mathbf{V}/\partial t)(\mathbf{r}, t)\delta t + O(\delta t^2)$.

Thus, by addition, $\delta \mathbf{V} = [\partial \mathbf{V}/\partial t(\mathbf{r}, t) + (\mathbf{V}\cdot \mathrm{grad})\mathbf{V}(\mathbf{r}, t + \delta t)]\delta t + O(\delta t^2)$. Dividing throughout by δt and letting δt go to zero, $\lim_{\delta t \to 0}(\delta \mathbf{V}/\delta t) = (\partial \mathbf{V}/\partial t) + (\mathbf{V}\cdot \mathrm{grad})\mathbf{V}$. By the definition of differentiation, the total rate of

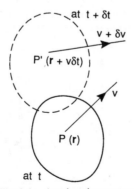

Fig. 2.4. Acceleration vector.

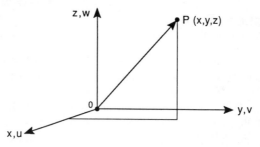

Fig. 2.5.   Cartesian coordinate axes.

change DV/Dt is defined as the rate of change of V following a particle of fluid:

$$\frac{DV}{Dt} = \frac{\partial V}{\partial t} + (V \cdot grad)V, \tag{2.11}$$

which is the sum of the local and convective rates of change of V.

Here, the operator $(D/Dt) = (\partial/\partial t) + (V \cdot grad)$ is the total differentiation following the motion of the fluid. Lighthill (1986) gives a very elegant description of this operator from the physical viewpoint.

The operator D/Dt is given below for three different sets of orthogonal axes.

(i) *In rectangular Cartesian coordinates* (Fig. 2.5)

The velocity vector is $V = (u, v, w)$. Elementary lengths are $(\delta x, \delta y, \delta z)$, $grad = \nabla = [(\partial/\partial x), (\partial/\partial y), (\partial/\partial z)]$. Therefore

$$\frac{D}{Dt} = \frac{\partial}{\partial t} + u\frac{\partial}{\partial x} + v\frac{\partial}{\partial y} + w\frac{\partial}{\partial z} \dots \tag{2.12}$$

(ii) *In cylindrical polar coordinates* (Fig. 2.6)

The relationship between Cartesian and cylindrical polar coordinates can be written as $x = r \cos\theta$, $y = r \sin\theta$, and $z = z$. The elementary

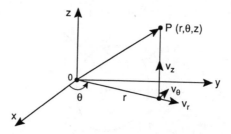

Fig. 2.6.   Cylindrical polar coordinate axes.

lengths are $(\delta r, r\delta\theta, \delta z)$. The velocity vector $\mathbf{V} = (v_r, v_\theta, v_z)$, and $\nabla = [(\partial/\partial r), (\partial/r\partial\theta), (\partial/\partial z)]$. Therefore

$$\frac{D}{Dt} = \frac{\partial}{\partial t} + v_r \frac{\partial}{\partial r} + \frac{v_\theta}{r} \frac{\partial}{\partial \theta} + v_z \frac{\partial}{\partial z}. \tag{2.13}$$

(iii) *In spherical polar coordinates* (Fig. 2.7)
The relationship between Cartesian and spherical polar coordinates can be written as $x = (r \sin \phi)\cos \theta$, $y = (r \sin \phi)\sin \theta$, and $z = r \cos \phi$. The elementary lengths are given by $(\delta r, r\delta\phi, r \sin \phi \, \delta\theta)$, the velocity vector $\mathbf{V} = (v_r, v_\phi, v_\theta)$ and $\nabla = [(\partial/\partial r), (\partial/r\partial\phi), (1/r \sin \phi)(\partial/\partial\theta)]$. Therefore

$$\frac{D}{Dt} = \frac{\partial}{\partial t} + v_r \frac{\partial}{\partial r} + \frac{v_\phi}{r} \frac{\partial}{\partial \phi} + \frac{v_\theta}{r \sin \phi} \frac{\partial}{\partial \theta}. \tag{2.14}$$

For a Cartesian resolution, the components of the velocity vector $\mathbf{V}$ would, at all times and at all points in space, be parallel to the Cartesian axes. However, consideration of a polar resolution of the velocity vector $\mathbf{V}$ indicates a rotation of the axes. At time t the position of P′ will be $\mathbf{r}(r, \phi, \theta)$ where the velocity will be $\mathbf{V}(v_r, v_\phi, v_\theta)$. After an interval of time, δt, the fluid will have moved to P″, where the components of velocity, v_r, v_ϕ, v_θ, are now parallel to the axes r', ϕ', θ'. Referring to Curle and Davies (1968), the acceleration of the fluid particle in the moving frame of axes can be defined as

$$\text{acceleration} = \frac{D\mathbf{V}}{Dt} + [\zeta \wedge \mathbf{V}],$$

where ζ is the angular velocity vector of axes and $\wedge$ is the vector cross-product.

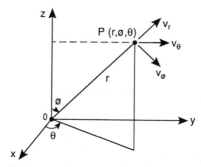

Fig. 2.7.   Spherical polar coordinate axes.

In the spherical polar frame of axes (see Fig. 2.7), the angular velocity vector $\zeta = [(v_\theta/r)\cot\phi, -(v_\theta/r), (v_\phi/r)]$ and the linear velocity vector $\mathbf{V} = (v_r, v_\phi, v_\theta)$. Then

$$[\zeta \wedge \mathbf{V}] = \begin{vmatrix} \mathbf{i} & \mathbf{j} & \mathbf{k} \\ \dfrac{v_\theta}{r}\cot\phi & -\dfrac{v_\theta}{r} & \dfrac{v_\phi}{r} \\ v_r & v_\phi & v_\theta \end{vmatrix} \tag{2.15}$$

from which we obtain the component

$$\left[-\frac{v_\theta^2 + v_\phi^2}{r}, \frac{v_r v_\phi}{r} - \frac{v_\theta^2}{r}\cot\phi, \frac{v_\theta v_\phi}{r}\cot\phi + \frac{v_r v_\theta}{r} \right].$$

Therefore the components of the acceleration vector in spherical polar coordinates can be written as

$$\frac{Dv_r}{Dt} - \frac{v_\theta^2 + v_\phi^2}{r}, \frac{Dv_\phi}{Dt} + \frac{v_r v_\phi}{r} - \frac{v_\theta^2}{r}\cot\phi, \frac{Dv_\theta}{Dt} + \frac{v_r v_\theta}{r} + \frac{v_\theta v_\phi}{r}\cot\phi. \tag{2.16}$$

In the cylindrical polar frame of axes (see Fig. 2.6), the angular velocity components are $\zeta = [0, 0, (v_\theta/r)]$.

Note that this result can be obtained from the angular velocity vector ζ with respect to the spherical polar frame of axes by changing ϕ to θ and then putting $v_\phi = 0$. The linear velocity vector is $\mathbf{V} = (v_r, v_\theta, v_z)$. Then

$$[\zeta \wedge \mathbf{V}] = \begin{vmatrix} \mathbf{i} & \mathbf{j} & \mathbf{k} \\ 0 & 0 & \dfrac{v_\theta}{r} \\ v_r & v_\theta & v_z \end{vmatrix} = \left(-\frac{v_\theta^2}{r}, \frac{v_r v_\theta}{r}, 0 \right).$$

Therefore the components of the acceleration vector in cylindrical polar coordinates can be written as

$$\frac{Dv_r}{Dt} - \frac{v_\theta^2}{r}, \frac{Dv_\theta}{Dt} + \frac{v_r v_\theta}{r}, \frac{Dv_z}{Dt}. \tag{2.17}$$

2.4 Equations of motion

We shall consider the mathematical development of the conservation of mass and Euler's equations of motion.

(I) Conservation of mass

Let $\rho(\mathbf{r}, t)$ be the density of the fluid occupying a volume V enclosed by a surface S (Fig. 2.8). Then $\Delta M = \rho(\mathbf{r}, t)\Delta V$, where ΔM is the elementary mass of the fluid which has occupied an elementary volume ΔV. Hence, the total mass can be obtained by integration to give

$$M = \int_V \rho(\mathbf{r}, t)\mathrm{d}V. \tag{2.18}$$

By definition

$$\frac{\mathrm{D}M}{\mathrm{D}t} = \lim_{\delta t \to 0} \frac{\int_{V+\delta V} \rho(\mathbf{r}, t + \delta t)\mathrm{d}V - \int_V \rho(\mathbf{r}, t)\mathrm{d}V}{\delta t}.$$

However,

$$\int_{V+\delta V} \rho(\mathbf{r}, t + \delta t)\mathrm{d}V = \int_V \rho(\mathbf{r}, t + \delta t)\mathrm{d}V + \int_{\delta V} \rho(\mathbf{r}, t + \delta t)\mathrm{d}V.$$

By Taylor's theorem

$$\int_V \rho(\mathbf{r}, t + \delta t)\mathrm{d}V = \int_V \rho(\mathbf{r}, t)\mathrm{d}V + \int_V \frac{\partial \rho}{\partial t} \delta t\, \mathrm{d}V + O(\delta t^2)$$

and

$$\int_{\delta V} \rho(\mathbf{r}, t + \delta t)\mathrm{d}V = \int_{\delta V} \rho(\mathbf{r}, t)\mathrm{d}V + \int_{\delta V} \frac{\partial \rho}{\partial t} \delta t\, \mathrm{d}V + O(\delta t^2)$$

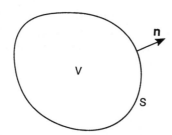

Fig. 2.8.   Fluid occupying volume V.

where ρ and $\partial\rho/\partial t$ are functions of $\mathbf{r}$ and t only. Hence

$$\frac{DM}{Dt} = \lim_{\delta t \to 0} \frac{\int_V \frac{\partial\rho}{\partial t}\,\delta t\,dV + \int_{\delta V}\rho(\mathbf{r},t)dV + \int_{\delta V}\frac{\partial\rho}{\partial t}\,\delta t\,dV}{\delta t} + O(\delta t).$$

As $\delta t \to 0$, consequently δV must go to zero. In that situation

$$\int_{\delta V}\frac{\partial\rho}{\partial t}\,dV = 0.$$

Thus

$$\frac{DM}{Dt} = \int_V \frac{\partial\rho}{\partial t}\,dV + \lim_{\delta t \to 0}\frac{\int_{\delta V}\rho(\mathbf{r},t)dV}{\delta t}.$$

Following the work of Curle and Davies (1968), it may be easily shown that

$$\lim_{\delta t \to 0}\int_{\delta V}\frac{\rho\,\delta V}{\delta t} = \lim_{\delta t \to 0}\int_S \frac{\rho(lu + mv + nw)\,\delta t\,\delta S}{\delta t} = \int_S \rho\mathbf{V}\cdot\mathbf{n}\,dS,$$

where $\mathbf{V}$ is the velocity vector and $\mathbf{n}$ is the unit normal vector. Using the divergence theorem, we obtain $\lim_{\delta t \to 0}\int_{\delta V}\rho(\delta V/\delta t) = \int_V \mathrm{div}(\rho\mathbf{V})dV$. Therefore

$$\frac{DM}{Dt} = \int_V\left(\frac{\partial\rho}{\partial t} + \mathrm{div}(\rho\mathbf{V})\right)dV. \tag{2.19}$$

Equation (2.19) can be deduced in the following manner also (see Aris (1962)). Let $\rho(\mathbf{r},t)$ be the mass per unit volume of a homogeneous fluid at a position $\mathbf{r}$ and time t. Then the mass of any finite volume $V(t)$ is

$$M = \iiint_{V(t)}\rho(\mathbf{r},t)dV.$$

We know that if the coordinate system is changed from coordinates $\mathbf{r_0}$ to coordinates $\mathbf{r}$, the element of volume changes by the formula $dV = J\,dV_0$, where

$$J = \frac{\partial(x,y,z)}{\partial(x_0,y_0,z_0)}$$

is called the Jacobian of transformations. Then we have the total derivative of M with respect to time t:

$$\frac{DM}{Dt} = \frac{D}{Dt}\iiint\limits_{V(t)} \rho(\mathbf{r},t)dV = \frac{D}{Dt}\iiint\limits_{V_0} \rho[\mathbf{r}(\mathbf{r}_0,t),t]J\,dV_0.$$

But we know $(DJ/Dt) = (\nabla\cdot\mathbf{V})J$. Thus,

$$\frac{DM}{Dt} = \iiint\limits_{V_0}\left[\frac{D\rho}{Dt} + \rho(\nabla\cdot\mathbf{V})\right]J\,dV_0 = \iiint\limits_{V(t)}\left[\frac{D\rho}{Dt} + \rho(\nabla\cdot\mathbf{V})\right]dV.$$

Since $(D/Dt) = (\partial/\partial t) + \mathbf{V}\cdot\nabla$, we can write this formulas as follows:

$$\frac{DM}{Dt} = \iiint\limits_{V(t)}\left[\frac{\partial\rho}{\partial t} + \nabla\cdot\rho\mathbf{V}\right]dV = \iiint\limits_{V(t)}\left[\frac{\partial\rho}{\partial t} + \text{div}(\rho\mathbf{V})\right]dV$$

which is identical to (2.19).

If fluid is being neither injected nor removed from the flow field, the total mass of the fluid body must remain constant and so $(DM/Dt) = 0$. Thus eqn (2.19) reduces to

$$\int_V\left[\frac{\partial\rho}{\partial t} + \text{div}(\rho\mathbf{V})\right]dV = 0.$$

But the volume V is arbitrary and consequently

$$\frac{\partial\rho}{\partial t} + \text{div}(\rho\mathbf{V}) = 0. \tag{2.20}$$

Equation (2.20) describes the basic assumption that the fluid is continuous, and is known as the *equation of continuity* for a viscous compressible flow. Equation (2.20) can be further simplified to yield

$$\frac{D\rho}{Dt} + \rho\,\text{div}(\mathbf{V}) = 0. \tag{2.21}$$

If the fluid is incompressible, then ρ is constant and in this situation $(D\rho/Dt) = 0$, and consequently eqn (2.21) reduces to

$$\text{div}\,\mathbf{V} = 0. \tag{2.22}$$

This is the *equation of continuity* when the fluid is incompressible.

(II) Euler's equation of motion

Consider the following momentum integral

$$I = \int_V \rho \mathbf{V}\,dV \tag{2.23}$$

where ρ is the density, $\mathbf{V}$ is the resultant velocity vector, δV is the elementary volume, and δS is the elementary surface area (Fig. 2.9).

Here $\rho\,dV$ is the elementary mass and I represents the total momentum.

The rate of change of momentum of the fluid bounded by the surface S, whether compressible or incompressible, real or ideal fluid, will be given by

$$\frac{DI}{Dt} = \frac{D}{Dt} \iiint_{V(t)} \rho \mathbf{V}\,dV$$

$$= \frac{D}{Dt} \iiint_{V_0} \rho \mathbf{V} J\,dV_0$$

$$= \iiint_{V_0} \left(\frac{D}{Dt}(\rho\mathbf{V})J + (\rho\mathbf{V})\frac{DJ}{Dt} \right) dV_0.$$

However, $(DJ/Dt) = (\nabla \cdot \mathbf{V})J$.

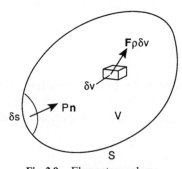

Fig. 2.9. Elementary volume.

Therefore,

$$\frac{DI}{Dt} = \iiint_{V_0} \left(\frac{D}{Dt}(\rho \mathbf{V})J + (\rho \mathbf{V})(\nabla \cdot \mathbf{V})J \right) dV_0$$

$$= \iiint_{V(t)} \left(\frac{D}{Dt}(\rho \mathbf{V}) + (\rho \mathbf{V})(\nabla \cdot \mathbf{V}) \right) dV$$

$$= \iiint_{V(t)} \left[\rho \frac{D\mathbf{V}}{Dt} + \mathbf{V} \left(\frac{D\rho}{Dt} + \rho(\nabla \cdot \mathbf{V}) \right) \right] dV.$$

Since $(D\rho/Dt) + \rho(\nabla \cdot \mathbf{V}) = 0$, we have

$$\frac{DI}{Dt} = \iiint_{V(t)} \rho \frac{D\mathbf{V}}{Dt} \, dV. \tag{2.24}$$

By Newton's laws we know that the rate of change of momentum is equal to the sum of the impressed forces. These forces are given by (a) the normal pressure thrust on the boundary S, and (b) the external force $\mathbf{F}$ per unit mass.

Therefore

$$\frac{DI}{Dt} = \int_V \rho \mathbf{F} \, dV - \int_S P\mathbf{n} \, dS, \tag{2.25}$$

where $\mathbf{n}$ is the unit normal vector to δS. By using Gauss' theorem, eqn (2.25) may be written as

$$\frac{DI}{Dt} = \int_V \rho \mathbf{F} \, dV - \int_V (\text{grad } P) dV. \tag{2.26}$$

From eqns (2.24) and (2.26) we obtain

$$\int_V \left(\rho \frac{D\mathbf{V}}{Dt} + \text{grad } P - \rho \mathbf{F} \right) dV = 0. \tag{2.27}$$

Equation (2.27) is true for any arbitrary volume V; it follows that

$$\frac{D\mathbf{V}}{Dt} = \mathbf{F} - \frac{1}{\rho} \text{grad } P,$$

which can be written as

$$\frac{\partial \mathbf{V}}{\partial t} + (\mathbf{V} \cdot \text{grad})\mathbf{V} = \mathbf{F} - \frac{1}{\rho} \text{grad } P. \tag{2.28}$$

Equation (2.28) is known as *Euler's equation of motion* for an ideal compressible or incompressible fluid.

(III) Bernoulli's equation

If the external force $\mathbf{F}$ is conservative, that is the forces which are single-valued functions of space coordinates, then $\mathbf{F}$ may be written as

$$\mathbf{F} = -\operatorname{grad} V_e, \tag{2.29}$$

where V_e is defined as the scalar potential. Also, from vector calculus

$$(\mathbf{V}\cdot\operatorname{grad})\mathbf{V} = \operatorname{grad}\left(\tfrac{1}{2}v^2\right) - \mathbf{V}\wedge\operatorname{curl}\mathbf{V}. \tag{2.30}$$

Assuming that the pressure is a function of density only, then Euler's equation of motion (2.28) can be written as

$$\frac{\partial \mathbf{V}}{\partial t} - \mathbf{V}\wedge\operatorname{curl}\mathbf{V} + \operatorname{grad}\left(\int \frac{\mathrm{d}P}{\rho} + \frac{1}{2}v^2 + V_e\right) = 0. \tag{2.31}$$

Defining

$$\operatorname{grad}\left(\int \frac{\mathrm{d}P}{\rho} + \frac{1}{2}v^2 + V_e\right) = \operatorname{grad} H, \tag{2.32}$$

and substituting eqn (2.32) into eqn (2.31) gives

$$\frac{\partial \mathbf{V}}{\partial t} - \mathbf{V}\wedge\operatorname{curl}\mathbf{V} + \operatorname{grad} H = 0. \tag{2.33}$$

Integration of eqn (2.32) yields

$$\int \frac{\mathrm{d}P}{\rho} + \frac{1}{2}v^2 + V_e = H, \tag{2.34}$$

where H is known as the *Bernoulli function*. Equation (2.34) is known as *Bernoulli's equation*.

Special case

We now consider irrotational flows which are defined as flows throughout which the vorticity is zero, i.e. $\operatorname{curl}\mathbf{V} = 0$. A particular property of irrotational flows is the existence of a velocity potential which greatly facilitates their calculation. This particular property allows the Bernoulli relationship between the velocity field and the pressure field which is normally restricted to steady flows to be modified so as to become applicable in the general case of unsteady fields when these are irrotational.

When the motion is *irrotational* curl $\mathbf{V} = 0$. For this case a scalar function $\phi(\mathbf{r}, t)$ exists such that

$$\mathbf{V} = \text{grad } \phi, \tag{2.35}$$

where ϕ is the velocity potential. Substituting eqn (2.35) into eqn (2.31) yields

$$\frac{\partial}{\partial t}(\text{grad } \phi) + \text{grad}\left(\int \frac{dP}{\rho} + \frac{1}{2}v^2 + V_e\right) = 0.$$

By integrating partially with respect to the space variables

$$\frac{\partial \phi}{\partial t} + \int \frac{dP}{\rho} + \frac{1}{2}v^2 + V_e = C(t), \tag{2.36}$$

where the integration constant C is a function of t.

For steady motion $\partial\phi/\partial t = 0$ and eqn (2.36) reduces to

$$\int \frac{dP}{\rho} + \frac{1}{2}v^2 + V_e = C, \tag{2.37}$$

where C is an absolute constant. Using $\mathbf{V}$ as defined by eqn (2.35) in the continuity eqn (2.22), it follows that the velocity potential ϕ satisfies Laplace's equation

$$\nabla^2\phi = 0. \tag{2.38}$$

Remark on the existence of irrotational motion

The vector quantity $\omega = \text{curl } \mathbf{V}$ is defined as the *vorticity vector* and if the fluid moves in such a way that the vorticity vector is zero, then the motion is said to be *irrotational*. That the irrotational motion exists in ideal fluids can be demonstrated by taking the curl of Euler's equation of motion (2.31), yielding $(\partial\omega/\partial t) - \text{curl}(\mathbf{V} \wedge \omega)$ which is identically satisfied if $\omega = \text{curl } \mathbf{V} = 0$. As a matter of fact such a motion exists for many physical problems including water wave mechanics in inviscid oceans. That the irrotational motion persists in inviscid fluids can be demonstrated below.

If C is a closed contour which moves with the fluid consisting of the same fluid particles, then the *circulation* can be defined as the integral

$\Gamma = \int_C \mathbf{V} \cdot \mathbf{dr}$ where $\mathbf{V}$ is the velocity vector and $\mathbf{dr}$ the elementary length of C. By Stokes' theorem (see Wylie and Barrett (1982)) this can be written as

$$\Gamma = \iint_S \operatorname{curl} \mathbf{V} \, dS = \iint_S \omega \, dS$$

where S is the surface bounded by the curve C every point of which lies within the fluid. Thus if $\omega = 0$ initially, so will Γ be zero. From another perspective, by differentiating with respect to time following the motion of the fluid we obtain

$$\frac{D\Gamma}{Dt} = \int_C \frac{D\mathbf{V}}{Dt} \cdot \mathbf{dr} + \int_C \mathbf{V} \cdot \mathbf{dV} = \int_C \frac{D\mathbf{V}}{Dt} \cdot \mathbf{dr} + \int_C d\left(\tfrac{1}{2}v^2\right).$$

The second integral around the closed contour C is zero. However, using Euler's equation (2.33) and after some simplification we can write

$$\frac{D\Gamma}{Dt} = -\int_C \operatorname{grad}\left(H - \tfrac{1}{2}v^2\right) \cdot \mathbf{dr} = -\iint_S \operatorname{curl}\operatorname{grad}\left(H - \tfrac{1}{2}v^2\right) dS$$

where in the surface integral we have used Stokes' theorem. It is now obvious that because curl grad of a scalar function is identically zero, we consequently have $D\Gamma/Dt = 0$ which expresses the fact that the circulation round a closed contour moving with the fluid is constant for all time. This result is usually known as Kelvin's theorem. In conclusion we state that the zero vorticity condition, namely the irrotational motion, exists for inviscid fluid flow motion.

2.5 Two-dimensional flow

It has been found experimentally that a large class of problems exists in which one of the velocity components, say w, is small when compared with the components u and v. Modelling such flows with the simplification obtained by setting $w = 0$ and allowing u and v to be functions of x and y, but not of z, leads to excellent agreement between theory and observation. The flow is defined as being two dimensional.

For incompressible flow in two dimensions the continuity equation, $\operatorname{div}\mathbf{V} = 0$, where $\mathbf{V} = (u, v, 0)$, becomes

$$\frac{\partial u}{\partial x} + \frac{\partial v}{\partial y} = 0. \tag{2.39}$$

Consider the following first-order ordinary differential equation:

$$u\,dy - v\,dx = 0. \tag{2.40}$$

From the theory of first-order ordinary differential equations we know that eqn (2.40) will be *exact* if the following condition is satisfied: $(\partial u/\partial x) + (\partial v/\partial y) = 0$, which is precisely the equation of continuity (2.39). Thus there exists a scalar function $\psi(x, y)$ such that

$$d\psi = u\,dy - v\,dx = 0. \tag{2.41}$$

Integrating eqn (2.41) gives

$$\psi(x, y) = \text{constant}. \tag{2.42}$$

Here $\psi(x, y)$ is a *stream function*, since by definition, the velocity is tangential to a streamline; therefore the differential equation of streamlines can be written as

$$\frac{dx}{u} = \frac{dy}{v}. \tag{2.43}$$

This equation can be derived from eqn (2.41). From eqn (2.41) we have

$$\frac{\partial \psi}{\partial x}\,dx + \frac{\partial \psi}{\partial y}\,dy = u\,dy - v\,dx.$$

Hence

$$u = \frac{\partial \psi}{\partial y}, \qquad v = -\frac{\partial \psi}{\partial x}. \tag{2.44}$$

It is noted that $\psi(x, y)$ is related to u and v, and also that this stream function exists only in two-dimensional flow. A good account of a velocity field and its streamlines can be found in the work of Lighthill (1986).

For the case where the motion is *irrotational* then we must have curl $\mathbf{V} = 0$, that is

$$\frac{\partial v}{\partial x} - \frac{\partial u}{\partial y} = 0. \tag{2.45}$$

Equation (2.45) can be recognized as the condition for the differential equation

$$u\,dx + v\,dy = 0 \tag{2.46}$$

to be *exact*. Thus there exists a scalar function $\phi(x, y)$ such that

$$d\phi = u\,dx + v\,dy, \tag{2.47}$$

such that $d\phi = 0$ and upon integration yields $\phi(x, y) = $ constant.
From eqn (2.47), $(\partial\phi/\partial x)dx + (\partial\phi/\partial y)dy = u\,dx + v\,dy$. Therefore

$$u = \frac{\partial\phi}{\partial x}, \qquad v = \frac{\partial\phi}{\partial y}, \tag{2.48}$$

and the velocity vector, $\mathbf{V}$, can be written as

$$\mathbf{V} = \text{grad } \phi. \tag{2.49}$$

Here $\phi(x, y)$ is called the *velocity potential*. This function exists in both
two- and three-dimensional flow. It is easily shown that both the stream
function $\psi(x, y)$ and the velocity potential $\phi(x, y)$ satisfy Laplace's equation. Using eqn (2.44) together with the irrotational flow condition (2.45)
yields

$$\frac{\partial^2\psi}{\partial x^2} + \frac{\partial^2\psi}{\partial y^2} = 0. \tag{2.50}$$

Similarly, using eqn (2.48) together with the continuity condition (2.39)
yields

$$\frac{\partial^2\phi}{\partial x^2} + \frac{\partial^2\phi}{\partial y^2} = 0. \tag{2.51}$$

Remark

(a) Physical interpretation of velocity potential

In practice the velocity potential ϕ is defined as the value of the line
integral of the velocity value $\mathbf{V} = (u, v, w)$ as

$$\phi = \int_C \mathbf{V}\cdot d\mathbf{r} = \int_C (u\,dx + v\,dy + w\,dz)$$

where C is the contour of integration. The quantity $\mathbf{V}\cdot d\mathbf{r}$ is a measure of
the fluid velocity in the direction of the contour at each point. Therefore ϕ
is related to the product of the velocity and length along the path between

two distinct points on C. For the value of ϕ to be independent of the path, i.e. for the flow rate between these two points to be the same no matter how the integration is carried out, the term in the integrand must be an exact differential $d\phi$, so that $d\phi = u\,dx + v\,dy + w\,dz$, and therefore, $\mathbf{V} = \text{grad } \phi$. To ensure that this scalar function ϕ exists, it is confirmed that the curl of the velocity vector $\mathbf{V}$ must be zero, which indeed is so because the vector calculus identity confirms that $\text{curl } \mathbf{V} = \text{curl grad } \phi = 0$ always. This curl of the velocity vector is referred to as the vorticity ω as described in the last section.

(b) Physical interpretation of stream function

For the velocity potential, we defined ϕ in three dimensions as the line integral of the velocity vector projected on the line element. Let us define in a similar manner the line integral composed of the velocity components perpendicular to the line element in two dimensions as $\psi = \int_C \mathbf{V} \cdot \mathbf{n}\,dl$ where $dl = |\mathbf{dr}|$. The integrand here will physically imply that ψ represents the amount of fluid crossing the line C between two distinct points of C. The unit normal vector $\mathbf{n}$ is perpendicular to the path of integration C. This vector can be obtained from the relation that $\mathbf{n} \cdot \mathbf{dr} = 0$ such that the normal unit vector components can be obtained as $n_x = dy/dl$ and $n_y = -(dx/dl)$. The integral then can be written as $\psi = \int_C (u\,dy - v\,dx)$. The value of this integral, i.e. the flow between these two distinct points, will be independent of the path of integration provided the integrand becomes an exact differential, $d\psi$. This requires that $u = \psi_y$ and $v = -\psi_x$, from which we deduce that $u_x + v_y = 0$ which is precisely the continuity equation in two dimensions. The ψ is defined as the stream function. It is to be noted from this mathematical analysis that there exists a stream function for two-dimensional incompressible flow. However, in general, there can be no stream function for three-dimensional flows with the exception of axisymmetric flows, whereas, as we have seen already, the velocity potential exists in any three-dimensional flow that is irrotational.

Complex potential

We have seen that the velocity components in two-dimensional flow can be related to $\psi(x, y)$ and $\phi(x, y)$ by the following equations:

$$u = \frac{\partial \psi}{\partial y} = \frac{\partial \phi}{\partial x}, \qquad v = \frac{\partial \phi}{\partial y} = -\frac{\partial \psi}{\partial x}. \tag{2.52}$$

The above equations are usually defined as the Cauchy–Riemann equations and enable hydrodynamicists to utilize the powerful techniques of functions of complex variables (see Milne-Thomson (1960)). Using the Cauchy–Riemann conditions it can be easily verified that the lines of constant stream function (ψ = constant) and the lines of constant velocity potential (ϕ = constant) are perpendicular. Also these conditions provide the necessary condition for the function

$$W = \phi + i\psi, \tag{2.53}$$

to be an analytic function of z, where $z = x + iy$.

The complex function W, whose real and imaginary parts are the velocity potential and stream function, respectively, is called the complex potential of the flow. W is an analytic function of z and hence

$$\frac{dW}{dz} = \frac{\partial\phi}{\partial x} + i\frac{\partial\psi}{\partial x} = u - iv = q\,e^{-i\theta}, \tag{2.54}$$

where q, the speed of the fluid, is given by

$$\left|\frac{dW}{dz}\right| = \sqrt{u^2 + v^2}, \tag{2.55}$$

and θ, the velocity direction relative to the real axis, by

$$\tan^{-1}\left(\frac{v}{u}\right) = \arg\left(\frac{dW}{dz}\right). \tag{2.56}$$

Also, at a stagnation point the fluid velocity is zero. Thus, if the complex potential W describing the motion is known, the stagnation points can be obtained from the equation $dW/dz = 0$.

2.6 Singularities in two-dimensional flow

In this section some basic singularities in two-dimensional flow are introduced. These are usually called source, sink and doublet. If the two-dimensional motion of a fluid consists of outward flow from a point, symmetrical in all directions in the reference plane, the point is called a *simple source* (Fig. 2.10). A *sink* is defined as a negative source. Thus a sink is a point of inward flow at which fluid is absorbed continuously. By combining a source and a sink, a further flow singularity may be defined; this is called a *doublet*.

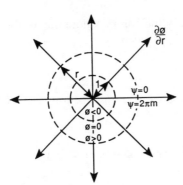

Fig. 2.10.   The source.

The strength (m) of a source is defined as a measure of the flux across any curve subtending unit radius at the source. As the flow is purely radial and symmetrical, then the flux across a circle of radius r, the centre of which is the source position, can be written as $2\pi r(\partial\phi/\partial r)$. The equation of continuity yields

$$2\pi r\,\frac{\partial\phi}{\partial r}=2\pi m. \tag{2.57}$$

The transverse velocity $(1/r)(\partial\phi/\partial\theta)$ is zero. Equation (2.57) can be written as $\partial\phi/\partial r=m/r$. Integrating this equation gives

$$\phi=m\ln r. \tag{2.58}$$

The Cauchy–Riemann equations in polar coordinates may be written as

$$v_r=\frac{\partial\phi}{\partial r}=\frac{1}{r}\frac{\partial\psi}{\partial\theta}\quad\text{and}\quad v_\theta=\frac{1}{r}\frac{\partial\phi}{\partial\theta}=-\frac{\partial\psi}{\partial r}. \tag{2.59}$$

Now from the first equation

$$\frac{\partial\phi}{\partial r}=\frac{m}{r}=\frac{1}{r}\frac{\partial\psi}{\partial\theta}.$$

Integrating this equation gives

$$\psi=m\theta. \tag{2.60}$$

Therefore the complex potential of the flow due to a source of strength m at the origin is given by

$$W=\phi+i\psi=m\ln z, \tag{2.61}$$

where $z = r\,e^{i\theta}$. Similarly the complex potential of the flow due to a sink of strength m at the origin is given by

$$W = -m \ln z. \tag{2.62}$$

In Cartesian coordinates, stream function (2.60) and velocity potential (2.58) can be written as

$$\psi = m \tan^{-1}\left(\frac{y}{x}\right) \quad \text{and} \quad \phi = m \ln(x^2 + y^2)^{\frac{1}{2}}. \tag{2.63}$$

If the source is at the point z_0, then by a shift of origin, the complex potential is

$$W = m \ln(z - z_0). \tag{2.64}$$

Reversing the roles of ψ and ϕ in (2.60) and (2.58) yields

$$\psi = -K \ln r \quad \text{and} \quad \phi = K\theta. \tag{2.65}$$

By direct differentiation of (2.65), we obtain the velocity pattern

$$v_r = 0 \tag{2.66}$$

and

$$v_\theta = \frac{K}{r}. \tag{2.67}$$

Thus the flow is purely circular with the tangential velocity dropping off as $1/r$.

The flow pattern is shown in Fig. 2.11, there being a singularity at the origin, where the velocity is infinite; ϕ and ψ are not defined. This type of singularity in the flow field is called the *line vortex* of strength K.

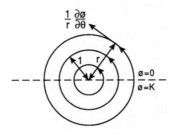

Fig. 2.11.   Line vortex.

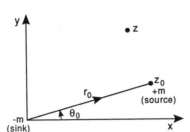

Fig. 2.12. Doublet.

By combining a source and sink in a particular way, a further flow singularity may be defined.

Consider a source of strength m situated at a point $z_0 = r_0 e^{i\theta_0}$ and a sink of equal strength situated at the origin. The complex potential of the flow due to this system is given by

$$W = m \ln(z - z_0) - m \ln z = m \ln\left(1 - \frac{z_0}{z}\right) = m \ln\left(1 - \frac{r_0 e^{i\theta_0}}{z}\right)$$

$$= -\frac{mr_0 e^{i\theta_0}}{z} + O\left(\frac{r_0^2}{z^2}\right).$$

Now if $\lim_{r_0 \to 0} mr_0 = \mu$, then $W = -(\mu e^{i\theta_0}/z)$. Such a combination of source and sink is said to be a *doublet* of strength μ (Fig. 2.12). Detailed discussion of flow singularities may be found in the standard textbooks, including Lamb (1945) and Lighthill (1986).

2.7 Navier–Stokes equations of motion

In this section the equation of continuity and the momentum equations for viscous incompressible fluids will be presented without detailed derivations. For such details, see the work of Lamb (1945), Rosenhead (1963), Phillips (1966), and Batchelor (1967). A more sophisticated derivation, using the full power of tensor properties, is given by Jeffreys (1931).

It will be assumed that throughout the motion of any element of fluid, its mass is conserved; hence, for incompressible flow, the volume of the fluid element must remain constant. This condition yields the *equation of continuity. The momentum equations*, which must be satisfied by the flow

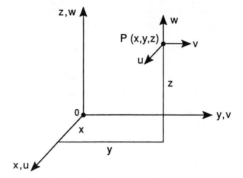

Fig. 2.13. Cartesian coordinate axes.

quantities at each point of the fluid, are deduced by applying Newton's second law of motion to the fluid, which occupies an elementary volume.

Thus, in a Cartesian coordinate set of axes (Fig. 2.13), the equation of continuity and the equations of motion can be written as

$$u_x + v_y + w_z = 0 \tag{2.68}$$

$$u_t + uu_x + vu_y + wu_z = -\frac{1}{\rho}P_x + X + \nu\nabla^2 u \tag{2.69}$$

$$v_t + uv_x + vv_y + wv_z = -\frac{1}{\rho}P_y + Y + \nu\nabla^2 v \tag{2.70}$$

$$w_t + uw_x + vw_y + ww_z = -\frac{1}{\rho}P_z + Z + \nu\nabla^2 w, \tag{2.71}$$

where ρ is the density, μ the kinematic viscosity and $\nu = \mu/\rho$ the dynamic viscosity of the fluid, and

$$\nabla^2 = \frac{\partial^2}{\partial x^2} + \frac{\partial^2}{\partial y^2} + \frac{\partial^2}{\partial z^2}. \tag{2.72}$$

Here u, v and w are the velocity components along the X, Y and Z directions, and given by $\dot{x} = u$, $\dot{y} = v$ and $\dot{z} = w$.

With cylindrical polar coordinates (r, θ, z) where $x = r\cos\theta$, $y = r\sin\theta$ and $z = z$ (Fig. 2.14), the equation of continuity and the equations of

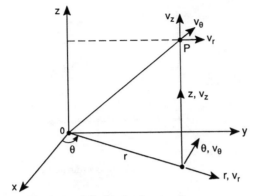

Fig. 2.14. Cylindrical polar coordinate axes.

motion can be written as

$$\frac{\partial v_r}{\partial r} + \frac{1}{r} v_r + \frac{1}{r} \frac{\partial v_\theta}{\partial \theta} + \frac{\partial v_z}{\partial z} = 0 \tag{2.73}$$

$$\frac{\partial v_r}{\partial t} + v_r \frac{\partial v_r}{\partial r} + \frac{v_\theta}{r} \frac{\partial v_r}{\partial \theta} + v_z \frac{\partial v_r}{\partial z} - \frac{v_\theta^2}{r}$$

$$= -\frac{1}{\rho} \frac{\partial P}{\partial r} + \nu \left(\nabla^2 v_r - \frac{v_r}{r^2} - \frac{2}{r^2} \frac{\partial v_\theta}{\partial \theta} \right) + X_r \tag{2.74}$$

$$\frac{\partial v_\theta}{\partial t} + v_r \frac{\partial v_\theta}{\partial r} + \frac{v_\theta}{r} \frac{\partial v_\theta}{\partial \theta} + v_z \frac{\partial v_\theta}{\partial z} + \frac{v_r v_\theta}{r}$$

$$= -\frac{1}{\rho} \frac{\partial P}{r \partial \theta} + \nu \left(\nabla^2 v_\theta + \frac{2}{r^2} \frac{\partial v_r}{\partial \theta} - \frac{v_\theta}{r^2} \right) + X_\theta \tag{2.75}$$

$$\frac{\partial v_z}{\partial t} + v_r \frac{\partial v_z}{\partial r} + \frac{v_\theta}{r} \frac{\partial v_z}{\partial \theta} + v_z \frac{\partial v_z}{\partial z} = -\frac{1}{\rho} \frac{\partial P}{\partial z} + \nu \nabla^2 v_z + X_z \tag{2.76}$$

where

$$\nabla^2 = \frac{\partial^2}{\partial r^2} + \frac{1}{r} \frac{\partial}{\partial r} + \frac{1}{r^2} \frac{\partial^2}{\partial \theta^2} + \frac{\partial^2}{\partial z^2}. \tag{2.77}$$

The velocity components are given by $\dot{r} = v_r$, $\dot{\theta} = v_\theta/r$ and $\dot{z} = v_z$.

With spherical polar coordinates (r, θ, ϕ) where $x = r \sin \theta \cos \phi$, $y = r \sin \theta \sin \phi$ and $z = r \cos \theta$ (Fig. 2.15), the equation of continuity and the

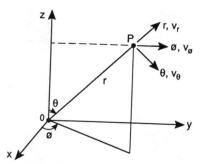

Fig. 2.15.   Spherical polar coordinate axes.

momentum equations can be written as

$$\frac{\partial v_r}{\partial r} + \frac{2}{r} v_r + \frac{1}{r} \frac{\partial v_\theta}{\partial \theta} + \left(\frac{1}{r} \cot \theta \right) v_\theta + \frac{1}{r \sin \theta} \frac{\partial v_\phi}{\partial \phi} = 0 \tag{2.78}$$

$$\frac{\partial v_r}{\partial t} + v_r \frac{\partial v_r}{\partial r} + \frac{v_\theta}{r} \frac{\partial v_r}{\partial \theta} + \frac{v_\phi}{r \sin \theta} \frac{\partial v_r}{\partial \phi} - \frac{v_\theta^2 + v_\phi^2}{r} = -\frac{1}{\rho} \frac{\partial P}{\partial r}$$

$$+ \nu \left(\nabla^2 v_r - \frac{2v_r}{r^2} - \frac{2}{r^2} \frac{\partial v_\theta}{\partial \theta} - \frac{2v_\theta \cot \theta}{r^2} - \frac{2}{r^2 \sin \theta} \frac{\partial v_\phi}{\partial \phi} \right) + X_r$$

$$\tag{2.79}$$

$$\frac{\partial v_\theta}{\partial t} + v_r \frac{\partial v_\theta}{\partial r} + \frac{v_\theta}{r} \frac{\partial v_\theta}{\partial \theta} + \frac{v_\phi}{r \sin \theta} \frac{\partial v_\theta}{\partial \phi} + \frac{v_r v_\theta}{r} - \frac{v_\phi^2 \cot \theta}{r} = -\frac{1}{\rho} \frac{\partial P}{r \partial \theta}$$

$$+ \nu \left(\nabla^2 v_\theta + \frac{2}{r^2} \frac{\partial v_r}{\partial \theta} - \frac{v_\theta}{r^2 \sin^2 \theta} - \frac{2 \cos \theta}{r^2 \sin^2 \theta} \frac{\partial v_\phi}{\partial \phi} \right) + X_\theta \tag{2.80}$$

$$\frac{\partial v_\phi}{\partial t} + v_r \frac{\partial v_\phi}{\partial r} + \frac{v_\theta}{r} \frac{\partial v_\phi}{\partial \theta} + \frac{v_\phi}{r \sin \theta} \frac{\partial v_\phi}{\partial \phi} + \frac{v_r v_\phi}{r} + \frac{v_\theta v_\phi \cot \theta}{r}$$

$$= -\frac{1}{\rho} \frac{1}{r \sin \theta} \frac{\partial P}{\partial \phi}$$

$$+ \nu \left(\nabla^2 v_\phi - \frac{v_\phi}{r^2 \sin^2 \theta} + \frac{2}{r^2 \sin \theta} \frac{\partial v_r}{\partial \phi} + \frac{2 \cos \theta}{r^2 \sin^2 \theta} \frac{\partial v_\theta}{\partial \phi} \right) + X_\phi$$

$$\tag{2.81}$$

where

$$\nabla^2 = \frac{1}{r^2}\frac{\partial}{\partial r}\left(r^2\frac{\partial}{\partial r}\right) + \frac{1}{r^2\sin\theta}\frac{\partial}{\partial\theta}\left(\sin\theta\frac{\partial}{\partial\theta}\right) + \frac{1}{r^2\sin^2\theta}\frac{\partial^2}{\partial\phi^2}.$$

(2.82)

The velocity components are given by $\dot{r} = v_r$, $\dot{\theta} = v_\theta/r$ and $\dot{\phi} = v_\theta/(r\sin\phi)$. The Navier–Stokes equations reduce to Euler's equation of motion when $v = 0$, that is when the fluid is inviscid.

2.8 Circulation around a curve

The *circulation* around a closed contour is defined as the integral taken around the contour of the tangential component of the velocity vector $\mathbf{V}$ (Fig. 2.16).

Thus the circulation is $\Gamma = \int_C \mathbf{V} \cdot d\mathbf{r} = \int_C (u\,dx + v\,dy + w\,dz)$. Using Stokes' theorem, this can be rewritten as

$$\Gamma = \int_C \mathbf{V}\cdot d\mathbf{r} = \iint_S \operatorname{curl}\mathbf{V}\cdot d\mathbf{s} = \iint_S \boldsymbol{\omega}\cdot d\mathbf{s}.$$

The vector $\boldsymbol{\omega} = \operatorname{curl}\mathbf{V}$ is called the *vorticity vector*. The rate of rotation of a fluid particle about the axes is defined as $\boldsymbol{\xi} = \frac{1}{2}\boldsymbol{\omega} = \frac{1}{2}\operatorname{curl}\mathbf{V}$. In a Cartesian coordinate system, the components of the rotation vector $\boldsymbol{\xi}$ may be written as

$$\xi_x = \frac{1}{2}\left(\frac{\partial w}{\partial y} - \frac{\partial v}{\partial z}\right), \qquad \xi_y = \frac{1}{2}\left(\frac{\partial u}{\partial z} - \frac{\partial w}{\partial x}\right)$$

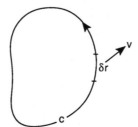

Fig. 2.16. Circulation.

and

$$\xi_z = \frac{1}{2}\left(\frac{\partial v}{\partial x} - \frac{\partial u}{\partial y}\right).$$

We now consider two-dimensional fluid motion such that the stream-lines of the flow are concentric circles. Assuming the flow is symmetrical, the transverse component of velocity will be constant for any given streamline and the radial component of velocity will be zero. The circulation around a streamline of a circuit of radius r will be

$$\int_C \mathbf{V}\cdot\mathbf{dr} = v_\theta\cdot\int_C \mathbf{dr} = v_\theta\cdot 2\pi r = \left(\frac{1}{r}\frac{\partial\phi}{\partial\theta}\right)\cdot 2\pi r,$$

and $v_r = \partial\phi/\partial r = 0$.

Kelvin's circulation theorem states that the rate of change of the circulation around C is zero, i.e. $(D/Dt)\int_C \mathbf{V}\cdot\mathbf{dr} = 0$. Equivalently, this statement implies that the circulation around the closed contour C moving with the fluid remains constant. Since the circulation around any stream-line is a constant, K, regarded as positive in an anticlockwise direction, then $(1/r)(\partial\phi/\partial\theta)\cdot 2\pi r = K$, which on integration yields $\Phi = (K/2\pi)\theta$, the velocity potential. The Cauchy–Riemann equations in polar form are

$$v_r = \frac{1}{r}\frac{\partial\Psi}{\partial\theta} = \frac{\partial\Phi}{\partial r}$$

$$v_\theta = \frac{\partial\Psi}{\partial r} = -\frac{1}{r}\frac{\partial\Phi}{\partial\theta}.$$

Therefore $\partial\Psi/\partial r = -(K/2\pi)(1/r)$, which integrates to give $\Psi = -(K/2\pi)\ln r$. Thus the complex potential due to a rectilinear vortex filament is given by

$$W = \Phi + i\Psi$$

$$= \frac{K}{2\pi}(\theta - i\ln r)$$

$$= -\frac{iK}{2\pi}(\ln r + i\theta)$$

$$= -\frac{iK}{2\pi}\ln z.$$

Using the integral $\int_C dW$, where C is a plane closed contour, a mixed flow can be produced. Writing

$$\int_C dW = \int_C \frac{dW}{dz} dz$$

$$= \int_C (u - iv)(dx + i\,dy)$$

$$= \int_C (u\,dx + v\,dy) + i\int_C (u\,dy - v\,dx)$$

$$= J + iQ,$$

where $J = \int_C u\,dx + v\,dy = \int_C \mathbf{V} \cdot d\mathbf{r}$, the circulation in circuit C, $Q = \int_C (u\,dy - v\,dx)$ is the total amount of fluid that flows outward across C. Using Green's lemma, the above integrals can be expressed as follows:

$$J = \int_C u\,dx + v\,dy = \iint_R \left(\frac{\partial v}{\partial x} - \frac{\partial u}{\partial y} \right) dx\,dy,$$

$$Q = \int_C (u\,dy - v\,dx) = \iint_R \left(\frac{\partial u}{\partial x} + \frac{\partial v}{\partial y} \right) dx\,dy.$$

If the integrands are analytic then the values of J and Q will be zero; conversely, if the integrands are non-analytic within the region R then J and Q are not zero. It is to be noted, however, that $Q = 0$ implies that the fluid is incompressible and if in addition $J = 0$, which implies that the flow is irrotational, then we are talking about the irrotational inviscid fluid flow. By combining one or more singularities with a uniform stream, a physical field may be obtained which will be bounded for the streamlines including the singularity from that region. The complex potential for a uniform stream may be written as $W = Uz$, where U is the uniform stream velocity. Combining this with the flow singularities such as source, sink, vortex or doublet, we can demonstrate the physical field by considering the following examples:

Example 2.1

We can obtain the flow past a semi-infinite body by the combination of potential due to a source in a uniform stream. The complex potential may be shown to be $W = Uz + m \ln z$. Then $\phi + i\psi = Ur\,e^{i\theta} + m \ln r + im\theta$.

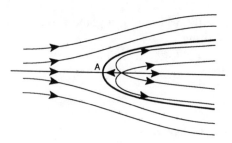

Fig. 2.17.   Flow past a semi-infinite body.

Therefore $\phi = Ur\cos\theta + m\ln r = Ux + (m/2)\ln(x^2 + y^2)$ and

$$\psi = Ur\sin\theta + m\theta = Uy + m\tan^{-1}\left(\frac{y}{x}\right).$$

The streamlines are shown in Fig. 2.17.

Example 2.2

Flow past an oval-shaped body may be obtained by the combination of a source and equal sink in a uniform stream. The complex potential is given by $W = Uz + m\ln(z + a) - m\ln(z - a)$. Then

$$\phi + i\psi = Uz + m\ln\left(\frac{z+a}{z-a}\right) = Ur\,e^{i\theta} + m\ln\frac{r_2\,e^{i\theta_2}}{r_1\,e^{i\theta_1}}$$

$$= Ur\,e^{i\theta} + m\ln\left(\frac{r_2}{r_1}\right) + im(\theta_2 - \theta_1).$$

Therefore $\phi = Ur\cos\theta + m\ln(r_2/r_1)$, $\psi = Ur\sin\theta + m(\theta_2 - \theta_1)$.   The streamlines are shown in Fig. 2.18.

Example 2.3

Flow past a circular cylindrical body may be obtained by the combination of a doublet in a uniform stream. The complex potential may be shown to

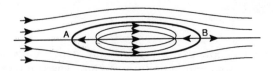

Fig. 2.18.   Flow past an oval-shaped body.

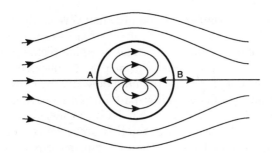

Fig. 2.19. Flow past a circular cylinder.

be $W = Uz + (\mu/z)$. Then $\phi + i\psi = Ur\,e^{i\theta} + (\mu/r)e^{-i\theta}$. Therefore $\phi = [Ur + (\mu/r)]\cos\theta$, $\psi = [Ur - (\mu/r)]\sin\theta$. The streamlines are shown in Fig. 2.19.

Example 2.4

Flow past a circular cylinder with circulation may be obtained by the combination of cylinder and vortex flow in a uniform stream. The complex potential may be written as $W = Uz + (Ua^2/z) + iK\ln z$. Thus

$$\phi + i\psi = Ur\,e^{i\theta} + \frac{Ua^2}{r}\,e^{-i\theta} + iK\ln r - K\theta.$$

Therefore $\phi = U[r + (a^2/r)]\cos\theta - K\theta$, $\psi = U[r - (a^2/r)]\sin\theta + K\ln r$. The streamlines are plotted in Fig. 2.20.

2.9 Conformal mapping

The difficulties in investigating two-dimensional irrotational motion of an incompressible inviscid fluid about a body largely depend upon the shape of the body. As already observed, the flow past a regular-shaped body, for

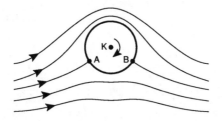

Fig. 2.20. Flow past a cylinder with circulation.

example a circular cylinder, can be obtained very easily from the complex potential. However, when the body geometry is not regular, recourse must be taken to other methods; conformal mapping techniques are useful in such situations. The main idea is to map a complicated geometry onto a simple configuration, replacing the fluid flow and the body geometry in the physical z-plane by a fictitious flow past a hypothetical configuration in the ξ-plane, where ξ is related to z by a known analytic function.

As an example, let us consider the flow around a corner. We know that a corner with included angle θ in the z-plane can be mapped onto a straight line in the ξ-plane by the transformation $\xi = z^{\pi/\theta}$. Figure 2.21 shows the mapping.

Another interesting example concerns the flow of an incompressible inviscid fluid past a fish-like profile with a sharp trailing edge, known as an aerofoil. An elegant account of the mathematical and physical developments leading to the conformal transformation can be found in standard textbooks including Lighthill (1986). We only describe here a brief summary leading to the transformation known as the Joukowski transformation. A simple mapping function $\xi = z + (a^2/z)$ where $a > 0$ can be shown to transform a circle $|z| = c$ on the z-plane into an ellipse of semi-major axis $c + (a^2/c)$ and semi-minor axis $c - (a^2/c)$ on the ξ-plane or vice versa. Thus it may be possible to establish a relationship between the complex variables ξ and z, say $\xi = f(z)$, such that, when a point on the z-plane traces the outline of the aerofoil, the corresponding point in the ξ-plane traces out a circle or some such simplified profile about which the flow can be calculated. Then, by means of the transformation $\xi = f(z)$ we can obtain the flow around the aerofoil by the much simpler process of calculating the flow past a circular cylinder. To make this possible, the transformation $\xi = f(z)$ must satisfy some conditions. Firstly, when we consider the flow around the aerofoil and flow lines, and the circle and the flow lines, it is seen that we must have a bilinear transformation, i.e. corresponding to one point on the z-plane there must exist one and only

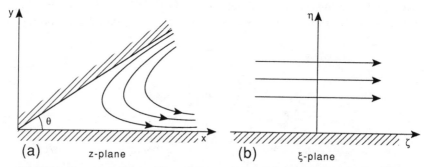

Fig. 2.21. The streamlines on the z-plane are shown as the straight lines in the ξ-plane.

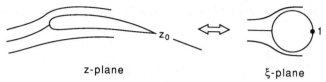

z-plane   ξ-plane

Fig. 2.22. Joukowski transformation.

one point on the ξ-plane and, conversely, corresponding to one point on the ξ-plane there must exist one and only one point on the z-plane. Secondly, there must also be a relationship between the velocities at corresponding points and as the increment in time is independent of the planes this means that corresponding infinitesimal lengths must be similar and angles related to corresponding directions must be preserved. Thirdly, there must be a relationship between corresponding elementary masses. As the fluid is incompressible, this amounts to saying that corresponding elementary masses must be similar. In the following we briefly present a mathematical deduction to obtain such a transformation initially derived by Joukowski which maps an aerofoil on the z-plane onto the ξ-plane. By using the Joukowski transformation, this aerofoil drawn in the z-plane can be transformed into the unit circle $|\xi| = 1$ in the ξ-plane such that the flow conditions at infinity are identical (Fig. 2.22).

A brief derivation of the Joukowski transformation is given below.

Let the position of the cusp $z = z_0$ on the aerofoil correspond to the point $\xi = 1$ on the unit circle. Consider the transformation

$$z = f(\xi). \tag{2.83}$$

The existence of a cusp at $z = z_0$ implies a singular point in the transformation. Therefore $d\xi/dz$ must be infinite at $\xi = 1$, and in addition the flow conditions at infinity need to be identical. Thus, mathematically, we have

$$\frac{dz}{d\xi} \simeq (\xi - 1) \quad \text{for} \quad |\xi - 1| \ll 1 \tag{2.84}$$

$$\frac{dz}{d\xi} \to 1 \quad \text{when} \quad |\xi| \to \infty. \tag{2.85}$$

The analytic function $f(\xi)$ which satisfies these two conditions may be found from the transformation

$$\frac{dz}{d\xi} = \frac{(\xi - 1)(\xi - c)}{(\xi - b)^2}, \tag{2.86}$$

where b and c are two arbitrary constants. Expanding the right-hand side for large ξ yields

$$\frac{\mathrm{d}z}{\mathrm{d}\xi} = 1 + \frac{2b - (1 + c)}{\xi} + \dots. \tag{2.87}$$

We know that, for the function to be analytic, its integral taken around a closed contour C in the ξ-plane must be zero. Thus

$$\int_C \frac{\mathrm{d}z}{\mathrm{d}\xi}\, \mathrm{d}\xi = \int_C \frac{(\xi - 1)(\xi - c)}{(\xi - b)^2}\, \mathrm{d}\xi = 0,$$

provided the residue at infinity is zero, which leads to $2b = 1 + c$.

Thus the required transformation is

$$\frac{\mathrm{d}z}{\mathrm{d}\xi} = 1 - \frac{\frac{1}{4}(1 - c)^2}{[\xi - \frac{1}{2}(1 + c)]^2} \tag{2.88}$$

which after integration can be written as

$$z = \xi + \frac{\frac{1}{4}(1 - c)^2}{\xi - \frac{1}{2}(1 + c)} \tag{2.89}$$

provided $|c| < 1$ and $|(1/c)(1 + c)| < 1$.

An application of the aerofoil theory can be illustrated by Blasius's theorem, which enables us to calculate the lift force and pitching moment acting on the aerofoil.

Consider a cylinder of arbitrary cross-section, which is in a steady irrotational flow of fluid. The aerodynamic force per unit length of the cylinder may be reduced to a force (X, Y) and a pitching moment M, as shown in Fig. 2.23.

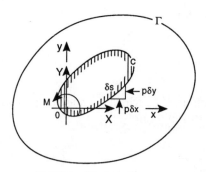

Fig. 2.23. Blasius's theorem.

Pressure forces on an element δs of the contour C may be written as δX and δY. A moment due to these forces about the origin is δM. Then $\delta X = -p\,\delta y$, $\delta Y = p\,\delta x$ and $\delta M = x\,\delta Y - y\,\delta X = p(x\,\delta x + y\,\delta y)$. Thus, $\delta X - i\,\delta Y = -p\,\delta y - ip\,\delta x = -ip(\delta x - i\,\delta y) = -ip\,\delta\bar{z}$ and therefore $\delta(X - iY) = -ip\,\delta\bar{z}$. Integrating along the contour C gives $X - iY = -i\int_C p\,d\bar{z}$ which, on using Bernoulli's equation, becomes $-i\int_C(\text{constant} - \frac{1}{2}\rho q^2)d\bar{z}$. Thus $X - iY = -i\int_C(\text{constant})d\bar{z} + \frac{1}{2}i\rho\int_C q^2\,d\bar{z}$, $\int_C d\bar{z} = 0$, and $q^2 = (dW/dz)\cdot(d\bar{W}/d\bar{z})$, where $q = (dW/dz) = u - iv$ and $q^2 = u^2 + v^2$ where W is the complex potential. Hence,

$$X - iY = i\frac{\rho}{2}\int_C \frac{dW}{dz}\,dW.$$

Since on C the stream function ψ is a constant, then $dW = d\bar{W}$. The above equation may be expressed as

$$X - iY = i\frac{\rho}{2}\int\left(\frac{dW}{dz}\right)^2 dz = i\frac{\rho}{2}\int_\Gamma\left(\frac{dW}{dz}\right)^2 dz, \tag{2.90}$$

where Γ is a contour reconcilable with C. To find the moment about the origin, note that $\delta M = p(x\,\delta x + y\,\delta y)$, since $\mathrm{Re}(z\,\delta\bar{z}) = x\,\delta x + y\,\delta y$, $\delta M = \mathrm{Re}(pz\,\delta\bar{z}) = \mathrm{Re}[(\text{constant} - \frac{1}{2}\rho q^2)z\,\delta\bar{z}]$. Integrating along the contour C

$$M = \mathrm{Re}\int_C\left[\text{constant}\tfrac{1}{2}\rho q^2\right]z\,\delta\bar{z} = \mathrm{Re}\left(-\tfrac{1}{2}\rho\int_C q^2 z\,d\bar{z}\right), \tag{2.91}$$

since

$$\int_C z\,d\bar{z} = 0 = \mathrm{Re}\left[-\tfrac{1}{2}\rho\int_C z\left(\frac{dW}{dz}\right)^2 dz\right] = \mathrm{Re}\left[-\frac{\rho}{2}\int_\Gamma z\left(\frac{dW}{dz}\right)^2 dz\right].$$

$$\tag{2.92}$$

Equations (2.90) and (2.92) are known as *Blasius's theorem*.

2.10 Lift force

Let dW/dz have no singularities outside the contour C. Then we may assume Γ to be a contour at infinity. Thus, it is reasonable to expand dW/dz in inverse powers of z as follows:

$$\frac{dW}{dz} = a_0 + \frac{a_1}{z} + \frac{a_2}{z^2} + \dots.$$

It is known that the fluid velocity must tend to a constant which is the velocity of the uniform stream.

Suppose the uniform stream velocity is U at an angle α to the real axis. Then $|z| \to \infty$, $dW/dz \to U e^{-i\alpha} = a_0$, where α is the incidence angle or angle of attack of the aerofoil. Thus $W = U e^{-i\alpha}z + a_1 \ln z - a_2/z + \dots$. The second term, being the only multiple-valued function, must represent the circulation $2\pi K$ around the aerofoil. Hence $a_1 = iK$. Thus

$$\frac{dW}{dz} = U e^{-i\alpha} + \frac{iK}{z} + \frac{a_2}{z^2} + \dots ,$$

$$\left(\frac{dW}{dz}\right)^2 = U^2 e^{-2i\alpha} + (2U e^{-i\alpha}) \cdot \frac{iK}{z} + O\left(\frac{1}{z^2}\right).$$

It is now obvious that the coefficient of $1/z = 2U e^{-i\alpha}(iK)$ is the residue. Therefore

$$X - iY = \int_\Gamma \frac{i}{2} \rho \left(\frac{dW}{dz}\right)^2 dz = \left(\frac{i}{2}\rho\right)(2\pi i) \times \text{Residue}$$

$$= \left(\frac{i}{2}\rho\right)(2\pi i)(2U e^{-i\alpha} \cdot iK) = -2\pi i \rho U e^{-i\alpha}K$$

$$= -2\pi\rho UK(i\cos\alpha + \sin\alpha).$$

On comparing real and imaginary parts, $X = -2\pi\rho UK \sin\alpha$, and $Y = 2\pi\rho UK \cos\alpha$. The lift force (Fig. 2.24) is given by

$$L = \sqrt{X^2 + Y^2} = 2\pi\rho UK = (2\pi) \times (\text{density})$$

$$\times (\text{main stream}) \times (\text{circulation}).$$

This lift force is perpendicular to the main stream U and is independent of the form of the aerofoil. This result is attributed to Kutta and Joukowski.

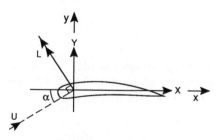

Fig. 2.24.   Lift force.

2.11 The Schwarz–Christoffel transformation

Consider a polygon in the ξ-plane having vertices at $\xi_1, \xi_2, \ldots, \xi_n$ with the corresponding interior angles $\theta_1, \theta_2, \ldots, \theta_n$, respectively. Let the points $\xi_1, \xi_2, \ldots, \xi_n$ map, respectively, onto the points $x_1, x_2, \ldots, x_n$ on the real axis of the plane as shown in Fig. 2.25.

A transformation which maps the interior D of the polygon of the ξ-plane onto the upper half D of the z-plane and the boundary of the polygon onto the real axis is given by

$$\frac{d\xi}{dz} = K(z - x_1)^{(\theta_1/\pi)-1}(z - x_2)^{(\theta_2/\pi)-1} \ldots (z - x_n)^{(\theta_n/\pi)-1}, \qquad (2.93)$$

where K is a complex constant.

To verify this, we begin with a point z on the x-axis to the left of the first of the given points $x_1, x_2, \ldots, x_n$ and investigate the locus of its image as it moves to the right along the x-axis as shown in Fig. 2.25. From eqn (2.93) we immediately obtain the relation

$$\arg(d\xi) = \arg(K) + \left(\frac{\theta_1}{\pi} - 1\right)\arg(z - x_1) + \left(\frac{\theta_2}{\pi} - 1\right)\arg(z - x_2)$$

$$+ \ldots + \left(\frac{\theta_n}{\pi} - 1\right)\arg(z - x_n) + \arg(dz).$$

It is apparent that until z reaches x_1, every term on the right remains constant since $z - x_1, z - x_2, \ldots, z - x_n$ are all negative real numbers and

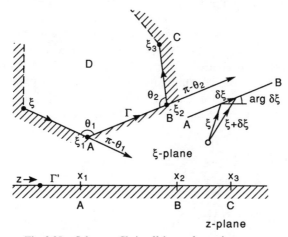

Fig. 2.25. Schwarz–Christoffel transformation.

hence have π for their respective arguments and since dz is positive and has 0 as its argument. Thus the image point ξ traces a straight line since the argument of the increment $d\xi$ remains constant. However, as z passes through x_1, the difference $z - x_1$ changes abruptly from negative to positive, and thus argument $z - x_1$ decreases abruptly from π to 0. Hence, $\arg d\xi$ changes by the amount $[(\theta_1/\pi) - 1](-\pi) = \pi - \theta_1$. This is the precise amount through which it is necessary to turn if ξ is to begin to move in the direction of the next side of the polygon. As z moves from x_1 to x_2, the same situation exists. Thus as z traverses the x-axis, it is evident that ξ moves along the boundary of a polygon whose interior angles are precisely the given angles $\theta_1, \theta_2, \ldots, \theta_n$. Now the mapping function ξ is obtained by integrating (2.93) which yields

$$\xi = K \int (z - x_1)^{(\theta_1/\pi) - 1} (z - x_2)^{(\theta_2/\pi) - 1} \ldots (z - x_n)^{(\theta_n/\pi) - 1} \, dz + L,$$

(2.94)

where K and L are complex constants. This formula was established by both Schwarz and Christoffel and therefore attributed to them as the Schwarz–Christoffel formula.

It is easily seen that when $z = 0$, eqn (2.94) becomes

$$\xi = \text{constant} \int_{C'} dz + L = L.$$

Here L is an arbitrarily chosen point in the ξ-plane to correspond to $z = 0$ such that one of the vertices of the polygon can be taken to this point. From eqn (2.94) $\delta\xi = Kf(z)\delta z$, where $f(z)$ is the integrand of (2.93). Then $\arg(\delta\xi) = \arg(K) + \arg(f(z)) + \arg(\delta z)$. The argument of K, being a constant, merely adds a constant to the argument of $\delta\xi$, which means that the value of any K defines the orientation of the polygon. The modulus of K determines the scale of the polygon. As L, $\arg K$ and $|K|$ are arbitrary, then any three consecutive numbers, x_1, x_2, x_3, may be chosen arbitrarily. The remainder of the constants in the integrand must then be chosen to make the polygon the right shape. It is often convenient to choose one of the constants, say x_1, to be at infinity. In this case,

$$K = B/(-x_1)^{\left(\frac{\theta_1}{\pi}\right) - 1}.$$

So eqn (2.94) becomes

$$\xi = B \int_{C'} \left(\frac{z - x_1}{-x_1}\right)^{\left(\frac{\theta_1}{\pi} - 1\right)} (z - x_2)^{\left(\frac{\theta_2}{\pi} - 1\right)} \ldots (z - x_n)^{\left(\frac{\theta_n}{\pi} - 1\right)} \, dz + L,$$

and when $x_1 \to \infty$, this becomes

$$\xi = B \int_{C'} (z - x_2)^{\left(\frac{\theta_2}{\pi} - 1\right)} (z - x_3)^{\left(\frac{\theta_3}{\pi} - 1\right)} \ldots (z - x_n)^{\left(\frac{\theta_n}{\pi} - 1\right)} dz + L.$$

As an application of the Schwarz–Christoffel transformation, consider a semi-infinite rectangle in the z-plane, with two vertices A_∞ and D_∞ at infinity where the transformation is chosen to map into the real axis in the ξ-plane such that A_∞ maps onto $\xi = -\infty$. Consider the two other arbitrary points to map B and C onto the points $\xi = -1$ and $\xi = 1$, respectively. The vertex D_∞ obviously must map onto $\xi = +\infty$. From Fig. 2.26 it is clear that the angles of B and C are right angles $\pi/2$. Then the transformation which transforms the semi-infinite rectangle onto the real axis is given by

$$z = K \int (\xi + 1)^{-1/2} (\xi - 1)^{-1/2} \, d\xi + L = K \int \frac{d\xi}{(\xi^2 - 1)^{1/2}} + L.$$

On integration this yields $z = K \cosh^{-1}(\xi) + L$.

At C, $\xi = 1$, $z = 0$, and hence $0 = K \cosh^{-1}(1) + L$. However,

$$\cosh^{-1}(1) = 0,$$

and therefore $L = 0$. At B, $\xi = -1$, $z = ib$. Therefore

$$ib = K \cosh^{-1}(-1).$$

But $\cosh^{-1}(-1) = \pi i$ and hence $K = b/\pi$. Thus the required transformation is

$$z = \frac{b}{\pi} \cosh^{-1}(\xi)$$

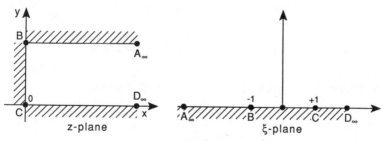

Fig. 2.26. Semi-infinite rectangle.

or

$$\xi = \cosh\left(\frac{\pi z}{b}\right).$$

2.12 Applications

Example 2.5

The complex potential of a fluid flow is given by $W(z) = V_0[z + (a^2/z)]$ where V_0 and a are positive constants.

(a) Obtain equations for streamlines and equipotential lines, represent them graphically and interpret them physically.
(b) Show that we can interpret the flow as that around a circular obstacle of radius a.
(c) Find the velocity at any point and determine its value far away from the obstacle.
(d) Find the stagnation points.

Solution

The complex potential is given by (a) $W = V_0[z + (a^2/z)]$. Therefore

$$\phi = i\psi = V_0\left(r\,e^{i\theta} + \frac{a^2}{r}\,e^{-i\theta}\right)$$

$$= V_0\left[r(\cos\theta + i\sin\theta) = \frac{a^2}{r}(\cos\theta - i\sin\theta)\right]$$

$$= V_0\left[\left(r + \frac{a^2}{r}\right)\cos\theta + i\left(r - \frac{a^2}{r}\right)\sin\theta\right].$$

Thus, equating real and imaginary parts we obtain

$$\phi = V_0\left(r + \frac{a^2}{r}\right)\cos\theta \quad \text{and} \quad \psi = V_0\left(r - \frac{a^2}{r}\right)\sin\theta.$$

The streamlines are given by $\psi = \text{constant} = \beta$. Therefore

$$V_0\left(r - \frac{a^2}{r}\right)\sin\theta = \beta.$$

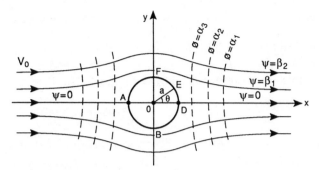

Fig. 2.27. Streamlines and equipotential lines.

The streamlines in Fig. 2.27 are indicated by the solid curves and show the actual paths taken by fluid particles. Note that $\psi = 0$ corresponds to $r = a$ and $\theta = 0$ or π. The equipotential lines are given by $\phi = $ constant $= \alpha$ or $V_0[r + (a^2/r)]\cos \alpha$. These are indicated by the dashed curves in Fig. 2.27 and are orthogonal to the family of streamlines.

(b) The circle $r = a$ represents a streamline and, since there can be no flow across a streamline, it can be considered as a circular obstacle of radius a placed in the path of the fluid.

(c) We have

$$\frac{dW}{dz} = W' = V_0\left(1 - \frac{a^2}{z^2}\right) = V_0\left(1 - \frac{a^2}{r^2}e^{-2i\theta}\right).$$

Therefore

$$u - iv = V_0\left(1 - \frac{a^2}{r^2}\cos 2\theta\right) + i\frac{V_0 a^2}{r^2}\sin 2\theta.$$

Equating real and imaginary parts we have

$$u = V_0\left(1 - \frac{a^2}{r^2}\cos 2\theta\right) \quad \text{and} \quad v = -V_0\left(\frac{a^2}{r^2}\sin 2\theta\right).$$

The absolute velocity is given by

$$q = V_0\sqrt{\left(1 - \frac{a^2}{r^2}\cos 2\theta\right)^2 + \frac{a^4}{r^4}\sin^2 2\theta} = V_0\sqrt{1 - \frac{2a^2\cos 2\theta}{r^2} + \frac{a^4}{r^4}}.$$

$$(2.95)$$

Far from the obstacle, as $r \to \infty$, we see that $q = V_0$ approximately, that is the fluid is travelling in the direction of positive x with constant speed V_0.

(d) The stagnation points are points at which the velocity is zero and are given by $dW/dz = 0$, $V_0[1 - (a^2/z^2)] = 0$ or $z = a$ and $z = -a$. The stagnation points are therefore A and D (see Fig. 2.27).

Example 2.6

(a) Find the complex potential due to a source at $z = -a$ and a sink at $z = a$ of equal strength m.

(b) Determine the equipotential lines and streamlines, and represent them graphically.

(c) Find the speed of the fluid at any point.

Solution

(a) The complex potential due to a source at $z = -a$ of strength m is $m \ln(z + a)$. The complex potential due to a sink at $z = a$ of strength m is $-m \ln(z - a)$. Then by superposition

$$W = m \ln(z + a) - m \ln(z - a) = m \ln\left(\frac{z + a}{z - a}\right).$$

(b) Let $z + a = r_1 e^{i\theta_1}$, $z - a = r_2 e^{i\theta_2}$. Then

$$W = \phi + i\psi = m \ln\left(\frac{r_1 e^{i\theta_1}}{r_2 e^{i\theta_2}}\right) = m \ln\left(\frac{r_1}{r_2} e^{i(\theta_1 - \theta_2)}\right)$$

$$= m \ln\left(\frac{r_1}{r_2}\right) + mi(\theta_1 - \theta_2).$$

On identifying real and imaginary parts, we have $\phi = m \ln(r_1/r_2)$ and $\psi = m(\theta_1 - \theta_2)$. The equipotential lines and the streamlines are thus $\phi = m \ln(r_1/r_2) = \alpha$, $\psi = m(\theta_1 - \theta_2) = \beta$. Using $r_1 = \sqrt{(x + a)^2 + y^2}$, $r_2 = \sqrt{(x - a)^2 + y^2}$, we have

$$\theta_1 = \tan^{-1}\left(\frac{y}{x + a}\right),$$

$$\theta_2 = \tan^{-1}\left(\frac{y}{x - a}\right).$$

The equipotential lines are given by $r_1/r_2 = e^{\alpha/m}$ which on substitution in the above expressions gives

$$\frac{\sqrt{(x+a)^2+y^2}}{\sqrt{(x-a)^2+y^2}} = e^{\alpha/m} = \lambda,$$

say. Thus $(x+a)^2+y^2 = \lambda^2[(x-a)^2+y^2]$, or

$$x^2 + 2ax + y^2 + a^2 = \lambda^2(x^2 - 2ax + a^2 + y^2),$$

or $(x^2 + a^2)(1-\lambda^2) + y^2(1-\lambda^2) + 2ax(1+\lambda^2) = 0$, or $x^2 + a^2 + y^2 + 2ax(1+\lambda^2)/(1-\lambda^2) = 0$, $\lambda \neq \pm 1$, or

$$x^2 - 2x\left(\frac{\lambda^2+1}{\lambda^2-1}a\right) + \left(\frac{\lambda^2+1}{\lambda^2-1}a\right)^2 + y^2 = a^2\left[\left(\frac{\lambda^2+1}{\lambda^2-1}\right)^2 - 1\right]$$

$$= a^2\frac{4\lambda^2}{(\lambda^2-1)^2}.$$

Therefore

$$\left(x - a\frac{\lambda^2+1}{\lambda^2-1}\right)^2 + y^2 = \frac{4a^2\lambda^2}{(\lambda^2-1)^2}$$

which represents circles having centres at $[a(\lambda^2+1)/(\lambda^2-1),0]$ and radii equal to $2a\lambda/(\lambda^2-1)$. When $\lambda = \pm 1$, the equipotential line is given by $x = 0$ (Fig. 2.28).

Streamlines are given by

$$\tan^{-1}\left(\frac{y}{x+a}\right) - \tan^{-1}\left(\frac{y}{x-a}\right) = \frac{\beta}{m}$$

and therefore

$$\tan^{-1}\left(\frac{\dfrac{y}{x+a} - \dfrac{y}{x-a}}{1 + \dfrac{y^2}{x^2-a^2}}\right) = \left(\frac{\beta}{m}\right),$$

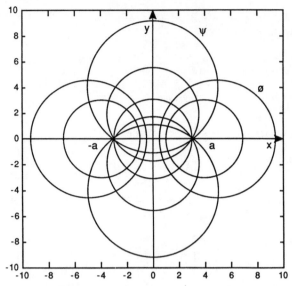

Fig. 2.28. Streamlines and equipotential lines.

$2ya/(x^2+y^2-a^2) = -\tan(\beta/m) = -(1/K)$ (say). Thus $x^2+y^2-a^2+2yKa = 0$, or $x^2+(y+aK)^2 = a^2+a^2K^2 = a^2(1+K^2)$, which for different values of K are circles having centres at $(0, -aK)$ and radii $a\sqrt{1+K^2}$.

(c) The fluid speed is given by

$$\frac{dW}{dz} = \frac{m}{z+a} - \frac{m}{z-a} = \frac{-2ma}{z^2-a^2} = \frac{-2ma}{r^2\,e^{2i\theta}-a^2} = \frac{2ma}{a^2-r^2\,e^{2i\theta}}$$

$$= \frac{2ma}{a^2-r^2\cos 2\theta - ir^2\sin 2\theta},$$

and therefore

$$\left|\frac{dW}{dz}\right| = \frac{2ma}{|a^2-r^2\cos 2\theta - ir^2\sin 2\theta|}$$

$$= \frac{2ma}{\sqrt{a^4+r^4\cos^2 2\theta - 2a^2r^2\cos 2\theta + r^4\sin^2 2\theta}}$$

$$= \frac{2ma}{\sqrt{a^4+r^4-2a^2r^2\cos 2\theta}}.$$

Example 2.7

Two sources each of strength m are placed at points $(-a, 0)$ and $(a, 0)$ and a sink of strength $2m$ is placed at the origin. Show that the streamlines are the curves $(x^2 + y^2)^2 = a^2(x^2 - y^2 + \lambda xy)$ where λ is a variable parameter.

Solution

The complex potential due to a source at $z = -a$ of strength m is $m \ln(z + a)$ (Fig. 2.29). The complex potential due to a source at $z = a$ of strength m is $m \ln(z - a)$. The complex potential due to a sink at $z = 0$ of strength $2m$ is $-2m \ln(z)$. Therefore by superposition, the total complex potential is

$$W = m \ln(z + a) + m \ln(z - a) - 2m \ln(z) = m \ln\left(\frac{(z + a)(z - a)}{z^2}\right).$$

Now let $z + a = r_1 e^{i\theta_1}$, $z - a = r_2 e^{i\theta_2}$, $z = r e^{i\theta}$, where

$$r_1 = \sqrt{(x + a)^2 + y^2}, \qquad r_2 = \sqrt{(x - a)^2 + y^2}, \qquad r = \sqrt{x^2 + y^2},$$

and $\theta_1 = \tan^{-1}[y/(x + a)]$, $\theta_2 = \tan^{-1}[y/(x - a)]$, $\theta = \tan^{-1}(y/x)$. Thus we have

$$W = \phi + i\psi = m \ln\left(\frac{r_1 r_2}{r^2} e^{i(\theta_1 + \theta_2 - 2\theta)}\right)$$

$$= m \ln\left(\frac{r_1 r_2}{r^2}\right) + im(\theta_1 + \theta_2 - 2\theta).$$

Therefore $\phi = m \ln(r_1 r_2 / r^2)$ is the velocity potential and $\psi = m(\theta_1 + \theta_2 - 2\theta)$ is the stream function.

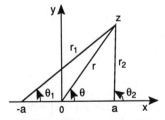

Fig. 2.29. Complex potential.

From the above expressions for θ_1 and θ_2 we have

$$\theta_1 + \theta_2 - 2\theta = \tan^{-1}\left(\frac{y}{x+a}\right) + \tan^{-1}\left(\frac{y}{x-a}\right) - 2\tan^{-1}\left(\frac{y}{x}\right)$$

or

$$\theta_1 + \theta_2 - 2\theta = \tan^{-1}\left(\frac{\dfrac{2xy}{x^2-a^2}}{1-\dfrac{y^2}{x^2-a^2}}\right) - \tan^{-1}\left(\frac{2y/x}{1-y^2/x^2}\right)$$

$$= \tan^{-1}\left(\frac{2xy}{x^2-y^2-a^2}\right) - \tan^{-1}\left(\frac{2xy}{x^2-y^2}\right)$$

$$= \tan^{-1}\left(\frac{2xya^2}{(x^2-y^2)(x^2-y^2-a^2)+4x^2y^2}\right).$$

The streamlines are given by

$$(x^2-y^2)(x^2-y^2-a^2) + 4x^2y^2 = 2a^2xy(\text{constant})$$

or

$$(x^2-y^2)^2 + 4x^2y^2 = a^2(x^2-y^2+\lambda xy)$$

or $(x^2+y^2)^2 = a^2(x^2-y^2+\lambda xy)$, where λ is a variable parameter.

Example 2.8

Prove that the pressure distribution p, around a circular cylinder fixed with its axis orthogonal to a uniform stream, V_0, is given by $p = \frac{1}{2}\rho V_0^2(1 - 4\sin^2\theta)$.

Solution

The complex velocity potential is $W(z) = V_0[z + (a^2/z)]$. Therefore

$$\phi + i\psi = V_0\left(r\,e^{i\theta} + \frac{a^2}{r}e^{-i\theta}\right)$$

$$= V_0\left[\left(r\cos\theta + \frac{a^2}{r}\cos\theta\right) + i\left(r\sin\theta - \frac{a^2}{r}\sin\theta\right)\right].$$

Thus $\phi = V_0[r + (a^2/r)]\cos\theta$ and $\psi = V_0[r - (a^2/r)]\sin\theta$.

By Cauchy–Riemann conditions, $V_r = \partial\phi/\partial r = (1/r)(\partial\psi/\partial\theta)$ and $V_\theta = (1/r)(\partial\phi/\partial\theta) = -(\partial\psi/\partial r)$, where V_r is the radial velocity component, V_θ the tangential velocity component.

But, on the surface of the cylinder, $V_r = 0$. Therefore

$$V_\theta = \left(\frac{1}{r}\frac{\partial\phi}{\partial\theta}\right)_{r=a} = -2V_0\sin\theta.$$

From Bernoulli's equation

$$p + \rho\frac{\partial\phi}{\partial t} + \frac{1}{2}\rho\left[\left(\frac{\partial\phi}{\partial x}\right)^2 + \left(\frac{\partial\phi}{\partial y}\right)^2\right] = C,$$

we obtain $p + \rho(\partial\phi/\partial t) + (\rho/2)q^2 = C$.

For steady flows $\partial\phi/\partial t = 0$, and hence $p + (\rho/2)q^2 = C$. We know that, at infinity, where $r\to\infty$, $q = V_0$ and $p = 0$. Therefore $C = (\rho/2)V_0^2$. Thus $p = \frac{1}{2}\rho(V_0^2 - q^2)$. However, $q^2 = V_\theta^2$ at $r = a$ and hence $p = \frac{1}{2}\rho V_0^2(1 - 4\sin^2\theta)$.

2.13 Exercises

1. Consider a region of space filled with a moving fluid. Let the density of the fluid at any point (x, y, z) at time t be $\rho(x, y, z, t)$ and let the particle instantaneously at the point (x, y, z) have velocity components u, v, and w, respectively, in the direction of the coordinate axes. By considering the flow through the boundaries of an infinitesimal region, of dimensions Δx, Δy, and Δz, show that the velocity components satisfy the equation of continuity

$$\frac{\partial\rho}{\partial t} + \frac{\partial}{\partial x}(\rho u) + \frac{\partial}{\partial y}(\rho v) + \frac{\partial}{\partial z}(\rho w) = 0.$$

2. Show that the acceleration components parallel to the cylindrical coordinate set (r, θ, z) are $(Dv_r/Dt) - (v_\theta^2/r)$, $(Dv_\theta/Dt) + (v_r v_\theta/r)$, Dv_z/Dt, where

$$\frac{D}{Dt} = \frac{\partial}{\partial t} + v_r\frac{\partial}{\partial r} + \frac{v_\theta}{r}\frac{\partial}{\partial\theta} + v_z\frac{\partial}{\partial z}.$$

3. Show that when Laplace's equation in Cartesian coordinates

$$\nabla^2 \phi = \phi_{xx} + \phi_{yy} + \phi_{zz} = 0$$

is transformed into cylindrical coordinates, by means of the substitutions $x = r \cos \theta$, $y = r \sin \theta$, $z = z$, it becomes

$$\frac{\partial^2 \phi}{\partial r^2} + \frac{1}{r} \frac{\partial \phi}{\partial r} + \frac{1}{r^2} \frac{\partial^2 \phi}{\partial \theta^2} + \frac{\partial^2 \phi}{\partial z^2} = 0,$$

where ϕ is defined to be the velocity potential.

4. Show that when Laplace's equation in Cartesian coordinates is transformed into spherical coordinates, by means of the substitutions $x = r \sin \theta \cos \phi$, $y = r \sin \theta \sin \phi$, $z = r \cos \theta$, it becomes

$$\frac{\partial^2 \Phi}{\partial r^2} + \frac{2}{r} \frac{\partial \Phi}{\partial r} + \frac{1}{r^2} \frac{\partial^2 \Phi}{\partial \theta^2} + \frac{\cot \theta}{r^2} \frac{\partial \Phi}{\partial \theta} + \frac{1}{r^2 \sin \theta} \frac{\partial^2 \Phi}{\partial \phi^2} = 0.$$

5. A family of surfaces $F(x, y, z) = \lambda$ is a family of equipotentials in free space. Show that $\nabla^2 \lambda / (\text{grad } \lambda)^2$ must be a function of λ only and determine the potential in terms of λ. If the family consists of the surfaces $r_2 - r_1 = \lambda$, where $r_1^2 = (x - c)^2 + y^2 + z^2$ and $r_2^2 = (x + c)^2 + y^2 + z^2$, c being a known constant, verify that the condition is satisfied and find the potential.

6. Find the velocity potential and stream function for the flow whose complex potential is given by $W = \cosh^{-1}(z)$. Show that this can be interpreted as the flow through an aperture of width $2b$ in an infinite plate.

7. Find the complex potential for the flow consisting of a source of strength m at the origin superimposed on a uniform flow of strength unity in the direction of the positive y-axis. Determine the velocity potential and the stream function and plot the streamlines.

8. Find the complex potential, the velocity potential, and the stream function for a source and a sink of strength $2m$ located, respectively, at $z = -1$ and $z = 1$, in a uniform stream of strength unity in the direction of the positive x-axis.

9. A source located at $z = 1$ and a sink at $z = -1$ of equal strength m

are between the parallel lines $y = \pm 1$. Show that the complex potential for the fluid motion is

$$W = m \ln\left(\frac{e^{\pi(z+1)} - 1}{e^{\pi(z-1)} - 1}\right).$$

10. Find the complex potential for the flow external to the cylinder $|z| = a$ due to a source of strength m at $z = 2a$.
(a) Prove that the force on the cylinder is $(\rho m^2)/(12\pi a)$; find its direction.
(b) Prove that $\int_C (dW/dz)^2 dz = 0$, where C is the contour $|z| = 3a$.
(c) Hence show that, in this flow, the equal and opposite force which acts on the source is given by Blasius's expression evaluated for a contour which encloses the source alone.

11. Prove that the transformation $\xi = e^{\pi z/2a}$ transforms a strip of breadth $2a$ into a half plane.
   Consider the case where incompressible fluid fills the space between two infinite parallel planes a distance $2a$ apart. Two line sources of strengths $+m$ and $-m$, respectively, are parallel to each other and lie midway between the two planes a distance $2b$ apart. Show that the complex potential function is

$$W = m \ln\left(\frac{\exp(\pi z/a) + \exp(\pi b/a)}{\exp(\pi z/a) + \exp(-\pi b/a)}\right).$$

12. Show that $W = K \ln(z^2 - a^2)$ gives the motion due to a two-dimensional source in the presence of a fixed wall, and, by using the transformation given by $d\xi/dz = [B/(z^2 - 1)^{1/2}]$, obtain the solution for such a source in a semi-infinite rectangle.

13. There is a source at A and a sink of equal strength at B. AB is the direction of a uniform stream. Determine the form of the streamlines. Given A is $(a, 0)$ and B is $(-a, 0)$ and the ratio of the flow issuing from A in unit time to the speed of the stream is $2\pi b$, show that the stream function is $\Psi = Vy - Vb \tan^{-1}[2ay/(x^2 + y^2 - a^2)]$. Also show that the length, $2l$, and the breadth, $2d$, of the closed wall that forms part of the dividing streamline is given by $l = \sqrt{a^2 + 2ab}$, $\tan(d/b) = 2ad/(d^2 - a^2)$ and the locus of the point at which the speed is equal to that of the stream is $x^2 - y^2 = a^2 + ab$.

14. A source is placed midway between two planes whose distance from one another is $2a$. Find the equation of the streamlines when the

motion is in two dimensions; show that those particles which are at an infinite distance from the source, and $\frac{1}{2}a$ from one of the boundaries, were issued from the source at an angle of $\pi/4$ with it.

15. The irrotational motion in two dimensions of a fluid bounded by the lines $y = \pm b$ is due to a doublet of strength μ at the origin, the axis of the doublet being in the positive direction of the x-axis. Prove that the motion is given by $W = (\pi\mu/2b)\coth(\pi z/2b)$. Show that the points where the fluid is moving parallel to the y-axis lie on the curve $\cosh(\pi x/b) = \sec(\pi y/b)$.

16. A wide stream of velocity U flows past a thin obstacle of length c which projects perpendicularly from a straight bank. Apply the Schwarz–Christoffel transformation to obtain the solution in the form $W^2 = U^2(z^2 + c^2)$. Find the pressure at any point of the obstacle and show that it becomes negative if $y > c(1+k)^{\frac{1}{2}}(1+2k)^{\frac{1}{2}}$, where $k = U^2/2p_0$, p_0 being the pressure at infinity.

17. Show that the complex potential $W = m\ln[\sinh(\pi z/(2a))]$ gives the flow from a large vase, of breadth $2a$, through a small hole in the centre of its base. Trace the general form of the streamlines, and prove that at a distance from the base greater than its breadth the flow is sensibly parallel to the walls of the vessel.

18. Show that the transformations $z = (a/\pi)(\sqrt{t^2-1} - \sec^{-1} t)$, $t = \exp[-(W\pi/aV)]$, where $z = x + iy$, $W = \phi + i\psi$, give the velocity potential ϕ and the stream function ψ for the flow of a straight river of breadth a, running with velocity V at right angles to the straight shore of an otherwise unlimited sheet of water into which it flows. By treating the motion as two dimensional, show that the real axis in the t-plane corresponds to the whole boundary of the liquid.

References

Aris, R. (1962). *Vectors, Tensors and the Basic Equations of Fluid Dynamics*, Prentice Hall, Englewood Cliffs, New Jersey.

Batchelor, G. K. (1967). *An Introduction to Fluid Dynamics*, Cambridge University Press, Cambridge.

Curle, N. and Davies, H. (1968). *Modern Fluid Dynamics*, Vol. 1, Van Nostrand, London.

Jeffreys, H. (1931). Tidal friction in shallow seas. *Philos. R. Soc.*, A 211, 239–264.

Lamb, H. (1945). *Hydrodynamics*, 6th edn, Dover, New York, Cambridge University Press, Cambridge.

Lighthill, J. (1986). *An Informal Introduction to Theoretical Fluid Mechanics*, Oxford University Press, Oxford.

Milne-Thomson, L. M. (1960). *Theoretical Hydrodynamics*, 4th edn, Macmillan, New York.

Phillips, O. M. (1966). *The Dynamics of the Upper Ocean*, Cambridge University Press, Cambridge.

Rosenhead, L. (ed.) (1963). *Laminar Boundary Layers*, Oxford University Press, Oxford.

Wylie, C. R. and Barrett, L. C. (1982). *Advanced Engineering Mathematics* McGraw-Hill, New York.

3
Solution techniques of partial differential equations

3.1 Introduction

We saw in the previous chapter that the motion of a fluid particle can be described by partial differential equations. In the study of potential wave theory, partial differential equations play a central role. We have seen, for example, how the velocity potential and the stream function, in fluid flow, are described by Laplace's equation. In the following chapters the main topics will be gravity waves and tidal waves. Here again, the physics of the problems will be described by a set of partial differential equations. Thus, for the proper study of the above phenomena, it is imperative that we master the solution techniques of partial differential equations.

In this chapter, our study will be directed to solution techniques which relate to physical situations described by the main types of partial differential equations, namely hyperbolic, parabolic and elliptic. From our elementary knowledge we know that the wave equation is of the hyperbolic type, the heat conduction equation is of the parabolic type and Laplace's equation is of the elliptic type. For surface wave problems we will solve Laplace's equation for the velocity potential; for one-dimensional tidal wave problems we will need d'Alembert's solution of the wave equation; for progressive waves, the method of characteristics will be used; and finally, for standing wave solutions, the method of separation of variables will be required.

3.2 Classifications of partial differential equations

In general, a second-order partial differential equation may be written as

$$A(x,y)u_{xx} + 2B(x,y)u_{xy} + C(x,y)u_{yy} = f(x,y,u,u_x,u_y) \qquad (3.1)$$

where $u(x, y)$ is the dependent variable and x, y are the independent variables. The coefficients A, B and C are, in general, functions of x and y.

The subscripts are used to define the partial derivatives, for example $u_{xx} = \partial^2 u / \partial x^2$, etc.

For simplicity, let us consider first the possibility of finding solutions of the form $u = F(x + \lambda y)$ with parameter λ for the equation

$$A u_{xx} + 2 B u_{xy} + C u_{yy} = 0 \tag{3.2}$$

where A, B and C are constants. Substituting our tentative solution into (3.2) we obtain $AF''(x + \lambda y) + 2BF''(x + \lambda y)\lambda + CF''(x + \lambda y)\lambda^2 = 0$ which will be an identity if, and only if

$$C\lambda^2 + 2B\lambda + A = 0. \tag{3.3}$$

Solving eqn (3.3) yields

$$\lambda = \frac{-B \pm \sqrt{B^2 - CA}}{C}. \tag{3.4}$$

Thus we see that the solutions of the form $F(x + \lambda y)$ exist, corresponding to the roots of λ, which may be real and distinct, or repeated and real, or complex.

From relation (3.4) we have

(i) λ has two real distinct roots if $B^2 - CA > 0$,
(ii) λ has two equal roots if $B^2 - CA = 0$,
(iii) λ has two complex roots if $B^2 - CA < 0$.

By analogy with the criterion for a conic to be hyperbolic, parabolic or elliptic, (3.2) is said to be of hyperbolic type if $B^2 - CA > 0$, parabolic type if $B^2 - CA = 0$, or elliptic type if $B^2 - CA < 0$.

In each case, the loci $x + \lambda_1 y = C_1$ and $x + \lambda_2 y = C_2$, where λ_1 and λ_2 are the roots of (3.3), are called the *characteristic curves*, or simply the *characteristics*, of (3.2). The characteristics define two families of parallel lines in the xy-plane.

Let us consider $x + \lambda y = C_1$ such that $dy/dx = -(1/\lambda)$. Now eliminating λ in (3.3) by the slope (dy/dx), we obtain the following differential equation:

$$A\left(\frac{dy}{dx}\right)^2 - 2B\left(\frac{dy}{dx}\right) + C = 0. \tag{3.5}$$

This equation is usually known as the characteristic equation.

More generally, (3.1) is said to be hyperbolic, parabolic or elliptic throughout a region D, respectively, according as $B^2(x, y) - A(x, y)C(x, y) > 0$, $B^2(x, y) - A(x, y)C(x, y) = 0$, or $B^2(x, y) - A(x, y)C(x, y) < 0$ at all points of D.

Furthermore, generalizing the property expressed in (3.5), we have

$$A(x,y)\left(\frac{dy}{dx}\right)^2 - 2B(x,y)\left(\frac{dy}{dx}\right) + C(x,y) = 0. \tag{3.6}$$

Solving (3.6) in terms of dy/dx yields

$$\frac{dy}{dx} = \frac{B \pm \sqrt{B^2 - AC}}{A}. \tag{3.7}$$

Equation (3.7) represents a pair of ordinary differential equations, the solution of which exists in the following forms: $\phi(x, y) = C_1$, and $\psi(x, y) = C_2$. These two solutions are usually said to be characteristic of the given partial differential equation (see Fig. 3.1).

The simplest examples of hyperbolic, parabolic and elliptic partial differential equations are, respectively, the wave equation $u_{tt} = a^2 u_{xx}$, the heat equation $u_t = a^2 u_{xx}$ and Laplace's equation $u_{xx} + u_{yy} = 0$.

Using the concept of characteristics, a second-order linear partial differential equation can be reduced to one of the above simple equations. The techniques to reduce a general second-order equation to standard simple form are as follows.

(I) Hyperbolic class

In this case, (3.1) will be hyperbolic if $B^2 - AC > 0$. The characteristics will be given by the solution of the following pair of ordinary differential equations: $dy/dx = (B \pm \sqrt{B^2 - AC})/A$. Solving these equations we obtain $\phi(x, y) = C_1$, and $\psi(x, y) = C_2$. Thus the characteristics are given by

$$\zeta = \phi(x, y), \qquad \eta = \psi(x, y). \tag{3.8}$$

Under this transformation, the original eqn (3.1) can be reduced to the standard form

$$u_{\zeta\eta} = g(\zeta, \eta, u, u_\zeta, u_\eta). \tag{3.9}$$

(II) Parabolic class

In this case, (3.1) will be parabolic if $B^2 - CA = 0$. The characteristics will be given by the solution of the following ordinary differential equation: $dy/dx = B/A$. Solving this equation gives $\phi(x, y) = C_1$. Thus the characteristics in this case are given by

$$\zeta = x, \qquad \eta = \phi(x, y). \tag{3.10}$$

Under the transformation, (3.1) can be transformed to the standard form

$$u_{\zeta\zeta} = g(\zeta, \eta, u, u_\zeta, u_\eta). \tag{3.11}$$

(III) Elliptic class

In this case, (3.1) will be elliptic if $B^2 - CA < 0$. The characteristics will be given by the solution of the following pair of ordinary differential equations: $dy/dx = (B \pm i\sqrt{CA - B^2})/A$. Taking the positive sign and solving, we get $\phi(x, y) = C_1$. Here $\phi(x, y)$ is a complex quantity. The characteristics in this case are given by

$$\zeta = \mathrm{Re}\{\phi(x, y)\}. \qquad \eta = \mathrm{Im}\{\phi(x, y)\} \tag{3.12}$$

Under this transformation, (3.1) reduces to the standard form

$$u_{\zeta\zeta} + u_{\eta\eta} = g(\zeta, \eta, u, u_\zeta, u_\eta). \tag{3.13}$$

Alternatively, considering both positive and negative signs and then solving the differential equations, we obtain $\phi(x, y) = C_1$, $\psi(x, y) = C_2$. These two functions are complex conjugates. The characteristics are given by

$$\zeta = \phi(x, y), \qquad \eta = \psi(x, y) = \overline{\phi}(x, y). \tag{3.14}$$

Both ζ and η are complex quantities. Under this transformation, (3.1) reduces to the standard form

$$u_{\zeta\eta} = g(\zeta, \eta, u, u_\zeta, u_\eta). \tag{3.15}$$

3.3 D'Alembert's solution of the wave equation

The characteristics of parabolic and elliptic equations are found to be of little importance in applied science problems. On the other hand, the two

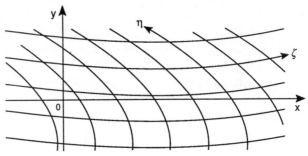

Fig. 3.1. Characteristics plane ($\zeta\eta$).

families of characteristics of hyperbolic equations, being real and distinct, are of considerable practical value. In one-dimensional progressive wave propagation in particular, consideration of the characteristics can give us a good deal of information about the propagation of wave fronts. Therefore we shall discuss the progressive wave solution of the one-dimensional wave equation. This solution was first obtained by d'Alembert, a French mathematician, and was subsequently attributed to him.

The one-dimensional wave equation is given by

$$u_{tt} = a^2 u_{xx}. \tag{3.16}$$

Comparing this equation with (3.1), we have $A = 1, B = 0, C = -a^2$. Then $B^2 - CA = 0 - (-a^2)(1) = a^2 > 0$. Hence (3.16) is of hyperbolic type.

The characteristics are given by the solution of the ordinary differential equation $A(dx/dt)^2 - 2B(dx/dt) + C = 0$ which reduces to $(dx/dt)^2 = a^2$ and hence $dx/dt = \pm a$. Integrating the above equations yields

$$x - at = C_1, \qquad x + at = C_2. \tag{3.17}$$

Hence the characteristics are defined by

$$\zeta = x - at, \eta = x + at. \tag{3.18}$$

Under this transformation, eqn (3.16) can be reduced to the following equation:

$$u_{\zeta\eta} = 0. \tag{3.19}$$

Integrating (3.19) partially with respect to η, we obtain $u_\zeta = f'(\zeta)$. Integrating partially again, with respect to ζ, we obtain

$$u = f(\zeta) + g(\eta) = f(x - at) + g(x + at) \tag{3.20}$$

where f and g are arbitrary functions. This solution is referred to as d'Alembert's solution of the wave equation.

The physical interpretation of these terms is quite interesting. The functions f and g represent two progressive waves travelling in opposite directions with speed a.

To see this, let us consider the solution $u = f(x - at)$. At $t = 0$, it defines the wave $u = f(x)$, and at any later time, $t = t_1$, it defines the curve $u = f(x - at_1)$. But these curves are identical (see Fig. 3.2) except that the latter is translated to the right a distance equal to at_1.

Thus the entire configuration moves along the positive direction of the x-axis a distance of at_1 in time t_1.

The velocity with which the wave is propagated is therefore $v = at_1/t_1 = a$. Similarly, the function $g(x + at)$ defines a wave which moves in the negative direction of the x-axis with constant velocity a.

The total displacement is, of course, the algebraic sum of these two travelling waves.

D'Alembert's solution (3.20) is a very convenient representation of progressive waves which travel large distances through a uniform medium.

Let us consider the following two initial conditions for a uniform medium extending over $-\infty < x < \infty$. The displacement and the velocity are, respectively, given by

$$u(x,0) = \phi(x) \qquad \text{and} \qquad u_t(x,0) = \psi(x), \tag{3.21}$$

where ϕ and ψ are arbitrarily assigned differentiable functions for $-\infty < x < \infty$. It can be easily confirmed from the form of the solution that will be subsequently given in eqn (3.25) that ϕ must be at least twice differentiable and ψ at least once differentiable. The initial condition (3.21) is known as Cauchy data in the theory of partial differential equations. Using these two conditions, we obtain from d'Alembert's solution (3.20)

$$f(x) + g(x) = \phi(x) \qquad \text{and} \qquad -af'(x) + ag'(x) = \psi(x) \tag{3.22}$$

for all values of x.

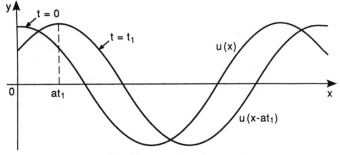

Fig. 3.2. A progressive wave.

Hence integrating the second equation, $-f(x) + g(x) = 1/a \int_{x_0}^x \psi(s)\mathrm{d}s + A$, and so

$$f(x) = \frac{1}{2}\left[\phi(x) - \frac{1}{a}\int_{x_0}^x \psi(s)\mathrm{d}s - A \right] \tag{3.23}$$

and

$$g(x) = \frac{1}{2}\left[\phi(x) + \frac{1}{a}\int_{x_0}^x \psi(s)\mathrm{d}s + A \right] \tag{3.24}$$

where A is an integration constant. Substituting these expressions into (3.20), we can write

$$u = \frac{1}{2}[\phi(x - at) + \phi(x + at)] + \frac{1}{2a}\int_{x-at}^{x+at} \psi(s)\mathrm{d}s. \tag{3.25}$$

This solution is usually called the *progressive wave* solution. The method of separation of variable yields the *standing wave* solution and will be considered later. The solution (3.25) which satisfies the Cauchy data is called *d'Alembert's formula* and is an important result since it provides valuable information about the nature of the solution u.

Remark

There are two conclusions which can be made from the wave equation (3.16) with its initial condition (3.21). The first conclusion that may be drawn from equation (3.25) is that the Cauchy data (3.21) is sufficient to specify a unique solution of the wave equation. This can be verified by considering the existence of two solutions u and v to eqn (3.16) such that both satisfy the same Cauchy data (3.21). Then because of the linearity of the wave equation the function $w = u - v$ will also be a solution of eqn (3.16), and it will have for its Cauchy data on the initial line $t = 0$ the homogeneous conditions $w(x,0) = 0$ and $w_t(x,0) = 0$. Thus d'Alembert's formula (3.25) shows that $w(x,t) = 0$ for all x, and $t \geq 0$, so that $u = v$ and hence the solution is unique.

The second conclusion is concerned with the concept of the *domain of dependence* and the *range of influence* of a point. These ideas are illustrated in Fig. 3.3(a) and 3.3(b). To understand these two concepts let us consider the solution which depends on the Cauchy data at a general point (x_0, t_0) on the (x,t) plane with $t > 0$.

Then from eqn (3.25) the solution at $Q(x_0, t_0)$ is seen to be determined only by Cauchy data given on the finite interval $x_0 - at_0 < x < x_0 + at_0$ of the initial line. More precisely, the solution is only influenced by the

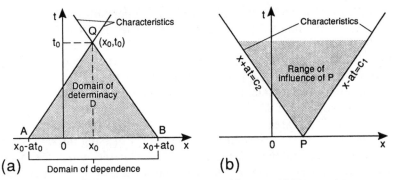

Fig. 3.3. (a) Domain of dependence of Q, (b) range of influence of P

functional values of ϕ at the end of this interval, and by the function ψ over the entire interval by virtue of the integral term in eqn (3.25). This interval $x_0 - at_0 < x < x_0 + at_0$ along $t = 0$ on the initial line is called the *domain of dependence* of the point $Q(x_0, t_0)$. The triangular region (shaded area) with the point $Q(x_0, t_0)$ as vertex and the domain of dependence of this point as base is called the *domain of determinacy* that is associated with the domain of dependence. An immediate consequence of eqn (3.25) is that two sets of different initial data that coincide only in some interval D of the initial line $t = 0$ define two different solutions which are, however, identical in their common domain of determinacy associated with D.

If a domain of determinacy on which initial data is given is reduced to a single point P of the initial line, then the solution will only be determined at that point. However, the data at P will influence, but not determine, the solution at points in the half plane $t > 0$ that lie in an open triangular region with P as vertex and with sides comprising the straight lines drawn through P in the direction of increasing time with gradients $dx/dt = \pm a$. It is for this region that the region so defined is called the *range of influence* of the point P.

As an example we consider the solution of the following initial value problem:

$$u_{tt} = a^2 u_{xx}, \qquad u(x,0) = \sin x, \qquad u_t(x,0) = \cos x.$$

Using d'Alembert's formula we find that the solution can be written as

$$u(x,t) = \frac{1}{2}[\sin(x + at) + \sin(x - at)] + \frac{1}{2a}\int_{x-at}^{x+at} \cos \lambda \, d\lambda$$

$$= \sin x \cos at + \frac{1}{a}\cos x \sin at.$$

According to eqn (3.25), the value of $u(x_0, t_0)$ depends on the initial data ϕ and ψ in the interval $[x_0 - at_0, x_0 + at_0]$ which is cut out of the initial line by the two characteristics $x - at = $ constant and $x + at = $ constant with slope $\pm 1/a$ passing through the point (x_0, t_0). This interval on the line $t = 0$ is the domain of dependence as shown in Fig. 3.3(a). On the other hand it follows from d'Alembert's solution that if an initial displacement or an initial velocity is located in a small neighbourhood (x_0, t_0) it can influence only the area $t > t_0$ bounded by the two characteristics $x - at = $ constant and $x + at = $ constant with slope $\pm 1/a$ passing through the point (x_0, t_0) as shown in Fig. 3.3(b). This means that the initial displacement propagates at the speed a whereas the effect of the initial velocity propagates at all speeds not exceeding a in the infinite sector, which is called the range of influence of the point (x_0, t_0).

3.4 Method of characteristics

We have seen how to classify a given second-order partial differential equation. General solutions of a wide range of partial differential equations can be obtained by using the method of characteristics, and we have already demonstrated d'Alembert's solution by using this method. It is a powerful technique as is shown by the following examples. These examples are selected to give the reader some insight of the technique.

Example 3.1

Find the characteristics of the following equation. Reduce it to the standard form and obtain its solution.

$$3u_{xx} + 10u_{xy} + 3u_{yy} = 0. \tag{3.26}$$

Solution

Comparing this equation with (3.1) we have $A = 3, B = 5, C = 3$. Then $B^2 - CA = 16 > 0$. Therefore the equation is of hyperbolic type.

The characteristics are given by the solution of the ordinary differential equation $A(dy/dx)^2 - 2B(dy/dx) + C = 0$. Hence $3(dy/dx)^2 - 10(dy/dx) + 3 = 0, dy/dx = 5 \pm 4/3 = 3, 1/3$. Integrating the above equations we obtain $3x - y = C_1$, and $x - 3y = C_2$. Hence the characteristics are defined by

$$\zeta = 3x - y \quad \text{and} \quad \eta = x - 3y. \tag{3.27}$$

Thus under this transformation

$$u_x = u_\zeta \zeta_x + u_\eta \eta_x = 3u_\zeta + u_\eta$$

$$u_y = u_\zeta \zeta_y + u_\eta \eta_y = -u_\zeta - 3u_\eta$$

$$u_{xx} = 3\{u_{\zeta\zeta}\zeta_x + u_{\zeta\eta}\eta_x\} + \{u_{\eta\zeta}\zeta_x + u_{\eta\eta}\eta_x\}$$

$$= 9u_{\zeta\zeta} + 6u_{\zeta\eta} + u_{\eta\eta}$$

$$u_{xy} = 3\{u_{\zeta\zeta}\zeta_y + u_{\zeta\eta}\eta_y\} + \{u_{\eta\zeta}\zeta_y + u_{\eta\eta}\eta_y\} = -3u_{\zeta\zeta} - 10u_{\zeta\eta} - 3u_{\eta\eta}$$

$$u_{yy} = -\{u_{\zeta\zeta}\zeta_y + u_{\zeta\eta}\eta_y\} - 3\{u_{\eta\zeta}\zeta_y + u_{\eta\eta}\eta_y\} = u_{\zeta\zeta} + 6u_{\zeta\eta} + 9u_{\eta\eta}.$$

Inserting the above into (3.26) yields

$$u_{\zeta\eta} = 0. \tag{3.28}$$

Integrating twice, as before, we obtain

$$u = f(\zeta) + g(\eta) = f(3x - y) + g(x - 3y) \tag{3.29}$$

where f and g are arbitrary functions.

Example 3.2

Find the characteristics of the equation

$$yu_{xx} + u_{yy} = 0 \tag{3.30}$$

in the lower half plane, $y < 0$. Then reduce it to the standard form.

Solution

Comparing this equation with (3.1), we have $A = y, B = 0, C = 1$. Then $B^2 - CA = 0^2 - y = -y > 0$ in the lower half plane, $y < 0$. Hence (3.30) is of hyperbolic type.

The characteristics are given by the solution of the ordinary differential equations $dy/dx = (B \pm \sqrt{B^2 - CA})/A = \pm(-y)^{-1/2}$. Integrating these equations gives

$$x + \tfrac{2}{3}(-y)^{3/2} = C_1, \qquad x - \tfrac{2}{3}(-y)^{3/2} = C_2, \tag{3.31}$$

and hence the characteristics are defined by

$$\zeta = x + \tfrac{2}{3}(-y)^{3/2} \qquad \text{and} \qquad \eta = x - \tfrac{2}{3}(-y)^{3/2}. \tag{3.32}$$

Thus under this transformation

$$u_x = u_\zeta \zeta_x + u_\eta \eta_x = u_\zeta + u_\eta$$

$$u_y = u_\zeta \zeta_y + u_\eta \eta_y = -(-y)^{1/2}(u_\zeta - u_\eta)$$

$$u_{xx} = u_{\zeta\zeta} + 2u_{\zeta\eta} + u_{\eta\eta}$$

$$u_{yy} = \tfrac{1}{2}(-y)^{-1/2}(u_\zeta - u_\eta) - (-y)^{1/2}\{u_{\zeta\zeta}\zeta_y + u_{\zeta\eta}\eta_y - u_{\zeta\eta}\zeta_y - u_{\eta\eta}\eta_y\}$$

$$= \tfrac{1}{2}(-y)^{-1/2}(u_\zeta - u_\eta) + (-y)\{u_{\zeta\zeta} - 2u_{\zeta\eta} + u_{\eta\eta}\}.$$

On substituting the above into (3.30), it can be reduced to

$$u_{\zeta\eta} - \frac{1}{6(\zeta - \eta)}(u_\zeta - u_\eta) = 0 \tag{3.33}$$

which is the required standard form.

It should be remembered that the given partial differential equation is due to Tricomi and is known as Tricomi's equation. This equation is elliptic for $y > 0$ ad hyperbolic for $y < 0$. For a detailed treatment, the reader is referred to Hellwig (1964).

Example 3.3

Determine the type of the following equation:

$$u_{xx} + u_{yy} = 0. \tag{3.34}$$

After reducing it to the hyperbolic form, deduce the formula $u(x, y)$ $= \tfrac{1}{2}\phi(x + iy) + \tfrac{1}{2}\overline{\phi(x + iy)}$ which expresses any harmonic function u as the real part of some analytic function ϕ of the complex variable $x + iy$.

Solution

In this case $A = 1$, $B = 0$, $C = 1$. Then $B^2 - CA = -1 < 0$. Therefore (3.34) is of elliptic type.

The characteristics are given by the solution of the ordinary differential equations $dy/dx = \pm i$ which on integration give $x + iy = C_1$, and $x - iy = C_2$. hence the characteristics are defined by

$$\zeta = x + iy \quad \text{and} \quad \eta = x - iy. \tag{3.35}$$

Thus under this transformation

$$u_x = u_\zeta \zeta_x + u_\eta \eta_x = u_\zeta + u_\eta, u_y = u_\zeta \zeta_y + u_\eta \eta_y = i(u_\zeta - u_\eta)$$

$$u_{xx} = u_{\zeta\zeta} + 2u_{\zeta\eta} + u_{\eta\eta}, u_{yy} = -(u_{\zeta\zeta} - 2u_{\zeta\eta} + u_{\eta\eta}).$$

After substituting the above into (3.34), it reduces to

$$u_{\zeta\eta} = 0. \tag{3.36}$$

Integrating partially with respect to η yields $u_\zeta = \frac{1}{2}\phi'(\zeta)$. Again integrating partially with respect to ζ gives $u = \frac{1}{2}\phi(\zeta) + \frac{1}{2}\psi(\eta)$ and therefore $u(x,y) = \frac{1}{2}\phi(x+iy) + \frac{1}{2}\psi(x-iy)$. But since $u(x,y)$ is real, then we must have $\overline{\psi(x-iy)} = \phi(x+iy)$ where $\overline{\phi}$ is the complex conjugate of ϕ.

Therefore

$$u(x,y) = \frac{1}{2}\phi(x+iy) + \frac{1}{2}\overline{\phi(x+iy)}. \tag{3.37}$$

Example 3.4

Using the method of characteristics find the general solutions of the following partial differential equation:

$$xu_{xy} + yu_{yy} = 0 \tag{3.38}$$

Solution

In this case $A = 0, B = x/2, C = y$. Then $B^2 - CA = x^2/4 > 0, x \neq 0$. Therefore, (3.38) is hyperbolic.

The characteristics are given by the solution of the following differential equation:

$$A\left(\frac{dy}{dx}\right)^2 - 2B\left(\frac{dy}{dx}\right) + C = 0. \tag{3.39}$$

Since $A = 0$, we rearrange this equation to give

$$C\left(\frac{dx}{dy}\right)^2 - 2B\left(\frac{dx}{dy}\right) + A = 0.$$

Therefore $(dx/dy)[y(dx/dy) - x] = 0$. Hence

$$\frac{dx}{dy} = 0, \qquad y\frac{dx}{dy} - x = 0. \tag{3.40}$$

On integration, we obtain $x = C_1$, and $y/x = C_2$. The characteristics are defined by

$$\zeta = x \qquad \text{and} \qquad \eta = \frac{y}{x}. \tag{3.41}$$

Thus under this transformation $u_x = u_\zeta - (y/x^2)u_\eta$, $u_{xy} = (1/x)u_{\zeta\eta} - (1/x^2)u_\eta - (y/x^3)u_{\eta\eta}$, $u_y = (1/x)u_\eta$, $u_{yy} = (1/x^2)u_{\eta\eta}$.
   On substituting the above into (3.38), it reduces to

$$u_{\zeta\eta} - \frac{1}{\zeta}u_\eta = 0. \tag{3.42}$$

Integrating (3.42) partially with respect to ζ yields $\ln(u_\eta) = \ln(\zeta) + \ln(g'(\eta))$. Therefore $u_\eta = \zeta g'(\eta)$.
   Again integrating partially with respect to η, $u = f(\zeta) + \zeta g(\eta)$, where f and g are arbitrary functions. In the original variables we have

$$u = f(x) + xg\left(\frac{y}{x}\right). \tag{3.43}$$

3.5 The method of separation of variables

In this section the method of separation of variables will be discussed; this is a powerful technique for solving boundary value problems which arise in engineering physics. This technique will be demonstrated with regard to the wave equation, the heat equation and Laplace's equation.

Example 3.5

The displacement of a vibrating string satisfies the wave equation as well as being a classical problem, the results of which can be applied to the problem of surface waves and long waves in the ocean. The progressive wave solution has already been obtained by the method of characteristics; here we will obtain the standing wave solution by the method of separation of variables.

The partial differential equation, together with the boundary and initial conditions, are, respectively, given by

$$u_{tt} = a^2 u_{xx}. \tag{3.44}$$

The boundary conditions are

$$x = 0: u(0,t) = 0; \; x = l: u(l,t) = 0 \tag{3.45}$$

and the initial conditions are at time $t = 0$:

$$u(x,0) = f(x) \quad \text{and} \quad u_t(x,0) = g(x). \tag{3.46}$$

Here u represents the displacement of the string which is fixed at its ends, $x = 0$ and $x = l$. u_t represents the velocity distribution of the string.

Solution

Let us consider the method of separation of variables. In this method we let

$$u(x,t) = X(x)T(t). \tag{3.47}$$

Equation (3.44) can be written as

$$\frac{T''}{T} = a^2 \frac{X''}{X} = k \tag{3.48}$$

where k is a separation constant. Therefore

$$T'' - kT = 0 \tag{3.49}$$

and

$$X'' - \frac{k}{a^2} X = 0. \tag{3.50}$$

The solution of the original partial differential equation will be determined by the solutions of the ordinary differential equations (3.49) and (3.50).

Assuming that we need to consider only physical solutions, there are three cases to investigate corresponding to $k = 0$, $k > 0$ and $k < 0$.

(I) If $k = 0$, the equations are $T'' = 0$ and $X'' = 0$. Therefore their solutions are $T = At + B$ and $X = Cx + D$, and hence $u(x,t) = X(x)T(t) = (Cx + D)(At + B)$. This solution is not periodic and therefore cannot describe the undamped vibration of a system. Hence, although product solutions exist for $k = 0$, they have no physical significance, and, therefore, we reject them. At the same time it can be noted that the given boundary conditions are satisfied only for the trivial case, i.e. $u(x,t) = 0$.

(II) If $k = \lambda^2 > 0$, the two differential equations and their solutions are

$$T'' - \lambda^2 T + 0, \qquad X'' - \left(\frac{\lambda}{a}\right)^2 X = 0$$

$$T = Ae^{\lambda t} + Be^{-\lambda t}, \qquad X = Ce^{(\lambda/a)x} + De^{-(\lambda/a)x}.$$

Hence

$$u(x,t) = X(x)T(t) = (Ce^{(\lambda/a)x} + De^{-(\lambda/a)x})(Ae^{\lambda t} + Be^{-\lambda t}).$$

This solution also is not acceptable from a physical point of view, since it cannot represent periodic motion. The case $k = \lambda^2$ is, therefore, rejected. At the same time the boundary conditions produce a trivial solution in which we are not interested.

(III) If $k = -\lambda^2 < 0$, the two differential equations and their solutions are $T'' + \lambda^2 T = 0$, $X'' + (\lambda/a)^2 X = 0$, and their solutions are $T = A \cos \lambda t + B \sin \lambda t$, $X = C \cos(\lambda x/a) + D \sin(\lambda x/a)$. From these solutions we obtain

$$u(x,t) = X(x)T(t) = \left(C \cos \frac{\lambda}{a} x + D \sin \frac{\lambda}{a} x\right)(A \cos \lambda t + B \sin \lambda t).$$

$$(3.51)$$

This solution is clearly periodic, repeating itself every time t increases by $2\pi/\lambda$. The solution represents a vibratory motion with period $2\pi/\lambda$, frequency $\lambda/2\pi$.

It now remains to find the value of λ and the constants A, B, C and D.

By using the boundary conditions (3.45), we find that $u(0,t) = 0 = C\{A \cos \lambda t + B \sin \lambda t\}$. This condition can obviously be satisfied if both A and B are zero. In this case we have a trivial solution, that is $u(x,t) = 0$ for all time and at all positions, and it is therefore of no interest to us. This leads us to the alternative, $C = 0$, which reduces (3.51) to the form $u(x,t) = D \sin(\lambda x/a)(A \cos \lambda t + B \sin \lambda t)$.

The second boundary condition of (3.45) requires that $u(l,t) = 0 = (D \sin \lambda l/a)(A \cos \lambda t + B \sin \lambda t)$. As before we reject the possibility that $A = 0 = B$. Moreover we cannot permit $D = 0$, since that too, with C already zero, leads to the trivial case. The only possibility which remains is that $\sin \lambda l/a = 0$, or $\lambda l/a = n\pi$, and therefore $\lambda_n = n\pi a/l, n = 1, 2, 3, \dots$. There is an infinity of values of λ; for each value a product solution of the wave equation exists.

Thus

$$u_n(x,t) = \left(\sin \frac{\lambda_n x}{a} \right) \{ A_n \cos \lambda_n t + B_n \sin \lambda_n t \}$$

$$= \sin \left(\frac{n\pi x}{l} \right) \left\{ A_n \cos \frac{n\pi at}{l} + B_n \sin \frac{n\pi at}{l} \right\}.$$

Since the given wave equation is linear, we may use the superposition principle to give us a Fourier expansion in the following way:

$$u(x,t) = \sum_{n=1}^{\infty} \sin \left(\frac{n\pi x}{l} \right) \left\{ A_n \cos \frac{n\pi at}{l} + B_n \sin \frac{n\pi at}{l} \right\} \tag{3.52}$$

where A_n and B_n are defined as the Fourier coefficients and are given in terms of the initial condition (3.46).

At $t = 0$, from the initial condition (3.46), we obtain

$$u(x,0) = f(x) = \sum_{n=1}^{\infty} A_n \sin \frac{n\pi x}{l}. \tag{3.53}$$

Now multiplying both sides of (3.53) by $\sin(n\pi x/l)$ and integrating with respect to x from 0 to 1, we have

$$A_n = \frac{2}{l} \int_0^l f(x) \sin \frac{n\pi x}{l} \, dx. \tag{3.54}$$

To determine the B_n, we note that

$$u_t(x,t) = \sum_{n=1}^{\infty} \sin \frac{n\pi x}{l} \left\{ -A_n \sin \frac{n\pi at}{l} + B_n \cos \frac{n\pi at}{l} \right\} \left(\frac{n\pi a}{l} \right).$$

Hence, putting $t = 0$, we have from the initial velocity condition (3.46)

$$u_t(x,0) = g(x) = \sum_{n=1}^{\infty} B_n\left(\frac{n\pi a}{l}\right)\sin\left(\frac{n\pi x}{l}\right). \tag{3.55}$$

Now multiplying both sides of (3.55) by $\sin(n\pi x/l)$ and integrating with respect to x from 0 to l, we obtain

$$B_n = \frac{2}{n\pi a}\int_0^l g(x)\sin\frac{n\pi x}{l}\,dx. \tag{3.56}$$

Therefore (3.52) is our solution where the Fourier coefficients A_n and B_n are given by (3.54) and (3.56).

It can be shown that when f and g vanish at $x = 0$, l and f and g are sufficiently differentiable, the infinite series solution (3.52) converges and satisfies (3.44), (3.45) and (3.46).

The solution presented in (3.52) can be regarded as the *standing wave* solution; the *progressive wave* solution (3.25) has already been derived. It is interesting to observe that the standing wave solution (3.52) can be expressed in the form of the progressive wave solution (3.25). We can show this as follows:

$$\begin{aligned}
u(x,t) &= \sum_{n=1}^{\infty}\left\{A_n\sin\frac{n\pi x}{l}\cos\frac{n\pi at}{l} + B_n\sin\frac{n\pi x}{l}\sin\frac{n\pi at}{l}\right\} \\
&= \sum_{n=1}^{\infty}\frac{A_n}{2}\left\{\sin\frac{n\pi(x+at)}{l} + \sin\frac{n\pi(x-at)}{l}\right\} \\
&\quad - \sum_{n=1}^{\infty}\frac{B_n}{2}\left\{\cos\frac{n\pi(x+at)}{l} - \cos\frac{n\pi(x-at)}{l}\right\}.
\end{aligned}$$

Thus

$$\begin{aligned}
u(x,t) &= \sum_{n=1}^{\infty}\left\{\frac{A_n}{2}\sin\frac{n\pi(x+at)}{l} - \frac{B_n}{2}\cos\frac{n\pi(x+at)}{l}\right\} \\
&\quad + \sum_{n=1}^{\infty}\left\{\frac{A_n}{2}\sin\frac{n\pi(x-at)}{l} + \frac{B_n}{2}\cos\frac{n\pi(x-at)}{l}\right\},
\end{aligned}$$

which can be expressed as

$$u(x,t) = \phi(x+at) + \psi(x-at). \tag{3.57}$$

Equation (3.57) is the progressive wave form of the solution.

Remark

Consider the three-dimensional wave equation

$$u_{tt} = a^2\{u_{xx} + u_{yy} + u_{zz}\}. \tag{3.58}$$

Solutions of (3.58) in Cartesian coordinates x, y, z and time t have a behaviour and properties which are very similar to those of the one-dimensional wave equation discussed above. This equation governs a wide variety of wave and vibration phenomena. Therefore it is natural to enquire whether the wave equation (3.58) possesses solutions similar to d'Alembert's solution (3.57) of the one-dimensional wave equation. It can be verified that $u = F(lx + my + nz - at)$ satisfies the wave equation (3.58) whenever $l^2 + m^2 + n^2 = 1$. So l, m and n may be interpreted as direction cosines. Here F is an arbitrary differentiable function and a is the speed of the wave front. Similarly, it may be confirmed that $u = G(lx + my + nz + at)$ also satisfies (3.58) where l, m and n are direction cosines. These solutions represent plane waves moving on opposite surfaces in the direction of the unit vector $\mathbf{n} = l\mathbf{i} + m\mathbf{j} + n\mathbf{k}$.

The equation of a plane with the unit normal $\mathbf{n} = l\mathbf{i} + m\mathbf{j} + n\mathbf{k}$ is given by $lx + my + nz - p = 0$. The line through the origin along the unit normal direction, $\mathbf{n}$, intersects the plane at a distance p from the origin. Suppose the point of intersection moves along the normal line with a speed a and passes through the origin at $t = 0$, so that $p = at$; then a plane moving along its fixed normal direction, $\mathbf{n}$, with speed a is given by the equation $lx + my + nx - at = 0$. The surface $lx + my + nz - at = \zeta$, where ζ is a constant, represents the parallel plane.

Example 3.6

Consider the one-dimensional heat conduction equation

$$u_t = a^2 u_{xx}, \tag{3.59}$$

with the boundary conditions

$$x = 0 : u(0, t) = 0; \ x = l : u(l, t) = 0 \tag{3.60}$$

and the initial condition at

$$t = 0 : u(x, 0) = f(x). \tag{3.61}$$

Here u represents the temperature at a given time of a thin metal bar, which is insulated except at its ends, $x = 0$ and $x = l$.

Solution

By using the method of separation of variables

$$u(x,t) = X(x)T(t) \tag{3.62}$$

eqn (3.59) can be written as

$$\frac{T'}{T} = a^2 \frac{X''}{X} = k, \tag{3.63}$$

that is

$$T' - kT = 0 \quad \text{and} \quad X'' - \frac{k}{a^2} = 0. \tag{3.64}$$

Because of the fact that T'/T is a function of t alone which equals $a^2 X''/X$, a function of x alone, the separation parameter k must be a constant.

Assuming that we need to consider only physical situations, there are three cases to investigate corresponding to $k = 0$, $k > 0$ and $k < 0$.

(I) If $k = 0$, the equations are $T' = 0$, $X'' = 0$ and their solutions are $T = A$, $X = Cx + D$ and hence $u(x,t) = A(Cx + D)$.

This solution cannot describe a damped oscillatory solution. Hence, although the product solution exists for $k = 0$, it should be rejected on physical grounds. However, it has a most definite physical significance as the steady-state solution for problems with nonhomogeneous boundary conditions. It is to be noted that in this case the homogeneous boundary conditions produce a trivial solution.

(II) If $k = \lambda^2 > 0$, for real λ, the two differential equations are $T' - \lambda^2 T = 0$, $X'' - (\lambda/a)^2 X = 0$ and their solutions are $T = Ae^{\lambda^2 t}$, $X = Ce^{(\lambda/a)x} + De^{-(\lambda/a)x}$ and hence $u(x,t) = X(x)T(t) = (Ce^{(\lambda/a)x} + De^{-(\lambda/a)x})Ae^{\lambda^2 t}$. This solution cannot describe damped oscillatory motion of a system because it is not periodic. Thus, from the physical point of view, the case $k = \lambda^2$ must be rejected. Also, it is worth mentioning here that the given homogeneous boundary conditions produce a trivial solution.

(III) If $k = -\lambda^2 < 0$, the two differential equations are $T' + \lambda^2 T = 0$, $X'' + (\lambda/a)^2 X = 0$ and their solutions are $T = Ae^{-\lambda^2 t}$, $X = C\cos(\lambda/a)x + D\sin(\lambda/a)x$. From these solutions

$$u(x,t) = X(x)T(t) = \left(C \cos \frac{\lambda}{a}x + D \sin \frac{\lambda}{a}x \right) Ae^{-\lambda^2 t}. \tag{3.65}$$

This is clearly a damped oscillatory solution, oscillating with respect to the x variable and damping with respect to time. Thus it describes a physical solution of the heat conduction problem considered here.

Now it remains to find the value of λ and the constants A, C and D. By using the boundary conditions (3.60) we find that $u(0,t) = 0 = CAe^{-\lambda^2 t}$. This condition can obviously be satisfied if either A or C is zero. If $A = 0$, we get the trivial solution, that is $u(x,t) = 0$ for all time and at all positions, and it is of no interest. Hence, C must be zero and therefore $u(x,t) = DA \sin(\lambda x/a)e^{-\lambda^2 t}$. The second boundary condition of (3.60) requires that $u(l,t) = 0 = DA \sin(\lambda l/a)e^{-\lambda^2 t}$. As before, we reject the possibility that $A = 0$; moreover, we cannot allow $D = 0$, since that too, with C already zero, leads to the trivial case. The only possibility which remains is that $\sin(\lambda l/a) = 0$. Hence $\lambda l/a = n\pi$ and therefore $\lambda_n = n\pi a/l$, $n = 1,2,3,\ldots$. There is an infinity of values of λ; for each value a product solution of the heat equation exists.

Thus $u_n(x,t) = A_n \sin(n\pi x/l)e^{-(n\pi a/l)^2 t}$. Since the heat equation is linear, we may use the superposition principle to give the Fourier expansion in the following way:

$$u(x,t) = \sum_{n=1}^{\infty} u_n(x,t) = \sum_{n=1}^{\infty} A_n \sin\left(\frac{n\pi x}{l}\right)e^{-(n\pi a/l)^2 t}. \tag{3.66}$$

Using the initial condition (3.61) we can obtain A_n. At $t = 0$,

$$u(x,0) = f(x) = \sum_{n=1}^{\infty} A_n \sin\frac{n\pi x}{l}. \tag{3.67}$$

Now multiplying both sides of (3.67) by $\sin(n\pi x/l)$ and integrating with respect to x from 0 to l, we obtain

$$A_n = \frac{2}{l} \int_0^l f(x)\sin\left(\frac{n\pi x}{l}\right)\mathrm{d}x. \tag{3.68}$$

Thus the solution is given by (3.66), where the Fourier coefficient A_n is given by (3.68).

Example 3.7

Consider Laplace's equation

$$u_{xx} + u_{yy} = 0 \tag{3.69}$$

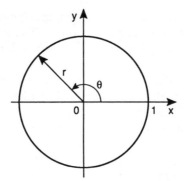

Fig. 3.4.   A unit circle as a domain.

where $u(x, y)$ represents the velocity potential of a fluid particle in a certain domain. For example, let us determine u inside a unit circle $x^2 + y^2 < 1$, when its values on the circumference, $x^2 + y^2 = 1$, are prescribed.

This boundary value problem for Laplace's equation is defined as the *Dirichlet problem*. This problem usually requires a different geometry and a lesser number of arbitrary functions in the boundary condition, which distinguishes it from the initial boundary value problems discussed in examples 3.5 and 3.6.

Solution

To obtain the solution of (3.69) we first make the following transformations $x = r \cos \theta$, and $y = r \sin \theta$ (Fig. 3.4). Then Laplace's equation (3.69) becomes

$$u_{rr} + u_r/r + u_{\theta\theta}/r^2 = 0 \tag{3.70}$$

where $u = u(r, \theta)$. The boundary condition for (3.70) is given by

$$u(1, \theta) = f(\theta). \tag{3.71}$$

By using the method of separation of variables $u(r, \theta) = R(r) \cdot \Theta(\theta)$ and on substituting this into (3.70), the following two ordinary differential equations result:

$$r^2 R'' + rR' - n^2 R = 0 \tag{3.72}$$

and

$$\Theta'' + n^2\Theta = 0, \tag{3.73}$$

where $n = 0, 1, 2, 3, \ldots$.

It is to be noted here that the separation constant $-n^2$ is chosen to obtain a circular solution to conform with the physical situation of the problem. Equation (3.72) can be readily recognized to be the Euler–Cauchy type of ordinary differential equation and the solution can be obtained by using the transformation $r = e^z$, where z is an independent variable.

The solutions of (3.72) and (3.73) are then given by

$$R = Ar^n + Br^{-n} \tag{3.74}$$

$$\Theta = C \sin n\theta + D \cos n\theta. \tag{3.75}$$

When $n = 0$, solutions of eqns (3.72) and (3.73) will be of the forms $A \ln(r) + B$ and $C\theta + D$ so that $u(r, \theta) = (A \ln(r) + B)(C\theta + D)$. It can be easily seen that a bounded solution of u will be produced provided we choose $C = 0 = A$. Hence corresponding to $n = 0$, the solution takes the form $u(r, \theta) = a_0/2$, where a_0 is a constant. From (3.74) it can be seen that the solution is bounded only if we choose $B = 0$.

Thus a bounded general solution can be written as

$$u(r, \theta) = \frac{a_0}{2} + \sum_{n=1}^{\infty} (a_n \cos n\theta + b_n \sin n\theta)r^n. \tag{3.76}$$

By applying the boundary condition (3.71) the Fourier coefficients a_n and b_n are given by

$$a_n = \frac{1}{\pi} \int_0^{2\pi} f(\theta)\cos n\theta \, d\theta, \qquad n = 0, 1, 2, \ldots ,$$

$$b_n = \frac{1}{\pi} \int_0^{2\pi} f(\theta)\sin n\theta \, d\theta, \qquad n = 1, 2, \ldots .$$

Remark

It has been noted that this result can be extended to obtain the potential inside a circular cylinder of radius a and the potential outside the cylinder very easily by the following expressions

$$u(r, \theta) = \frac{a_0}{2} + \sum_{n=1}^{\infty} (a_n \cos n\theta + b_n \sin n\theta)\left(\frac{r}{a}\right)^n, \qquad r \leq a$$

$$= \frac{a_0}{2} + \sum_{n=1}^{\infty} (a_n \cos n\theta + b_n \sin n\theta)\left(\frac{a}{r}\right)^n, \qquad r \geq a$$

where a_n and b_n are given by the above.

Example 3.8

Let us consider Laplace's equation in cylindrical coordinates (r, θ, z) which has the form

$$\nabla^2 u = \frac{1}{r} \frac{\partial}{\partial r}\left(r \frac{\partial u}{\partial r}\right) + \frac{1}{r^2} \frac{\partial^2 u}{\partial \theta^2} + \frac{\partial^2 u}{\partial z^2} = 0. \tag{3.77}$$

Solution

We apply the method of separation of variables by assuming that u is a product of the functions r, θ and z, that is $u = R(r)\Theta(\theta)Z(z)$. On substituting this into the partial differential equation (3.77), we obtain

$$\frac{\Theta Z}{r} \frac{d}{dr}\left(r \frac{dR}{dr}\right) + \frac{RZ}{r^2} \frac{d^2\Theta}{d\theta^2} + R\Theta \frac{d^2 Z}{dz^2} = 0.$$

Dividing throughout by $R\Theta Z/r^2$ yields

$$\frac{r}{R} \frac{d}{dr}\left(r \frac{dR}{dr}\right) + \frac{r^2}{Z} \frac{d^2 Z}{dz^2} = -\frac{1}{\Theta} \frac{d^2\Theta}{d\theta^2}.$$

Since the left-hand side of the above equation is independent of θ, the equation can be satisfied only if both sides are constant. It follows that $-(\Theta''/\Theta) = v^2$, and a second separation yields $(1/rR)(rR')' - (v^2/r^2) = -(Z''/Z) = -\lambda^2$, where a prime represents differentiation with respect to the argument. We have called the first separation constant v^2 because this will force Θ (and u) to be periodic of period 2π in θ. This is due to the physical situation of some applied problems. We have called the second separation constant λ^2 because we do not want Z (and u) to be periodic in z.

The values of the separation constants ν and λ are actually found by the nature of the boundary conditions that u must satisfy. Therefore, by the separation of variables, we have reduced Laplace's equation to the following three ordinary differential equations:

$$Z'' - \lambda^2 Z = 0 \tag{3.78}$$

$$\Theta'' + \nu^2 \Theta = 0 \tag{3.79}$$

$$R'' + \frac{1}{r} R' + \left(\lambda^2 - \frac{\nu^2}{r^2} \right) R = 0. \tag{3.80}$$

The solutions to the first two equations are straightforward and are given by

$$Z = A e^{\lambda z} + B e^{-\lambda z} \tag{3.81}$$

and

$$\Theta = C \cos \nu\theta + B \sin \nu\theta, \tag{3.82}$$

respectively. Equation (3.80) is a Bessel equation and has a solution of the form

$$R = E J_\nu(\lambda r) + F Y_\nu(\lambda r) \tag{3.83}$$

where $J_\nu(\lambda r)$ and $Y_\nu(\lambda r)$ are Bessel functions of the first and second kind of order ν, respectively.

Therefore the solution u can be written as

$$u = (A e^{\lambda z} + B e^{-\lambda z})(C \cos \nu\theta + D \sin \nu\theta)[E J_\nu(\lambda r) + F Y_\nu(\lambda r)]. \tag{3.84}$$

The function u satisfies Laplace's equation in cylindrical coordinates which are referred to as cylindrical harmonics. From the physics of the problem, the values of the six constants can be determined.

Example 3.9

In this example, we consider Laplace's equation in spherical coordinates, namely

$$\nabla^2 u = u_{rr} + \frac{2}{r} u_r + \frac{1}{r^2 \sin^2 \theta} u_{\phi\phi} + \frac{1}{r^2} u_{\theta\theta} + \frac{\cot \theta}{r^2} u_\theta = 0. \tag{3.85}$$

An equivalent form of this equation is

$$\frac{1}{r^2}(r^2 u_r)_r + \frac{1}{r^2 \sin\theta}(\sin(\theta)u_\theta)_\theta + \frac{1}{r^2 \sin^2\theta} u_{\phi\phi} = 0. \tag{3.86}$$

Solution

We seek a solution by the method of separation of variables by assuming a solution of the form $u(r, \theta, \phi) = R(r)\Theta(\theta)\Phi(\phi)$ and substituting this into (3.86) yields

$$\frac{\Theta\Phi}{r^2}(r^2 R')' + \frac{R\Phi}{r^2 \sin\theta}(\sin(\theta)\Theta')' + \frac{R\Theta}{r^2 \sin^2\theta}\Phi'' = 0.$$

Dividing throughout by $R\Theta\Phi/(r^2 \sin^2\theta)$, we obtain $(\sin^2\theta/R)(r^2 R')' + (\sin\theta/\Theta)(\sin(\theta)\Theta')' = -(\Phi''/\Phi)$. Since the left-hand side is independent of ϕ, we have

$$-\frac{\Phi''}{\Phi} = m^2, \qquad m = 0,1,2,\dots \tag{3.87}$$

where the first separation constant m is chosen to be a non-negative integer in order that the function Φ (and also u) be periodic of period 2π in ϕ.

On separating the variables again, we obtain

$$\frac{1}{R}(r^2 R')' = -\left(\frac{1}{\Theta \sin\theta}(\sin(\theta)\Theta')' - \frac{m^2}{\sin^2\theta}\right) = \lambda$$

where λ is the second separation constant. At this stage nothing further is known about λ.

We have reduced Laplace's equation to the following three second-order linear differential equations:

$$\Phi'' + m^2\Phi = 0 \tag{3.88}$$

$$(r^2 R')' - \lambda R = 0 \tag{3.89}$$

$$\frac{1}{\sin\theta}(\sin(\theta)\Theta')' + \left(\lambda - \frac{m^2}{\sin^2\theta}\right)\Theta = 0. \tag{3.90}$$

Solving (3.88), we obtain

$$\Phi = A_m \cos m\phi + B_m \sin m\phi, \qquad m = 0,1,2,3,\dots. \tag{3.91}$$

Equation (3.89) can be written as

$$r^2 R'' + 2rR' - \lambda R = 0, \tag{3.92}$$

and is known as the Cauchy–Euler equation. It can be solved by making the substitution $R = r^k$, in which case the differential equation (3.92) becomes $r^2 k(k-1)r^{k-2} + 2rkr^{k-1} - \lambda r^k = 0$, that is $(k^2 + k - \lambda)r^k = 0$. Hence $R = r^k$ is a solution provided that $k^2 + k - \lambda = 0$.

To find a second, linearly independent solution, consider the following. If we choose $k = n$, a non-negative integer, then $\lambda = n(n+1)$. If we choose $k = -(n+1)$, then $\lambda = n(n+1)$ again. Thus with $\lambda = n(n+1)$, (3.89) has two linearly independent solutions, r^n and $r^{-(n+1)}$, so that the general solution can be written as

$$R_n(r) = C_n r^n + D_n r^{-(n+1)}. \tag{3.93}$$

In order to solve (3.90), we make the following substitution: $x = \cos\theta$, $\Theta(\theta) = y(x)$, $d/d\theta = (d/dx)(dx/d\theta) = -\sin\theta(d/dx)$. Then

$$\frac{d}{d\theta}\left(\sin\theta \frac{d\Theta}{d\theta}\right) = -\sin\theta \frac{d}{dx}\left(\sin\theta \frac{d\Theta}{dx}\frac{dx}{d\theta}\right)$$

$$= -\sin\theta \frac{d}{dx}\left(-\sin^2\theta \frac{d\Theta}{dx}\right)$$

$$= -\sqrt{1-x^2}\,\frac{d}{dx}\left(-(1-x^2)\frac{dy}{dx}\right)$$

$$= \sqrt{1-x^2}\,\frac{d}{dx}\left((1-x^2)\frac{dy}{dx}\right).$$

Equation (3.90) can now be written as

$$\frac{d}{dx}\left\{(1-x^2)\frac{dy}{dx}\right\} + \left\{n(n+1) - \frac{m^2}{1-x^2}\right\}y = 0, \tag{3.94}$$

or equivalently

$$(1-x^2)y'' - 2xy' + \left\{n(n+1) - \frac{m^2}{1-x^2}\right\}y = 0. \tag{3.95}$$

Equation (3.95) is Legendre's associated differential equation. Its general solution, found by the series method, is

$$y(x) = EP_n^m(x) + FQ_n^m(x), \tag{3.96}$$

where $P_n^m(x)$ and $Q_n^m(x)$ are called Legendre functions of the first and second kind, respectively.

The use of subscripts and superscripts indicates that these functions depend on m and n as well as upon the argument x.

Example 3.10

A thin sheet of metal coincides with a square in the xy-plane whose vertices are points $(0,0)$, $(1,0)$, $(1,1)$ and $(0,1)$. The upper and lower faces of the sheet are perfectly insulated, so that heat flow in the sheet is purely two dimensional. Initially, the temperature distribution in the sheet is $u(x, y, 0) = f(x, y)$, where $f(x, y)$ is a known function. If there are no sources of heat in the sheet, find the temperature at any point at any subsequent time, given that the edges parallel to the x-axis are perfectly insulated and that the edges parallel to the y-axis are maintained at the constant temperature, 0°C.

Solution

The heat equation is given by

$$\frac{\partial^2 u}{\partial x^2} + \frac{\partial^2 u}{\partial y^2} = a^2 \frac{\partial u}{\partial t}. \tag{3.97}$$

The boundary conditions are (Fig. 3.5).

$$x = 0: \quad u(0, y, t) = 0 \tag{3.98}$$

$$x = 1: \quad u(1, y, t) = 0 \tag{3.99}$$

$$y = 0: \quad \frac{\partial u}{\partial y}(x, 0, t) = 0 \tag{3.100}$$

$$y = 1: \quad \frac{\partial u}{\partial y}(x, 1, t) = 0. \tag{3.101}$$

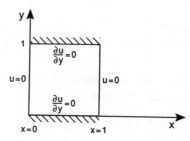

Fig. 3.5. A square sheet of metal.

The initial condition is

$$t = 0: \qquad u(x,y,0) = f(x,y). \tag{3.102}$$

Using the method of separation of variables, let $u(x,y,t) = X(x)Y(y)T(t)$. Substituting this into (3.97) yields $X''YT + XY''T = a^2 XYT'$, which on dividing throughout by XYT gives

$$\frac{X''}{X} + \frac{Y''}{Y} = a^2 \frac{T'}{T}. \tag{3.103}$$

Now assume

$$\frac{X''}{X} = -\lambda^2 \qquad and \qquad \frac{Y''}{Y} = -\nu^2. \tag{3.104}$$

Then on substituting into (3.103)

$$\frac{T'}{T} = -\left(\frac{\lambda^2 + \nu^2}{a^2}\right) \qquad and \qquad T' + \left(\frac{\lambda^2 + \nu^2}{a^2}\right)T = 0 \tag{3.105}$$

The solutions can be written as $X = A\cos\lambda x + B\sin\lambda x$, $Y = C\cos\nu y + D\sin\nu y$, and

$$T = E \exp\left(-\frac{\lambda^2 + \nu^2}{a^2}t\right).$$

Therefore,

$$u(x,y,t) = (A\cos\lambda x + B\sin\lambda x)$$

$$\times (C\cos\nu y + D\sin\nu y)Ee^{-[(\lambda^2+\nu^2)/a^2]t}. \tag{3.106}$$

The constants A, B, C, E, λ and ν must be chosen to satisfy the boundary conditions. Now

$$u(0,y,t) = 0 = (A)(C\cos\nu y + D\sin\nu y)(Ee^{-[(\lambda^2+\nu^2)/a^2]t}).$$

Therefore A must be zero.
   Also we have

$$u(1,y,t) = 0 = (B\sin\lambda)(C\cos\nu y + D\sin\nu y)(Ee^{-[(\lambda^2+\nu^2)/a^2]t}).$$

However $B \neq 0$. Hence $\sin\lambda = 0$ and therefore $\lambda_n = n\pi$ for $n = 1,2,3,\ldots$.

The solution becomes

$$u(x,y,t) = E(B \sin \lambda x)(C \cos \nu y + D \sin \nu y)e^{-[(\lambda^2+\nu^2)/a^2]t}.$$

Therefore, differentiating,

$$\frac{\partial u}{\partial y}(x,y,t) = (B \sin \lambda x)(-C\nu \sin \nu y + D\nu \cos \nu y)Ee^{-[(\lambda^2+\nu^2)/a^2]t},$$

and so

$$\frac{\partial u}{\partial y}(x,0,t) = (B \sin \lambda x)(D\nu)(Ee^{-[(\lambda^2+\nu^2)/a^2]t}) = 0.$$

Therefore $D = 0$.
   Also we have

$$\frac{\partial u}{\partial y}(x,1,t) = (B \sin \lambda x)(-C\nu)\sin \nu(Ee^{-[(\lambda^2+\nu^2)/a^2]t}) = 0,$$

$C \neq 0$; therefore $\sin \nu = 0$, or $\nu_m = m\pi$, $m = 1,2,3,\ldots$. Thus we obtain

$$u_{mn}(x,y,t) = (B \sin n\pi x)(C \cos m\pi y)\left(E \exp\left[-\frac{(m^2+n^2)\pi^2}{a^2} t \right] \right)$$

$$= E_{mn} \sin n\pi x \cos m\pi y e^{-((m^2+n^2)\pi^2/a^2)t}.$$

The general solution for u is a double Fourier series

$$u(x,y,t) = \sum_{m,n} u_{mn}(x,y,t)$$

$$= \sum_{n=0}^{\infty} \sum_{m=1}^{\infty} E_{mn} \sin m\pi x \cos n\pi y \exp\left[-\frac{m^2+n^2}{a^2} \pi^2 t \right].$$

By using the initial condition at $t = 0$, we have

$$f(x,y) = \sum_{n=0}^{\infty} \left\{ \sum_{m=1}^{\infty} E_{mn} \sin m\pi x \right\} \cos n\pi y = \sum_{n=0}^{\infty} G_n(x)\cos n\pi y,$$

where $G_n(x) = \sum_{m=1}^{\infty} E_{mn} \sin m\pi x$.

Now $G_n(x) = 2\int_0^1 f(x,y)\cos n\pi y\,dy$. Also we get

$$E_{mn} = 2\int_0^1 G_n(x)\sin m\pi x\,dx$$

$$= 2\int_0^1 \left(2\int_0^1 f(x,y)\cos n\pi y\,dy\right)\sin m\pi x\,dx$$

$$= 4\int_0^1 \int_0^1 f(x,y)\cos n\pi y\,\sin m\pi x\,dx\,dy.$$

With E_{mn} determined for all values of m and n, the formal solution of the problem is now complete.

3.6 Application: motion in an axially symmetric 3-D body

In two-dimensional motion, we saw that the velocity potential ϕ and the stream function ψ satisfy the two-dimensional Laplace's equation. These two variables can be obtained as the real and imaginary parts of an analytic function, $W(z)$, which is known as the complex potential. Although no such development can be discussed in the case of general three-dimensional flow, a stream function may be defined for the case of axially symmetric flow, which represents the flux across any surface of revolution about the axis of symmetry. This stream function is attributed to Stokes (1847). Although the relationship between the stream function and the velocity potential does not satisfy the Cauchy–Riemann conditions, nevertheless the velocity potential ϕ satisfies the three-dimensional Laplace's equation. In spherical polar coordinates (r, θ, ω), Laplace's equation can be written as

$$\frac{\partial}{\partial r}\left(r^2\frac{\partial\phi}{\partial r}\right) + \frac{1}{\sin\theta}\frac{\partial}{\partial\theta}\left(\sin\theta\frac{\partial\phi}{\partial\theta}\right) + \frac{1}{\sin^2\theta}\frac{\partial^2\phi}{\partial\omega} = 0. \qquad (3.107)$$

By using the method of separation of variables, the solution of ϕ can be written as

$$\phi = \sum_{n=0}^{\infty}\sum_{m=0}^{\infty}(A_n r^n + B_n r^{-n-1})P_n^m(\mu)\{C_m\cos m\omega + D_m\sin m\omega\}$$

$$(3.108)$$

where $P_n^m(\mu)$ is the solution of the ordinary differential equation known as the associated Legendre equation, namely

$$\frac{d}{d\mu}\left((1-\mu^2)\frac{dP}{d\mu}\right) + \left(n(n+1) - \frac{m^2}{1-\mu^2}\right)P = 0 \tag{3.109}$$

where $\mu = \cos\theta$. For simplicity the superscript m and subscript n in the symbol P_n^m have been omitted.

For axisymmetric flow about $\theta = 0$, the flow configuration in all axial planes, ω being constant, the velocity potential ϕ is independent of ω. This is equivalent to equating m to zero in (3.109) which then becomes Legendre's equation

$$\frac{d}{d\mu}\left((1-\mu^2)\frac{dP}{d\mu}\right) + n(n+1)P = 0. \tag{3.110}$$

The solution of $P_n(\mu)$ can be obtained by means of Frobenius' method and exists in the following form:

$$P_n(\mu) = \sum_{r=0}^{p} \frac{(-1)^r(2n-2r)!}{2^n r!(n-r)!(n-2r)!}\mu^{n-2r} \tag{3.111}$$

where the integer p is $\frac{1}{2}n$ or $\frac{1}{2}(n-1)$, according as n is even or odd. Then the velocity potential ϕ in the axisymmetric case is

$$\phi = \sum_{n=0}^{\infty} (A_n r^n + B_n r^{-n-1})P_n(\cos\theta). \tag{3.112}$$

For instance, the motion induced by a sphere moving at a speed of $U(t)$ through an infinite fluid at rest can be regarded as the axisymmetric flow (see Fig. 3.6).

The velocity potential is given by (3.112). To determine the two constants, A_n and B_n, we need two boundary conditions, which are given by

$$\left(\frac{\partial\phi}{\partial r}\right)_{r=a} = U\cos\theta \tag{3.113}$$

and

$$(\phi)_{r\to\infty} = 0. \tag{3.114}$$

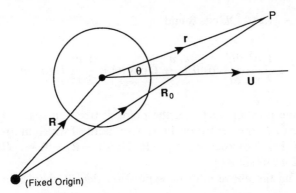

Fig. 3.6. Moving sphere referred to fixed origin.

Using these two boundary conditions we see that (3.114) requires $A_n = 0$ and condition (3.113) yields

$$\left(\frac{\partial \phi}{\partial r}\right)_{r=a} = \sum_{n=0}^{\infty} \left[B_n(-n-1)r^{-n-2}P_n(\cos\theta)\right]_{r=a} = U\cos\theta. \quad (3.115)$$

The first few Legendre polynomials (see Curle and Davies (1968), Wylie and Barrett (1982), and Rahman (1991)) are

$$P_0(\cos\theta) = 1$$

$$P_1(\cos\theta) = \cos\theta$$

$$P_2(\cos\theta) = \tfrac{1}{2}(3\cos^2\theta - 1)$$

$$P_3(\cos\theta) = \tfrac{1}{2}(5\cos^3\theta - 3\cos\theta)$$

$$P_4(\cos\theta) = \tfrac{1}{8}(35\cos^4\theta - 30\cos^2\theta + 3).$$

Corresponding to $n = 1$, from (3.115) we have $-2B_1 a^{-3} = U$, giving $B_1 = -\tfrac{1}{2}Ua^3$. Hence

$$\phi = -\frac{1}{2}Ua\left(\frac{a}{r}\right)^2\cos\theta. \quad (3.116)$$

The pressure distribution exerted on the surface of the sphere can be obtained from Bernoulli's equation

$$\frac{P}{\rho} + \frac{1}{2}(\nabla\phi)^2 + \frac{\partial\phi}{\partial t} = C(t) \quad (3.117)$$

where $\phi = -\frac{1}{2}(a^3/r^2)U\cos\theta$ and

$$\frac{\partial\phi}{\partial t} = -\frac{1}{2}\frac{a^3}{r^2}\frac{dU}{dt}\cos\theta + \frac{a^3}{r^3}U\cos\theta\,\dot{r} + \frac{1}{2}\frac{a^3}{r^2}U\sin\theta\,\dot{\theta}. \qquad (3.118)$$

If $\mathbf{R}_0$ is the position vector of the point P (fixed in space) and $\mathbf{R}$ is the position vector of the centre of the sphere, both of which are referred to a fixed origin (Fig. 3.6), then $\mathbf{r} = \mathbf{R}_0 - \mathbf{R}$. Thus $\dot{\mathbf{r}} = \dot{\mathbf{R}}_0 - \dot{\mathbf{R}} = -U\cos\theta$, $r\dot{\theta} = U\sin\theta$ and $\dot{\theta} = U\sin\theta/r$.

Substituting the above into the expression $\partial\phi/\partial t$ gives

$$\frac{\partial\phi}{\partial t} = -\frac{1}{2}\frac{a^3}{r^2}\cos\theta\frac{dU}{dt} + \frac{1}{2}\frac{a^3}{r^3}U^2 - \frac{3}{2}\frac{a^3}{r^3}U^2\cos^2\theta. \qquad (3.119)$$

The speed at any point is given by $(\nabla\phi)^2 = (\partial\phi/\partial r)^2 + (\partial\phi/r\partial\theta)^2 = (U^2 a^6/r^6)(\cos^2\theta + \frac{1}{4}\sin^2\theta)$.

From (3.117), the expression for the pressure is

$$\frac{P}{\rho} = C(t) + \frac{1}{2}\frac{a^3}{r^2}\cos\theta\frac{dU}{dt} - \frac{1}{2}\frac{a^3}{r^3}U^2$$

$$+ \frac{3}{2}\frac{a^3}{r^3}U^2\cos^2\theta - \frac{U^2 a^6}{2r^6}\left(\cos^2\theta + \frac{1}{4}\sin^2\theta\right).$$

When $r \to \infty$, $p = p_\infty$ and hence $C(t) = P_\infty/\rho$. Thus the pressure force on the body of the sphere, $r = a$, is given by

$$\frac{P - P_\infty}{\rho} = \frac{1}{2}a\cos\theta\frac{dU}{dt} + \frac{1}{8}U^2(9\cos^2\theta - 5). \qquad (3.120)$$

The drag force on the sphere can be obtained by integrating the resolved pressure force over the surface of the sphere. Alternatively, this result may be obtained by equating the rate of change of the kinetic energy of the fluid to the work done by the fluid forces. We know that the kinetic energy is given by $T = \frac{1}{2}\rho\int_s \phi(\partial\phi/\partial n)\,ds$, where $\partial\phi/\partial n$ is the specified velocity of the boundary along the outward normal from the fluid.

Referring to Fig 3.7 we have $\delta s = (2\pi a\sin\theta)(a\,\delta\theta)$ and $\partial\phi/\partial n = -(\partial\phi/\partial r) = -(U_a^3/r^3)\cos\theta$. Therefore $[\phi(\partial\phi/\partial n)]_{r=a} = \frac{1}{2}aU^2\cos^2\theta$.

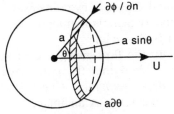

Fig. 3.7. Drag force.

The kinetic energy of the fluid is

$$T_1 = \frac{\rho}{2} \int_s \phi \frac{\partial \phi}{\partial n}\, \mathrm{d}s = \frac{\rho}{2} \int_0^\pi \left(\frac{1}{2} aU^2 \cos^2 \theta \right)(2\pi a^2 \sin \theta\, \mathrm{d}\theta)$$

$$= \frac{\rho}{2} \pi a^3 U^2 \int_0^\pi \sin \theta \cos^2 \theta\, \mathrm{d}\theta$$

$$= \frac{\rho}{2} \pi a^3 U^2 \frac{1}{3}(-\cos^3 \theta)_0^\pi = \frac{\rho}{3} \pi a^3 U^2 = \frac{1}{4} M'U^2,$$

where $M' = \frac{4}{3} \rho \pi a^3$ is the mass of the liquid displaced by the sphere. Usually $\frac{1}{2} M'$ is defined as the added mass of the sphere due to the motion.

However, the kinetic energy produced by the moving sphere is equal to $\frac{1}{2} MU^2$, where M is the mass of the sphere. Therefore, the total kinetic energy is given by $T = \frac{1}{2} MU^2 + \frac{1}{4} M'U^2 = \frac{1}{2}(M + \frac{1}{2} M')U^2$. The virtual mass of the sphere is $M + \frac{1}{2} M'$; thus the effect of the liquid is to increase the inertia of the sphere by half the mass of the liquid displaced.

If the total force (including the inertial field) component is X in the direction of U, then $\mathrm{d}T/\mathrm{d}t = \mathrm{d}/\mathrm{d}t[\frac{1}{2} MU^2 + \frac{1}{4} M'U^2]$ is the rate at which work is being done, i.e. XU.

Therefore $M(\mathrm{d}U/\mathrm{d}t) + \frac{1}{2} M'(\mathrm{d}U/\mathrm{d}t) = X$.

3.7 Exercises

1. The velocity potential, $\phi(r, \theta)$, for the two-dimensional irrotational flow of an ideal fluid satisfies Laplace's equation. The velocity of the fluid is $\mathbf{V} = \mathrm{grad}\, \phi$, and the radial velocity is $\partial \phi/\partial r$. If the velocity of the fluid at infinity is parallel to $\theta = 0$ and is $\mathbf{V} = U\mathbf{i}$, and if the flow passes around a circular cylinder $r = a$ on which the boundary condition is $\partial \phi/\partial r = 0$, confirm that the velocity potential is $\phi = Ur \cos \theta + U(a^2/r)\cos \theta$.

2. The velocity potential due to a two-dimensional point source (a line source normal to the xy-plane) of strength m at the point $x = 0, y = d$ in an infinite fluid is $\phi = m \ln\{x^2 + (y - d)^2\}^{1/2}$. If a fixed boundary, $y = 0$, is inserted into the flow giving rise to the boundary condition $\partial\phi/\partial y = 0$ on $y = 0$, show that the potential in the region $y > 0$ is $\phi = m \ln\{x^2 + (y - d)^2\}^{1/2} + m \ln\{x^2 + (y + d)^2\}^{1/2}$.

   Interpret the second term as an image source at $x = 0$, $y = -d$. Determine the flow in the quadrant $x \geq 0,: y \geq 0$, with the boundary conditions $\partial\phi/\partial x = 0$ on $x = 0$, $\partial\phi/\partial y = 0$ on $y = 0$, due to a fluid source of strength m at the point m at the point $x = a, y = b$ of the quadrant.

3. If the function T satisfies Laplace's equation in the cylindrical region $z \geq 0$, $0 \leq r \leq a$, subject to the boundary conditions $T = 1$ on $z = 0$, $0 \leq r \leq a$, $T = 0$ on $r = a$, $z \geq 0$, $T \to 0$ as $z \to \infty$, show that $T = \sum_{n=1}^{\infty} A_n J_0(\alpha_n r/a)\exp[-(\alpha_n z/a)]$, where the coefficients A_n must be chosen to satisfy $\sum_{n=1}^{\infty} A_n J_0(\alpha_n r/a) = 1$, $0 \leq r \leq a$, and where α_n are the roots of the equation $J_0(\alpha_n) = 0$, $n = 1, 2, 3, \ldots$.

4. The Legendre polynomials P_0, P_1, P_2 are $P_0(\cos \theta) = 1$, $P_1(\cos \theta) = \cos \theta$, $P_2(\cos \theta) = \frac{1}{2}(3 \cos^2 \theta - 1)$. Hence, the simplest axially symmetric solutions of Laplace's equation in spherical polar coordinates are $A_0 + B_0/r$, $(A_1 r + B_1/r^2)\cos \theta$, and $(A_2 r^2 + B_2/r^3)(3 \cos^2 \theta - 1)$. The irrotational flow of an ideal fluid has the velocity field $\mathbf{V} = \text{grad } \phi$, where ϕ is the velocity potential which satisfies $\nabla^2\phi = 0$. Confirm that the uniform flow field, $\mathbf{V} = U\mathbf{k}$, corresponds to the case $\phi = Ur \cos \theta$ where U is a constant. Similarly show that if a rigid sphere of radius $r = a$ is placed in this uniform flow field, then the potential becomes

$$\phi = U\left(r + \frac{a^3}{2r^2}\right)\cos \theta,$$

   where the boundary condition of zero normal flow on the sphere is $\partial\phi/\partial r = 0$ on $r = a$.

5. Confirm that the solution of the one-dimensional wave equation $\partial^2 u/\partial t^2 = a^2(\partial^2 u/\partial x^2)$ on $0 \leq x \leq l$, $t \geq 0$, subject to the boundary conditions $\partial u/\partial x = 0$ on both $x = 0$ and $x = l$ and the initial conditions $u = f(x)$, $\partial u/\partial t = 0$ on $0 \leq x \leq l$, when $t = 0$, is

$$u = \frac{1}{2}[f_e(x - at) + f_e(x + at)].$$

   Here $f_e(x)$ denotes the even periodic extension of $f(x)$ of period $2l$.

6. Find the characteristics of each of the following equations and reduce each equation to the appropriate standard form. Where possible,

obtain real solutions for the equation. (a) $u_{xx} - 4x^2 u_{yy} = (1/x)u_x$; (b) $xu_{xy} + yu_{yy} = 0$; (c) $x^2 u_{xx} + 2xyu_{xy} + y^2 u_{yy} = 0$; (d) $x^2 u_{xx} - 2xyu_{xy} + y^2 u_{yy} = 0$.

7. Show that the axially symmetric solution $u(r, z)$ of Laplace's equation in the infinite cylinder $-\infty < z < \infty$, $0 \le r \le a$, subject to the boundary condition $u = \cos \lambda z$ on $r = a$, is given by $u = [I_0(\lambda r)/I_0(\lambda a)]\cos \lambda z$. What is the solution which corresponds to the boundary condition $u = 1$ on $r = a$?

8. By determining a particular interval, solve the inhomogeneous heat equation $(1/\alpha)u_t = u_{xx} + h$, in the region $0 \le x \le a$, given that h is a constant and u is subject to the boundary and initial conditions $u(0, t) = 0$, $u(a, t) = 0$; $u(x, 0) = 0$. Show that the temperature tends to $1/2 hx(a - x)$ for large values of t.

9. Solve Laplace's equation, $\nabla^2 u = 0$, in the rectangular box $0 \le x \le a$, $0 \le y \le b$, $0 \le z \le c$, subject to the boundary conditions $u = 0$ on the faces $x = 0$, $x = a$, $y = 0$, $y = b$ and $z = c$, and $u = f(x, y)$ on $z = 0$. Show that $u = \sum_{n=1}^{\infty} \sum_{m=1}^{\infty} A_{mn} \sin(n\pi x/a)\sin(m\pi y/b)\sin[(n^2/a^2 + m^2/b^2)^{1/2}\pi(c - z)]$ where

$$A_{mn} = \left(4/ab\sinh\left(\frac{n^2}{a^2} + \frac{m^2}{b^2}\right)^{1/2} \pi c\right)$$

$$\times \int_0^b \int_0^a f(x, y)\sin(n\pi x/a)\sin(m\pi y/b)dx\,dy.$$

10. The solution of Laplace's equation in plane polar coordinates in the circle $r \le a$ subject to the boundary condition $u(a, \theta) = f(\theta)$, $0 \le \theta \le 2\pi$, is

$$u(r, \theta) = \frac{1}{2\pi} \int_0^{2\pi} f(s)ds$$

$$+ \frac{1}{\pi} \int_0^{2\pi} f(s) \sum_{n=0}^{\infty} (r/a)^n \cos n(s - \theta)ds.$$

By expressing the cosine in terms of the exponential function, the sum can be expressed as a pair of geometric series. By summing these series show that

$$u(r, \theta) = \frac{a^2 - r^2}{2\pi} \int_0^{2\pi} \frac{f(s)ds}{a^2 - 2ar\cos(s - \theta) + r^2}.$$

This is known as Poisson's formula.

11. A function ϕ satisfies Laplace's equation; it is continuous, single valued and finite at all points outside a sphere of radius a, and on this sphere its normal derivative can be expressed in the form $\sum_{n=0}^{\infty} A_n(a/d)^n P_n(\cos \theta)$. Show that the solution of ϕ is given by

$$\phi = - \sum_{n+0}^{\infty} \frac{A_n}{n+1} \left(\frac{a^2}{d}\right)^n \frac{a^2}{r^{n+1}} P_n(\cos \theta).$$

12. If $\phi = r^n s$ is a spherical harmonic, prove that, s being independent of r,

$$\frac{1}{\sin \theta} \frac{\partial}{\partial \theta} \left(\sin \theta \frac{\partial s}{\partial \theta}\right) + \frac{1}{\sin^2 \theta} \frac{\partial^2 s}{\partial \omega^2} + n(n+1)s = 0.$$

Deduce that s/r^{n+1} is also a spherical harmonic. (Note: ϕ is defined to be spherical harmonic provided it satisfies Laplace's equation in spherical coordinates.)

13. If $\phi = r^n s$ is a spherical harmonic symmetrical about the x-axis, and s is independent of r, show that $d/d\mu[(1 - \mu^2)ds/d\mu] + n(n+1)s = 0$ where $\mu = \cos \theta$. Show that the solutions of this equation corresponding to $n = 0$ and $n = 1$ are $P_0(\mu)$, $P_1(\mu)$, and show that

$$Q_0(\mu) = \tfrac{1}{2} \ln\left(\frac{1+\mu}{1-\mu}\right)$$

$$Q_1(\mu) = \tfrac{1}{2}\mu \ln\left(\frac{1+\mu}{1-\mu}\right) - 1.$$

14. The motion of a fluid is given by the velocity potential

$$\phi = C\left[\left(1 + \frac{1}{n}\right) \frac{r^n}{a^{n-1}} + \frac{a^{n+2}}{r^{n+1}}\right] P_n(\cos \theta),$$

in which C is a constant, and r and θ are spherical polar coordinates. Determine the stream function.

15. A sphere of radius a is surrounded by a concentric spherical shell of radius b, and the space between is filled with liquid. If the sphere is moving with velocity V, show that

$$\phi = \frac{Va^3}{b^3 - a^3}\left(r + \frac{b^3}{2r^2}\right) \cos \theta$$

and find the current function.

16. Incompressible fluid is in steady irrotational motion in the interior of a sphere of radius b and centre O. There is a source at a point A and a sink of equal strength at B, A and B being points on a diameter at a distance a on either side of O. Find the velocity potential for $r < a$ and for $a < r < b$ as a series in Legendre polynomials and determine the velocity at O.

17. Prove that the velocity potential due to the image of a source of strength m, in a sphere of radius a, is the same as that due to a distribution of doublets over the surface of the sphere, the axes being normal to the surface and strength per unit area being

$$m\left[(2a/cR) - (1/a)\ln\left(\frac{a + R + a^2/c}{a + R - a^2/c}\right)\right];$$

c is the distance of the source from the centre, and R is the distance from the inverse point.

18. A solid sphere of radius a moves in a fluid which far from the sphere remains at rest at pressure P_0. At time t the centre of the sphere is at $(c \sin \sigma t, 0, 0)$. Show that the pressure at the point $\mathbf{r}(x, \dot{y}, z)$ at time $t = (\pi/2\sigma)$, when the sphere is instantaneously at rest, is $p = P_0 - \frac{1}{2}ca^3\sigma^2(x/r^3)$.

19. Find the values of A and B for which $(Ar + B/r^2)\cos\theta$ is the velocity function of the motion of an incompressible fluid which fills the space between a solid sphere of radius a and a concentric spherical shell of radius $2a$. The sphere has a velocity U and the shell is at rest. Prove that the kinetic energy is of the fluid of density ρ is $10\pi\rho a^3 U^2/21$.

20. A sphere, of mass M and radius a, is at rest with its centre at a distance h from a plane boundary. Show that the magnitude of the impulse necessary to start the sphere with a velocity V directly towards the boundary is, very nearly,

$$V\left[M + \tfrac{1}{2}M'\left(1 + \frac{3a^3}{8h^3}\right)\right],$$

where M' is the mass of the displaced fluid. Find also the impulse on the plane boundary.

21. A sphere of radius a moves in a semi-infinite liquid of density ρ bounded by a plane wall, its centre being at a great distance h from the wall. Show that the approximate kinetic energy of the fluid is

$$1/3\pi\rho a^3 V^2 \left[\left(1 + \frac{3}{16}\frac{a^3}{h^3}(1 + \sin^2 \alpha)\right)\right].$$

The sphere is moving at an angle α with the wall, at a speed V.

22. The ellipsoid $x^2/a^2 + y^2/b^2 + z^2/c^2 = 1$ is placed in a uniform stream parallel to the x-axis. Prove that the lines of equal pressure on the ellipsoid are its curves of intersection with the cones $y^2/b^2 + z^2/c^2 = x^2/h^2$, where h is an arbitrary constant.

23. Find the transformation to give the two-dimensional flow of a stream of velocity $2U$ at infinity past a right-angled bend in a river bounded by the positive halves of the x- and y-axes and the straight lines $x = a$, $y > a$, and $y = a$, $x > a$.

References

Curle, N. and Davies, H. (1968). *Modern Fluid Dynamics*, Vol. 1, Van Nostrand, London.

Hellwig, G. (1964). *Partial Differential Equations*, Blaisdell, Waltham, Massachusetts.

Milne-Thomson, L. M. (1968). *Theoretical Hydrodynamics*, 5th ed, Macmillan, New York.

Rahman, M. (1991). *Applied Differential Equations for Scientists and Engineers, Vol. 2, Partial Differential Equations*, Computational Mechanics Publications, Southampton UK and Boston USA.

Stokes, G. G. (1847). On the theory of oscillatory waves. Trans. Cambridge Philos. Soc., 8, 441–455.

Wylie, C. R. and Barrett, L. C. (1982). *Advanced Engineering Mathematics*, McGraw-Hill, New York.

Part II

Water waves

4
Theory of surface waves

4.1 Introduction

As we know in practice, real water waves propagate in viscous oceans over an irregular rough bottom of varying permeability. However, a striking feature of the water waves in oceans is that in most cases the main body of the fluid motion is nearly irrotational such that the viscosity which causes the rotation may be negligible. Another remarkable fact is that water can be reasonably regarded as incompressible and, consequently, a velocity potential and a stream function should exist for wave propagation in oceans. We have already remarked in Chapter 2 about these two mathematical entities and their importance in the solution process of water wave mechanics.

The problems associated with water wave propagation are difficult owing to the complex nature of the process. The difficulties arise from the irregularity of the wave motion, wave breaking and energy dissipation due to friction, turbulence, etc. Any mathematical model that we attempt will require some simplifications. The application of a mathematical description often depends on the number of space dimensions involved with the problem. In the case of the one-dimensional model it is possible to introduce some nonlinear effects onto the method of solution. However, in the case of two- and three-dimensional models the mathematical formulations are often restricted to those solvable by linear harmonic wave theory.

In this chapter the analytical basis of elementary surface waves is presented and a sound physical and mathematical understanding of water wave motion in its simplest form is developed. The small-amplitude wave theory of this chapter might be best described in practical terms as 'a first approximation' to the complete theoretical description of wave behaviour. Since it is an approximation of the true behaviour, naturally there exists some error. In many engineering applications the magnitude of the error is negligible.

4.2 Equations of motion

Euler's equations of motion and Bernoulli's equation have been given in their three-dimensional forms in Chapter 2. For the treatment of wave motion as discussed here, only two-dimensional motion will be considered where the horizontal acceleration component along the x-axis and the vertical acceleration component along the z-axis will be retained in the equations of motion. The horizontal component of acceleration along the y-axis is assumed to be smaller than the other two components, and is neglected.

For incompressible, two-dimensional motion in the (x, z) plane, the continuity equation (2.22) becomes

$$\frac{\partial u}{\partial x} + \frac{\partial w}{\partial z} = 0. \tag{4.1}$$

If, in addition, the flow is irrotational, then a velocity potential exists which by definition gives $\mathbf{V} = \pm \text{grad } \phi$. Some authors' convention is to use the negative gradient and thus we have

$$u = -\frac{\partial \phi}{\partial x} \quad \text{and} \quad w = -\frac{\partial \phi}{\partial z}. \tag{4.2}$$

Substituting (4.2) into (4.1), we obtain

$$\nabla^2 \phi(x, z) = 0 \tag{4.3}$$

which is Laplace's equation, where $\nabla^2 = \partial^2/\partial x^2 + \partial^2/\partial z^2$. It is this partial differential equation which must be solved, subject to the appropriate boundary conditions, in order to investigate the behaviour of water waves. In the following analysis, we present a classical method in deriving Bernoulli's equation again applicable to water wave mechanics in the vertical (x, z) plane for the convenience of some readers who are still acquiring the requisite skills of vector calculus.

Euler's equations of motion (2.28) in the x- and z-directions can be explicitly written as

$$\frac{\partial u}{\partial t} + u\frac{\partial u}{\partial x} + w\frac{\partial u}{\partial z} = -\frac{1}{\rho}\frac{\partial P}{\partial x} + X \tag{4.4}$$

$$\frac{\partial w}{\partial t} + u\frac{\partial w}{\partial x} + w\frac{\partial w}{\partial z} = -\frac{1}{\rho}\frac{\partial P}{\partial z} + Z. \tag{4.5}$$

Assuming that the only external force acting on the fluid particle is the gravitational body force per unit mass, then we may write $X = 0$, $Z = -g$, where g is the gravitational constant.

Since we know the gravitational force to be derivable from a potential, we can write $Z = -g = -[\partial(gz)/\partial z]$. Under the assumption that the motion is irrotational, we have $\partial u/\partial z = \partial w/\partial x$, and, from the definition of velocity potentials (4.2), $-(\partial u/\partial t) = \partial^2\phi/\partial x \partial t$, $-(\partial w/\partial t) = \partial^2\phi/\partial z \partial t$. The equations of motion (4.4) and (4.5) may be written as

$$-\frac{\partial^2\phi}{\partial x \partial t} + u\frac{\partial u}{\partial x} + w\frac{\partial w}{\partial x} = -\frac{1}{\rho}\frac{\partial P}{\partial x} \tag{4.6}$$

$$-\frac{\partial^2\phi}{\partial z \partial t} + u\frac{\partial u}{\partial z} + w\frac{\partial w}{\partial z} = -\frac{1}{\rho}\frac{\partial P}{\partial z} - \frac{\partial}{\partial z}(gz). \tag{4.7}$$

Integrating (4.6) and (4.7) with respect to x and z respectively, we obtain

$$-\frac{\partial\phi}{\partial t} + \frac{1}{2}(u^2 + w^2) + \frac{P}{\rho} = F_1(z,t) \tag{4.8}$$

$$-\frac{\partial\phi}{\partial t} + \frac{1}{2}(u^2 + w^2) + \frac{P}{\rho} + gz = F_2(x,t). \tag{4.9}$$

Subtracting (4.8) from (4.9) yields $gz = F_2(x,t) - F_1(z,t)$, which suggests that F_2 is a function of t alone. Thus $F_2 = F_2(t)$, and hence $F_1 = F_2(t) - gz$. Therefore, (4.8) and (4.9) reduce to the single equation

$$-\frac{\partial\phi}{\partial t} + \frac{1}{2}(u^2 + w^2) + \frac{P}{\rho} + gz = F_2(t), \tag{4.10}$$

which is Bernoulli's equation, in two-dimensional fluid flow. This explicit form of Bernoulli's equation obtained through elementary procedures can be recovered from eqn (2.36) (obtained through powerful vector analysis) provided we redefine eqn (2.35) as $\mathbf{V} = -\text{grad }\phi$ which is exactly (4.2) in two dimensions.

For the steady-state case, $\partial\phi/\partial t = 0$, (4.10) reduces to

$$\frac{1}{2}(u^2 + w^2) + \frac{P}{\rho} + gz = \text{constant}. \tag{4.11}$$

Without loss of generality, $F_2(t)$ can be combined with the velocity potential, $\phi(x,z,t)$, so that Bernoulli's equation (4.10) becomes

$$-\frac{\partial\phi}{\partial t} + \frac{1}{2}(u^2 + w^2) + \frac{P}{\rho} + gz = 0. \tag{4.12}$$

Small-amplitude wave theory is based upon the further assumption that all motions are infinitely small, which enables us to neglect the square of the velocity components (u^2 and w^2). Hence (4.12) can be reduced to

$$-\frac{\partial \phi}{\partial t} + \frac{P}{\rho} + gz = 0, \tag{4.13}$$

which is the general equation of motion applied in the development of the small-amplitude theory of water waves.

4.3 Wave terminology

Water wave terminology is defined and illustrated in Fig. 4.1. The coordinate system is x, z. A simple harmonic progressive wave is shown moving in the positive x-direction.

With reference to Fig. 4.1, the symbols are defined as follows: h = water depth (from mean water level to the bottom of the sea), $\eta(x,t)$ = vertical displacement of water surface above mean water level, A = wave amplitude, $H = 2A$ = total amplitude, L = wavelength, T = wave period, C = velocity of wave propagation = phase velocity = L/T, k = wave number = $2\pi/L$, σ = wave angular frequency = $2\pi/T$.

The wave in Fig. 4.1 is moving to the right with a wave velocity C. The differential equation to be satisfied in the region $-h \leq z \leq \eta$ and $-\infty < x < \infty$ is Laplace's equation

$$\frac{\partial^2 \phi}{\partial x^2} + \frac{\partial^2 \phi}{\partial z^2} = 0. \tag{4.14}$$

The boundary conditions applied to (4.14) must be consistent with the type of wave motion illustrated in Fig. 4.1. In this problem, it is interesting to note that the surface boundary is not only moving, but its position also is not specified and depends on the solution of the problem.

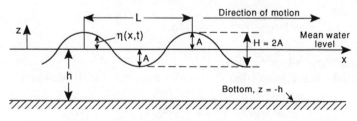

Fig. 4.1.   Schematic diagram of propagation of a progressive wave.

It is clear that if the bottom is regarded as a fixed, impermeable, horizontal boundary, then the boundary condition to be satisfied at the bottom is

$$w = - \frac{\partial \phi}{\partial z} = 0 \quad \text{on} \quad z = -h. \tag{4.15}$$

The boundary condition to be satisfied on the surface can be obtained by applying Bernoulli's equation (4.12) to the surface, $z = \eta$, and by taking the pressure P on this surface to be equal to zero. If in addition, we linearize as before by neglecting the second-order terms, namely u^2 and w^2, (4.12) yields the surface boundary condition

$$\eta = \frac{1}{g} \left(\frac{\partial \phi}{\partial t} \right)_{z = \eta}. \tag{4.16}$$

In order to use this boundary condition, it is necessary to make the further assumption that the amplitude of the waves is sufficiently small so that (4.16) may be simplified to the boundary condition

$$\eta = \frac{1}{g} \left(\frac{\partial \phi}{\partial t} \right)_{z = 0} \tag{4.17}$$

which is known as the dynamic boundary condition at the free surface. There is, however, another free surface boundary condition to be satisfied, known as the kinematic boundary condition. This condition is described in section 4.5 in connection with the evaluation of the wave dispersion relation.

In addition to the boundary conditions (4.15) and (4.17) above, it is noted that the physical solution must be periodic both in time t and in space dimension x.

4.4 Analytic solution of the wave problem

The complete boundary value wave problem can be restated as follows. The differential equation

$$\frac{\partial^2 \phi}{\partial x^2} + \frac{\partial^2 \phi}{\partial z^2} = 0 \tag{4.18}$$

is subject to the boundary conditions

$$\frac{\partial \phi}{\partial z} = 0 \quad \text{on} \quad z = -h \tag{4.19}$$

$$\eta = \frac{1}{g}\left(\frac{\partial \phi}{\partial t}\right) \quad \text{on} \quad z = 0. \tag{4.20}$$

In Chapter 3 the method of separation of variables was shown to be a powerful technique for solving Laplace's equation. This method usually leads to the standing wave solution. For the above problem, we assume

$$\phi(x,z,t) = X(x)Z(z)T(t), \tag{4.21}$$

where X is a function of x alone, Z is a function of z alone, and T is a function of t alone.

Substituting (4.21) into (4.18), we obtain

$$\frac{X''}{X} = -\frac{Z''}{Z} = -k^2, \tag{4.22}$$

where k^2 is a separation constant. The resulting two ordinary differential equations are

$$X'' + k^2 X = 0 \tag{4.23}$$

$$Z'' - k^2 Z = 0. \tag{4.24}$$

The solutions of (4.23) and (4.24) are $X = B \cos kx + D \sin kx$, and $Z = Ee^{kz} + Ge^{-kz}$, where B, D, E and G are arbitrary constants. Thus, the solution can be written as

$$\phi(x,z,t) = (B \cos kx + D \sin kx)(Ee^{kz} + Ge^{-kz})T(t). \tag{4.25}$$

From a physical viewpoint, solutions are required which are simple harmonic in time. It is therefore reasonable to express $T(t)$ as $\cos \sigma t$ or $\sin \sigma t$, where σ is defined as the frequency and is given by $\sigma = 2\pi/T$.

There are four independent elementary combinations of terms, periodic in x and t, which are solutions of Laplace's equation, namely

$$\phi_1 = A_1 Z(z)\cos kx \cos \sigma t \tag{4.26}$$

$$\phi_2 = A_2 Z(z)\sin kx \sin \sigma t \tag{4.27}$$

$$\phi_3 = A_3 Z(z)\sin kx \cos \sigma t \tag{4.28}$$

$$\phi_4 = A_4 Z(z)\cos x \sin \sigma t. \tag{4.29}$$

Decomposing the solution in this manner helps in the evaluation of the unknown constants. Since Laplace's equation is linear, a proper combination of these solutions will therefore satisfy both Laplace's equation and its boundary conditions.

The boundary conditions (4.19) and (4.20) will now be applied to solution (4.26). From (4.26), $\partial\phi_1\partial z = kA_1(Ee^{kz} - Ge^{-kz})\cos kx \cos \sigma t$ at $z = -h$.

Applying the condition $\partial\phi_1/\partial z = 0$ at $z = -h$ gives $Ee^{-kh} = Ge^{kh}$. Hence

$$E = Ge^{2kh}. \tag{4.30}$$

Then

$$\phi_1 = 2A_1Ge^{kh}\left(\frac{e^{k(z+h)} + e^{-k(z+h)}}{2}\right)\cos kx \cos \sigma t$$

$$= 2A_1Ge^{kh} \cosh k(z + h)\cos kx \cos \sigma t. \tag{4.31}$$

Applying the surface boundary condition $\eta_1 = 1/g(\partial\phi_1/\partial t)_{z=0}$ yields $\eta_1 = -2A_1Ge^{kh}/g(\sigma) \cosh kh \cos kx \sin \sigma t$. The maximum value of η is the wave amplitude A, and it will occur when $\cos kx \sin \sigma t = 1$. Therefore,

$$A_1Ge^{kh} = -\frac{Ag}{2\sigma \cosh kh} \tag{4.32}$$

and subsequently

$$\eta_1 = A \cos kx \sin \sigma t. \tag{4.33}$$

This represents a system of 'standing waves', the wavelength being $2\pi/k$ and the vertical amplitude A. The velocity potential ϕ_1 is now

$$\phi_1 = -\frac{Ag \cosh k(z + h)}{\sigma \cosh kh}\cos kx \cos \sigma t. \tag{4.34}$$

ϕ_1 is to be periodic in x with wavelength L. It is necessary that k is defined as $k = 2\pi/L$, which is commonly known as the *wave number*.

The constant in the other elementary ϕ can be evaluated in a similar manner. Thus

$$\phi_1 = -\frac{Ag \cosh k(z+h)}{\sigma \cosh kh} \cos kx \cos \sigma t$$

$$\phi_2 = \frac{Ag \cosh k(z+h)}{\sigma \cosh kh} \sin kx \sin \sigma t$$

$$\phi_3 = -\frac{Ag \cosh k(z+h)}{\sigma \cosh kh} \sin kx \cos \sigma t$$

$$\phi_4 = \frac{Ag \cosh k(z+h)}{\sigma \cosh kh} \cos kx \sin \sigma t.$$

Owing to the linearity of Laplace's equation, any linear combination will also be a solution to the problem. Thus

$$\phi = \phi_2 - \phi_1 = \frac{Ag \cosh k(z+h)}{\sigma \cosh kh} \cos(kx - \sigma t). \tag{4.35}$$

This velocity potential is due to a progressive wave travelling in the positive x-direction given below. From (4.20) we obtain the water elevation

$$\eta = \frac{1}{g}\left(\frac{\partial \phi}{\partial t}\right)_{z=0} = A \sin(kx - \sigma t), \tag{4.36}$$

which is periodic in x and t. This solution is usually called the progressive wave solution.

If we move along with the wave such that at all time t our position relative to the wave front remains fixed, then $kx - \sigma t = $ constant. The speed at which we must move to accomplish this is given by

$$\frac{dx}{dt} = \frac{\sigma}{k} = \frac{L}{T} = C, \tag{4.37}$$

which is known as the wave celerity, C, or speed of wave propagation. Equation (4.35) thus represents the velocity potential for a progressive wave travelling in the positive x-direction.

It is noted that from (4.35) the complete velocity field beneath a wave may be described, and also, from Bernoulli's equation, the pressure field may be determined.

In a similar manner the velocity potential for a progressive wave travelling in the negative x-direction is obtained from the combination $(\phi_1 + \phi_2)$. That is

$$\phi = (\phi_1 + \phi_2) = -\frac{Ag \cosh k(z+h)}{\sigma \cosh kh} \cos(kx + \sigma t), \qquad (4.38)$$

and the associated water elevation is

$$\eta = A \sin(kx + \sigma t). \qquad (4.39)$$

Similarly we obtain

$$\phi = -(\phi_3 + \phi_4) = \frac{Ag \cosh k(z+h)}{\sigma \cosh kh} \sin(kx - \sigma t) \qquad (4.40)$$

$$\eta = -A \cos(kx - \sigma t) \qquad (4.41)$$

and

$$\phi = -(\phi_4 - \phi_3) = -\frac{Ag \cosh k(z+h)}{\sigma \cosh kh} \sin(kx + \sigma t) \qquad (4.42)$$

$$\eta = -A \cos(kx + \sigma t). \qquad (4.43)$$

Velocity potentials (4.40) and (4.42) are identical to (4.35) and (4.38) except for a substitution which amounts only to a shift in wave phase with respect to the origin.

From the expression for velocity potential describing wave motion we are in a position to investigate a number of wave characteristics, the most important of which is wave dispersion. Before we proceed to find the wave dispersion relation, we present below some more insights of the velocity potential together with some physical characteristics.

As an alternative 'shortcut' derivation, consider the following approach. Assuming a progressive wave solution of the form $\phi \sim e^{i(kx - \sigma t)}$ we can obtain a solution of the two-dimensional Laplace's equation satisfied by the velocity potential ϕ.

For a progressive wave of amplitude A travelling in the positive x-direction, the velocity potential may be written as

$$\phi = Z(z)\mathrm{Re}\{e^{i(kx - \sigma t)}\} \qquad (4.44)$$

where Re represents the real part of the complex solution.

Therefore, the real form of the solution is

$$\phi = Z(z)\cos(kx - \sigma t). \tag{4.45}$$

Using this solution in Laplace's equation, we have

$$Z'' - k^2 Z = 0 \tag{4.46}$$

the solution of which is

$$Z = B \cosh kz + D \sinh kz \tag{4.47}$$

where B and D are arbitrary constants.
Therefore

$$\phi = (B \cosh kz + D \sinh kz)\cos(kx - \sigma t). \tag{4.48}$$

The boundary conditions to be satisfied by (4.48) are

$$\frac{\partial \phi}{\partial z} = 0 \qquad z = -h \tag{4.49}$$

$$\eta = \frac{1}{g}\left(\frac{\partial \phi}{\partial t}\right)_{z=0} \qquad z = 0. \tag{4.50}$$

Using (4.49), we obtain $B \sinh kh - D \cosh kh = 0$. Therefore

$$D = B \tanh kh. \tag{4.51}$$

Then (4.48) yields

$$\phi = B \frac{\cosh k(z + h)}{\cosh kh} \cos(kx - \sigma t). \tag{4.52}$$

Using (4.50), we have

$$\eta = \frac{B\sigma}{g} \sin(kx - \sigma t). \tag{4.53}$$

Defining

$$A = \frac{B\sigma}{g}, \tag{4.54}$$

where A is the amplitude of the wave, then

$$\eta = A \sin(kx - \sigma t). \tag{4.55}$$

Consequently

$$\phi = \frac{Ag}{\sigma} \frac{\cosh k(z+h)}{\cosh kh} \cos(kx - \sigma t). \tag{4.56}$$

Similarly, we can arrive at the other three solution forms of ϕ using the appropriate product solutions.

Table 4.1 Analytical Expressions of Physical Quantities Corresponding to Sine Progressive Waves

(a) Velocity potential: $\phi = (Ag/\sigma)\dfrac{\cosh k(z+h)}{\cosh kh} \cos(kx - \sigma t)$

Associated physical quantities:

Wave elevation: $\eta = \dfrac{1}{g}\left(\dfrac{\partial\phi}{\partial t}\right)_{z=0} = A \sin(kx - \sigma t)$

Horizontal velocity component: $u = -\dfrac{\partial\phi}{\partial x} = \dfrac{Agk}{\sigma} \dfrac{\cosh k(z+h)}{\cosh kh} \sin(kx - \sigma t)$

Vertical velocity component: $w = -\dfrac{\partial\phi}{\partial z} = -\dfrac{Agk}{\sigma} \dfrac{\sinh k(z+h)}{\cosh kh} \cos(kx - \sigma t)$

Horizontal acceleration: $\dfrac{\partial u}{\partial t} = -Agk \dfrac{\cosh k(z+h)}{\cosh kh} \cos(kx - \sigma t)$

Vertical acceleration: $\dfrac{\partial w}{\partial t} = -Agk \dfrac{\sinh k(z+h)}{\cosh kh} \sin(kx - \sigma t)$

Pressure: $P = \rho\dfrac{\partial\phi}{\partial t} - \rho gz = \rho Ag \dfrac{\cosh k(z+h)}{\cosh kh} \sin(kx - \sigma t) - \rho gz$

(b) Velocity potential: $\phi = -\dfrac{Ag}{\sigma} \dfrac{\cosh k(z+h)}{\cosh kh} \cos(kx - \sigma t)$

Associated physical quantities:

Wave elevation: $\eta = -\dfrac{1}{g}\left(\dfrac{\partial\phi}{\partial t}\right)_{z=0} = A \sin(kx - \sigma t)$

Horizontal velocity component: $u = \dfrac{\partial\phi}{\partial x} = \dfrac{Agk}{\sigma} \dfrac{\cosh k(z+h)}{\cosh kh} \sin(kx - \sigma t)$

Vertical velocity component: $w = \dfrac{\partial\phi}{\partial z} = -\dfrac{Agk}{\sigma} \dfrac{\sinh k(z+h)}{\cosh kh} \cos(kx - \sigma t)$

Horizontal acceleration: $\dfrac{\partial u}{\partial t} = -Agk \dfrac{\cosh k(z+h)}{\cosh kh} \cos(kx - \sigma t)$

Vertical acceleration: $\dfrac{\partial w}{\partial t} = -Agk \dfrac{\sinh k(z+h)}{\cosh kh} \sin(kx - \sigma t)$

Pressure: $P = -\rho\dfrac{\partial\phi}{\partial t} - \rho gz = \rho Ag \dfrac{\cosh k(z+h)}{\cosh kh} \sin(kx - \sigma t) - \rho gz$

If the progressive wave is travelling from $-\infty$ to ∞, making an angle δ with the x-axis, then the forms of ϕ and η must be modified to yield

$$\phi = \frac{Ag}{\sigma} \frac{\cosh k(z+h)}{\cosh kh} \cos(kx \cos \delta + ky \sin \delta - \sigma t) \tag{4.57}$$

$$\eta = A \sin(kx \cos \delta + ky \sin \delta - \sigma t). \tag{4.58}$$

Tables 4.1 and 4.2 summarize the analytic expressions of physical quantities corresponding to sine and cosine progressive waves, respectively, using $\mathbf{V} = \pm \operatorname{grad} \phi$.

Table 4.2 Analytical Expressions of Physical Quantities Corresponding to Cosine Progressive Waves

(a) Velocity potential: $\phi = -\dfrac{Ag}{\sigma} \dfrac{\cosh k(z+h)}{\cosh kh} \sin(kx - \sigma t)$

Associated physical quantities:

Wave elevation: $\eta = \dfrac{1}{g} \left(\dfrac{\partial \phi}{\partial t} \right)_{z=0} = A \cos(kx - \sigma t)$

Horizontal velocity component: $u = -\dfrac{\partial \phi}{\partial x} = \dfrac{Agk}{\sigma} \dfrac{\cosh k(z+h)}{\cosh kh} \cos(kx - \sigma t)$

Vertical velocity component: $w = -\dfrac{\partial \phi}{\partial z} = \dfrac{Agk}{\sigma} \dfrac{\sinh k(z+h)}{\cosh kh} \sin(kx - \sigma t)$

Horizontal acceleration: $\dfrac{\partial u}{\partial t} = Agk \dfrac{\cosh k(z+h)}{\cosh kh} \sin(kx - \sigma t)$

Vertical acceleration: $\dfrac{\partial w}{\partial t} = -Agk \dfrac{\cosh k(z+h)}{\cosh kh} \cos(kx - \sigma t)$

Pressure: $P = \rho \dfrac{\partial \phi}{\partial t} - \rho gz = \rho gA \dfrac{\cosh k(z+h)}{\cosh kh} \cos(kx - \sigma t) - \rho gz$

(b) Velocity potential: $\phi = \dfrac{Ag}{\sigma} \dfrac{\cosh k(z+h)}{\cosh kh} \sin(kx - \sigma t)$

Associated physical quantities:

Wave elevation: $\eta = -\dfrac{1}{g} \left(\dfrac{\partial \phi}{\partial t} \right)_{z=0} = A \cos(kx - \sigma t)$

Horizontal velocity component: $u = \dfrac{\partial \phi}{\partial x} = \dfrac{Agk}{\sigma} \dfrac{\cosh k(z+h)}{\cosh kh} \cos(kx - \sigma t)$

Vertical velocity component: $w = \dfrac{\partial \phi}{\partial z} = \dfrac{Agk}{\sigma} \dfrac{\sinh k(z+h)}{\cosh kh} \sin(kx - \sigma t)$

Vertical acceleration: $\dfrac{\partial w}{\partial t} = -Agk \dfrac{\sinh k(z+h)}{\cosh kh} \cos(kx - \sigma t)$

Pressure: $P = -\rho \dfrac{\partial \phi}{\partial t} - \rho gz = \rho gA \dfrac{\cosh k(z+h)}{\cosh kh} \cos(kx - \sigma t) - \rho gz.$

4.5 Dispersion relation of wave motion

For small-amplitude waves, the vertical velocity component w is equal to the rate of rise of the water surface at any point. Thus $w = d\eta/dt = \partial\eta/\partial t + u(\partial\eta/\partial x)$. On neglecting the second-order term on the right, we obtain $\partial\eta/\partial t \simeq w$ on $z = 0$. Since $w = -\partial\phi/\partial z$, hence

$$\frac{\partial\eta}{\partial t} = -\frac{\partial\phi}{\partial z} \tag{4.59}$$

which is known as the kinematic boundary condition. We know that the dynamic boundary condition is given by

$$\eta = \frac{1}{g}\left(\frac{\partial\phi}{\partial t}\right)_{z=0}. \tag{4.60}$$

Then combining these two boundary conditions at the free surface, we obtain a single free-surface boundary condition

$$\frac{\partial^2\phi}{\partial t^2} + g\frac{\partial\phi}{\partial z} = 0 \quad \text{at} \quad z = 0. \tag{4.61}$$

Considering a progressive wave moving in the positive x-direction, where the velocity potential is given by

$$\phi = \frac{Ag \cosh k(z+h)}{\sigma \cosh kh} \cos(kx - \sigma t),$$

we have

$$\frac{\partial^2\phi}{\partial t^2} = -Ag\sigma\frac{\cosh k(z+h)}{\cosh kh} \cos(kx - \sigma t) \tag{4.62}$$

$$g\frac{\partial\phi}{\partial z} = \frac{Ag^2 k \sinh k(z+h)}{\sigma \cosh kh} \cos(kx - \sigma t). \tag{4.63}$$

Substituting these values into (4.61) on $z = 0$ yields

$$\sigma^2 = gk \tanh kh. \tag{4.64}$$

This relation is known as the dispersion relation. The same result can be obtained for a progressive wave travelling in the negative x-direction.

Since $\sigma = kC$, (4.64) may be written as

$$C^2 = \frac{g}{k} \tanh kh. \tag{4.65}$$

Equation (4.65) expresses the rate of propagation of a surface wave as a function of water depth h and wavelength L. To find the wavelength, the dispersion relation (4.64) may be rearranged to give

$$L = \frac{gT^2}{2\pi} \tanh\left(\frac{2\pi h}{L}\right). \tag{4.66}$$

It is noted that for given water depth h and wave period T, the wavelength L can be determined from (4.66) by a trial and error procedure. Equations (4.64), (4.65) and (4.66) are the dispersion relations of water waves.

4.6 Classification of water waves

Water waves are classified into three main categories according to the relative depth of the ocean. This relative depth is defined as h/L, where h is the depth of the ocean and L is the wavelength. If the relative depth is below $\frac{1}{20}$, then the depth is considered small in comparison with the wavelength and the waves are termed 'shallow water' waves (or long waves). If the ratio is greater than $\frac{1}{2}$, the waves are called 'deep water' waves (or short waves). For $\frac{1}{20} < h/L < \frac{1}{2}$, the waves are called 'intermediate depth' waves, and usually in this range the wave equations do not simplify; however, in many circumstances the waves belong to either the 'shallow water' or 'deep water' categories.

The dispersion relation obtained in (4.64), (4.65) and (4.66) can now be simplified.

For shallow water waves, we know that $\tanh kh \simeq kh$ and, therefore, the dispersion relation (4.65) reduces to

$$C^2 = gh, \tag{4.67}$$

which is the well-known equation for the propagation of a small-amplitude surge. In this case, the wave celerity is seen to be independent of the wavelength.

For deep water waves, we know that $\tanh kh \simeq 1$, and, therefore, the dispersion relations (4.65) and (4.66) can be expressed as

$$C^2 = \frac{gL}{2\pi} \quad \text{or} \quad L = \frac{gT^2}{2\pi}. \tag{4.68}$$

Thus the celerity and wavelength are independent of water depth. When $g = 9.81$ m/s^2, then

$$L = 1.56T^2 \tag{4.69}$$

where the dimension of L is in metres.

4.7 Particle motion and pressure

As observed the velocity potential for a small-amplitude progressive wave travelling in the positive x-direction is

$$\phi = \frac{Ag \cosh k(z+h)}{\sigma \cosh kh} \cos(kx - \sigma t).$$

Using the definition of velocity components of the fluid particle we may obtain the horizontal and vertical components of velocity as follows:

$$\frac{dx}{dt} = u = -\frac{\partial \phi}{\partial x} = \frac{Agk \cosh k(z+h)}{\sigma \cosh kh} \sin(kx - \sigma t) \tag{4.70}$$

$$\frac{dz}{dt} = w = -\frac{\partial \phi}{\partial z} = -\frac{Agk \sinh k(z+h)}{\sigma \cosh kh} \cos(kx - \sigma t). \tag{4.71}$$

These equations express the velocity components within the wave at any depth z. At a given depth the velocities are seen to be harmonic in x and t. At a given phase angle $\theta = kx - \sigma t$, the hyperbolic function of z causes an exponential decay of the velocity components with distance below the free surface.

Experimental data confirms that at $z = -L/2$ the velocity becomes negligible, and below this depth there is practically no motion (see Fig. 4.2).

The local accelerations are easily obtained from (4.70) and (4.71), and are

$$\frac{\partial u}{\partial t} = -Agk \frac{\cosh k(z+h)}{\cosh kh} \cos(kx - \sigma t) \tag{4.72}$$

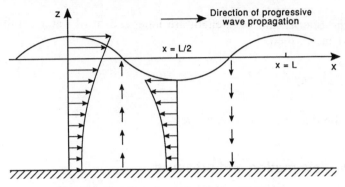

Fig. 4.2. Variation of particle velocity with depth.

and

$$\frac{\partial w}{\partial t} = -Agk\frac{\sinh k(z+h)}{\cosh kh}\sin(kx - \sigma t). \tag{4.73}$$

The vertical particle displacement cannot exceed the wave amplitude A; therefore, it is assumed that the displacement of any fluid particle from its mean position is small. We can compute the horizontal and vertical displacements of the fluid particle from its mean position by using

$x = \xi =$ horizontal displacement from mean position

$$= \int u\,\mathrm{d}t = \frac{Agk}{\sigma}\frac{\cosh k(z+h)}{\cosh kh}\int \sin(kx - \sigma t)\mathrm{d}t$$

$$= \frac{Agk}{\sigma^2}\frac{\cosh k(z+h)}{\cosh kh}\cos(kx - \sigma t), \tag{4.74}$$

and

$z = \zeta =$ vertical displacement from mean position

$$= \int w\,\mathrm{d}t = -\frac{Agk}{\sigma}\frac{\sinh k(z+h)}{\cosh kh}\int \cos(kx - \sigma t)\mathrm{d}t$$

$$= \frac{Agk}{\sigma^2}\frac{\sinh k(z+h)}{\cosh kh}\sin(kx - \sigma t). \tag{4.75}$$

By using the dispersion relation, $\sigma^2 = gk\tanh kh$, (4.74) and (4.75) may be further simplified to give the following expressions:

$$\xi = A\frac{\cosh k(z+h)}{\sinh kh}\cos(kx - \sigma t) \tag{4.76}$$

and

$$\zeta = A \frac{\sinh k(z+h)}{\sinh kh} \sin(kx - \sigma t). \tag{4.77}$$

These two equations can be combined to give

$$\frac{\xi^2}{\alpha^2} + \frac{\zeta^2}{\beta^2} = 1, \tag{4.78}$$

where

$$\alpha = \frac{A \cosh k(z+h)}{\sinh kh} \tag{4.79}$$

and

$$\beta = \frac{A \sinh k(z+h)}{\sinh kh}. \tag{4.80}$$

Equation (4.78) represents an ellipse with a major semi-axis (horizontal) of α and a minor semi-axis (vertical) of β. The particle paths are, therefore, generally elliptic in shape. The specific form of the particle paths for shallow water and deep water can easily be determined by examining the values of α and β.

For shallow water waves, it may be readily seen that

$$\alpha = \frac{A}{kh} \quad \text{and} \quad \beta = \frac{Ak(z+h)}{kh}. \tag{4.81}$$

The maximum horizontal displacement is constant from the surface to the bottom of the ocean. The maximum vertical displacement varies from zero at the bottom to the wave amplitude A at the surface.

For deep water waves, the values of α and β are given by

$$\alpha = A \frac{e^{kz}e^{kh} + e^{-kz}e^{-kh}}{e^{kh} - e^{-kh}} \tag{4.82}$$

$$\beta = A \frac{e^{kz}e^{kh} - e^{-kz}e^{-kh}}{e^{kh} - e^{-kh}}. \tag{4.83}$$

Thus when $h \to \infty$,

$$\alpha = A e^{kz} \quad \text{and} \quad \beta = A e^{kz}. \tag{4.84}$$

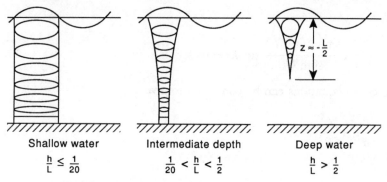

Shallow water $\qquad$ Intermediate depth $\qquad$ Deep water

$$\frac{h}{L} \leq \frac{1}{20} \qquad\qquad \frac{1}{20} < \frac{h}{L} < \frac{1}{2} \qquad\qquad \frac{h}{L} > \frac{1}{2}$$

Fig. 4.3. Schematic diagram of water particle trajectories.

The major and minor axes for this case are equal so that each particle describes a circular path. The radii of these circles are given by the formula Ae^{kz}, and, therefore, diminish rapidly downwards. Again at the surface, the vertical displacement is equal to the wave amplitude A. Figure 4.3 is a schematic representation of water particle trajectories.

The pressure field within a progressive wave can be obtained from the linearized Bernoulli's equation

$$\frac{P}{\rho} = \frac{\partial \phi}{\partial t} - gz. \tag{4.85}$$

Using the velocity potential ϕ for a progressive wave moving in the positive x-direction, (4.85) becomes

$$\frac{P}{\rho} = Ag \frac{\cosh k(z+h)}{\cosh kh} \sin(kx - \sigma t) - gz$$

or

$$P = A\rho g \frac{\cosh k(z+h)}{\cosh kh} \sin(kx - \sigma t) - \rho gz. \tag{4.86}$$

From the pressure expression (4.86), a number of quantities of physical interest can be derived such as wave forces and moments.

4.8 Superposition of waves

We know that Laplace's equation is linear. Therefore, the velocity potential of the wave system, ϕ_T, is given by the sum of the potentials of the individual waves.

Thus

$$\phi_T = \phi_1 + \phi_2 + \ldots + \phi_n,$$ (4.87)

where

$$\phi_n = \frac{gA_n}{\sigma_n} \frac{\cosh k_n(z+h)}{\cosh k_n h} \cos(k_n x \pm \sigma_n t + \delta_n).$$ (4.88)

Equation (4.88) is the velocity potential for a progressive small-amplitude wave travelling in the positive $(-\sigma_n t)$ or negative $(+\sigma_n t)$ x-direction. δ_n is the phase difference between various waves and is measured from the origin $(kx \pm \sigma t)$.

We know that at the free surface $z = \eta$,

$$\eta = \frac{1}{g}\left(\frac{\partial \phi}{\partial t}\right)_{z=0}$$ (4.89)

and hence, applying (4.88), we obtain

$$\eta_T = \eta_1 + \eta_2 + \ldots + \eta_n.$$ (4.90)

Similarly, the fluid particle velocity components can be written as

$$u_T = u_1 + u_2 + \ldots + u_n, \qquad w_T = w_1 + w_2 + \ldots + w_n.$$ (4.91)

Also, the pressure field can be evaluated as

$$\frac{P_T}{\rho} = \frac{\partial \phi_1}{\partial t} + \frac{\partial \phi_2}{\partial t} + \ldots + \frac{\partial \phi_n}{\partial t} - gz.$$ (4.92)

Using this superposition technique, the equation for waves travelling in the positive x-direction, but of variable amplitudes, is

$$\eta_T = \sum_n A_n \sin(k_n x - \sigma_n t + \delta_n).$$ (4.93)

We shall assume that all the waves are moving in water of the same depth, and also that the period of all wave components is the same. Then by the dispersion relationship we know that all the σ and the k are identical. Under these special condition (4.93) reduces to

$$\eta_T = r \sin(kx - \sigma t + \lambda).$$ (4.94)

Here r is such that $r \cos \lambda = \Sigma A_n \cos \delta_n$ and $r \sin \lambda = \Sigma A_n \sin \delta_n$.

Hence

$$r = \left[\left(\sum A_n \cos \delta_n \right)^2 + \left(\sum A_n \sin \delta_n \right)^2 \right]^{1/2} \tag{4.95}$$

and

$$\lambda = \tan^{-1} \left(\frac{\sum A_n \sin \delta_n}{\sum A_n \cos \lambda_n} \right). \tag{4.96}$$

For the special case in which there are only two components, both of the same period,

$$\eta_T = A_1 \sin(kx - \sigma t + \delta_1) + A_2 \sin(kx - \sigma t + \delta_2)$$

$$= \sin(kx - \sigma t) \sum_{n=1}^{2} A_n \cos \delta_n + \cos(kx - \sigma t) \sum_{n=1}^{2} A_n \sin \delta_n$$

$$= r \sin(kx - \sigma t + \lambda) \tag{4.97}$$

where

$$r^2 = A_1^2 + A_2^2 + 2A_1 A_2 \cos(\delta_1 - \delta_2)$$

$$\lambda = \tan^{-1} \left(\frac{A_1 \sin \delta_1 + A_2 \sin \delta_2}{A_1 \cos \delta_1 + A_2 \cos \delta_2} \right). \tag{4.98}$$

If the two components are in phase, that is if $\delta_1 = \delta_2$, then

$$\eta_T = (A_1 + A_2)\sin(kx - \sigma t + \delta_1). \tag{4.99}$$

If the two components are 180° out of phase, i.e. $\delta_1 - \delta_2 = \pi$, then

$$\eta_T = (A_1 - A_2)\sin(kx - \sigma t + \delta_1). \tag{4.100}$$

If the two components differ in phase by $\delta_1 - \delta_2 = \pi/2$, then

$$\eta_T = A_1 \sin(kx - \sigma t + \delta_1) - A_2 \cos(kx - \sigma t + \delta_1). \tag{4.101}$$

4.9 Wave reflection and standing wave

This section deals with the reflection of waves and consequently the concept of standing waves. Suppose the reflection is caused by a plane vertical barrier located at $x = b$. It is assumed that the outgoing reflective wave has the same amplitude as the incoming incident wave. Also, the frequency and wave number of the reflected wave are identical with those of the incident wave.

By definition we know that the reflection coefficient K_r is given by

$$K_r = \frac{\text{amplitude of reflected wave}}{\text{amplitude of incident wave}}$$

and must be unity for perfect reflection.

For this system of waves the velocity potential may be written

$$\phi_T = \frac{Ag}{\sigma} \frac{\cosh k(z+h)}{\cosh kh} [\cos(kx - \sigma t) - \cos(kx + \sigma t + \delta_2)].$$

$$(4.102)$$

If the barrier is impermeable, then the horizontal component of velocity must be zero at the barrier site, at $x = b$. Thus, the boundary condition is

$$u_T = -\frac{\partial \phi_T}{\partial x} = 0 \quad \text{at} \quad x = b. \tag{4.103}$$

Applying this boundary condition to (4.102), we obtain $\sin(kb - \sigma t) = \sin(kb + \sigma t + \delta_2)$. Expanding and equating the coefficients of $\sin \sigma t$ and $\cos \sigma t$, we obtain $\sin kb = \sin(kb + \delta_2)$ and $\cos kb = -\cos(kb + \delta_2)$. The solution of these equations yields $\delta_2 = (2n + 1)\pi - 2kb$, $n = 0, 1, 2, \ldots$.

For two progressive waves moving in opposite directions and having the same amplitudes

$$\eta_T = A \sin(kx - \sigma t) + A \sin(kx + \sigma t + \delta_2)$$

$$= A \sin(kx - \sigma t) + A \sin(kx + \sigma t)\cos \delta_2 + A \cos(kx + \sigma t)\sin \delta_2.$$

$$(4.104)$$

Substituting the values of δ_2 in this expression for η_T we obtain

$$\eta_T = A \sin kx \cos \sigma t - A \sin \sigma t \cos kx - A \cos 2kb \sin kx \cos \sigma t$$

$$- A \cos 2kb \cos kx \sin \sigma t + A \sin 2kb \cos kx \cos \sigma t$$

$$- A \sin 2kb \sin kx \sin \sigma t$$

which can be written in the form

$$\eta_T = 2A \sin(kb - \sigma t)\cos(kx - kb). \tag{4.105}$$

Equation (4.105) is the product of two terms, one independent of x and the other independent of t. Thus, there are certain times when $\eta_T = 0$ for all x and there are certain x of which $\eta_T = 0$ for all time. These latter points are called the nodes of the system and are located by the condition $\cos(kx - kb) = 0$, which on solving yields $x = b + [(2n + 1)\pi]/2k$, $n = 0, 1, 2, 3, 4, \ldots$. Such a condition of stationary nodes defines a 'standing wave'. Figure 4.4 shows the composition of standing waves.

It is noted that the slopes of the free surface of η_1 and η_2 are always equal and opposite at $x = b$. This yields the condition

$$\frac{\partial \eta_T}{\partial x} = 0 \quad \text{at} \quad x = b \quad \text{for all time} \tag{4.106}$$

which is evident from (4.105). It is convenient to take the origin of x at the barrier. This avoids having to locate the origin of x in order to determine b. Setting $b = 0$, we find that $\delta_2 = (2n + 1)\pi$. Taking $x = 0$ we see that the reflection process preserves the phase of the incident waves.

Equation (4.105) then becomes

$$\eta_T = -2A \sin \sigma t \cos kx. \tag{4.107}$$

In contrast to the stationary nodes of the standing wave, the 'progressive' wave

$$\eta = A \sin(kx - \sigma t) \tag{4.108}$$

has nodes which 'progress' according to the relation $\sin(kx - \sigma t) = 0$ which on solving yields $x_{\text{node}} = (n\pi + \sigma t)/k$, $n = 0, 1, 2, \ldots$. Using the

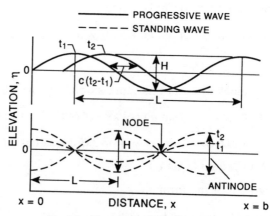

Fig. 4.4. Formation of standing waves.

value of δ_2 in (4.102), the velocity potential for the standing wave may be obtained and is

$$\phi_T = \frac{2Ag}{\sigma} \frac{\cosh k(z+h)}{\cosh kh} \cos(kb - \sigma t)\cos(kx - kb). \tag{4.109}$$

Using (4.109), the velocity components of standing waves are given by

$$u = -\frac{\partial \phi_T}{\partial x} = \frac{2Akg}{\sigma} \frac{\cosh k(z+h)}{\cosh kh} \cos(kb - \sigma t)\sin(kx - kb) \tag{4.110}$$

$$w = -\frac{\partial \phi_T}{\partial z} = -\frac{2Akg}{\sigma} \frac{\sinh k(z+h)}{\cosh kh} \cos(kb - \sigma t)\cos(kx - kb). \tag{4.111}$$

We have seen that the nodes occur where $\cos(kx - kb) = 0$; therefore, only horizontal motion occurs under the nodes and only vertical motion occurs under the antinodes. The velocity field for the standing wave is thus cellular.

4.10 Wave energy and group velocity

Consider the case of a wave group within an infinite series of disturbances. For simplicity we consider two waves moving in the positive x-direction, having equal amplitudes and in phase at the origin of $(kx - \sigma t)$.
Thus

$$\eta_T = A\sin(k_1 x - \sigma_1 t) + A\sin(k_2 x - \sigma_2 t), \tag{4.112}$$

which can be rewritten as

$$\eta_T = 2A\cos[\tfrac{1}{2}(k_1 - k_2)x - \tfrac{1}{2}(\sigma_1 - \sigma_2)t]$$
$$\times \sin[\tfrac{1}{2}(k_1 + k_2)x - \tfrac{1}{2}(\sigma_1 + \sigma_2)t]. \tag{4.113}$$

The point of zero amplitude of the wave envelope thus separates groups of individual waves. These nodal points are located by finding the zeros of the cosine factor in (4.113).

Now $\eta_{T\,max} = 0$ gives

$$\tfrac{1}{2}(k_1 - k_2)x - \tfrac{1}{2}(\sigma_1 - \sigma_2)t = (2n+1)\frac{\pi}{2}$$

or

$$x_{node} = \frac{\sigma_1 - \sigma_2}{k_1 - k_2}t + \frac{(2n+1)\pi}{k_1 - k_2}.$$

Since the position of all nodes is a function of time, they are not stationary. At $t=0$ there will be nodes at $x = [(2n+1)\pi]/(k_1 - k_2)$, $n = 0,1,2,3,\ldots$. Thus the distance between two consecutive nodes is

$$(x_2 - x_1) = \Delta x = \frac{2\pi}{k_1 - k_2} = \frac{L_1 L_2}{L_2 - L_1}.$$

The speed of propagation of these nodes, and hence of the wave group, is called the *group velocity* and is given by

$$\frac{\mathrm{d}x_{node}}{\mathrm{d}t} = C_g = \left(\frac{\sigma_1 - \sigma_2}{k_1 - k_2}\right). \tag{4.114}$$

Here C_g is given by the limit as σ_1 approaches σ_2, that is $C_g = \mathrm{d}\sigma/\mathrm{d}k$.

We know that $\sigma = Ck$ and thus in terms of the wavelength L and wave phase velocity C, (4.114) can be written as

$$C_g = \frac{\mathrm{d}(Ck)}{\mathrm{d}k} = C + k\frac{\mathrm{d}C}{\mathrm{d}k} = C + k\frac{\mathrm{d}C}{\mathrm{d}L}\frac{\mathrm{d}L}{\mathrm{d}k} = C - L\frac{\mathrm{d}C}{\mathrm{d}L}. \tag{4.115}$$

However, using the relation $C^2 = (g/k)\tanh kh$, the group velocity is given by

$$C_g = \frac{\mathrm{d}(Ck)}{\mathrm{d}k} = C\frac{1}{2}\left(1 + \frac{2kh}{\sinh 2kh}\right). \tag{4.116}$$

For deep water waves $C_g/C \sim \tfrac{1}{2}$, and for shallow water $C_g/C \sim 1$. From this it is seen that individual waves, since they always move faster than the group, appear to originate at the rear node of the group and move towards the head of the group, at which point the pressure of the front node forces their apparent disappearance.

Example 4.1

Establish the result given in eqn (4.116), and find the limiting conditions for deep and shallow water cases.

Solution

We know that the group velocity is given by

$$C_g = \frac{d(Ck)}{dk} = C + k\frac{dC}{dk},$$

where $C = \sqrt{g/k}\,(\tanh kh)^{1/2}$. Thus

$$\frac{dC}{dk} = \sqrt{g}\left(-\frac{1}{2}\right)k^{-3/2}(\tanh kh)^{1/2} + \sqrt{\frac{g}{k}}\,\frac{1}{2}(\tanh kh)^{-1/2}\,\text{sech}^2 kh(h).$$

So

$$k\frac{dC}{dk} = -\frac{1}{2}\sqrt{\frac{g}{k}}\,\tanh^{1/2} kh + \frac{1}{2}\sqrt{\frac{g}{k}}\,\tanh^{1/2} kh\,\frac{\text{sech}^2 kh}{\tanh kh}(kh)$$

$$= -\frac{1}{2}C + \frac{1}{2}Ckh\frac{1}{\sinh kh\cosh kh}$$

$$= -\frac{1}{2}C + \frac{C}{2}\frac{2kh}{\sinh 2kh}.$$

Thus

$$C_g = C + k\frac{dC}{dk}$$

$$= \frac{C}{2}\left(1 + \frac{2kh}{\sinh 2kh}\right),$$

where C_g = group velocity, C = phase velocity and

$$\frac{C_g}{C} = \frac{1}{2}\left(1 + \frac{2kh}{\sinh 2kh}\right)$$

for the intermediate depth case. For shallow water, $kh \to 0$, $C_g/C = 1$, and for deep water, $kh \to \infty$, $C_g/C = \frac{1}{2}$.

We shall calculate the potential energy due to the presence of the progressive wave form on the free surface. In order to determine this, we shall first find the potential energy of the wave above $z = -h$ where there is a wave form present; then we shall subtract from this the potential energy of the water where there is no wave form present.

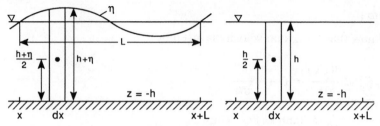

Fig. 4.5.   Definition sketches for potential energy.

The potential energy (with respect to $z = -h$) of a column of water $h + \eta$ high, dx long and of 1 unit width (Fig. 4.5) is

$$\Delta(PE_1) = \text{(height to centre of gravity)} \times g \, \Delta M$$

$$\Delta(PE_1) = \left(\frac{h + \eta}{2}\right)(h + \eta)\rho g \, \Delta x$$

$$= \frac{(h + \eta)^2}{2} \rho g \, \Delta x.$$

Therefore, the average potential energy per unit surface area is

$$PE_1 = \frac{\rho g}{2LT} \int_t^{t+T} \int_x^{x+L} (h + \eta)^2 \, dx \, dt. \tag{4.117}$$

Using $\eta = A \sin(kx - \sigma t)$, we obtain

$$PE_1 = \frac{\rho g h^2}{2} + \frac{\rho g A^2}{4}. \tag{4.118}$$

The potential energy in the absence of a wave form would be

$$PE_2 = \frac{\rho g}{2LT} \int_t^{t+T} \int_x^{x+L} h^2 \, dx \, dt = \frac{\rho g h^2}{2}. \tag{4.119}$$

Thus, the average potential energy which is attributed to the presence of the progressive wave on the free surface is

$$PE = PE_1 - PE_2 = \frac{\rho g A^2}{4}. \tag{4.120}$$

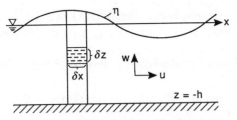

Fig. 4.6. Definition sketch for kinetic energy.

The kinetic energy of a small element δx long, δz high, of unit width and with velocity components u and w (Fig. 4.6), is given by the formula

$$\delta(KE) = \tfrac{1}{2}(u^2 + w^2)\delta M = \tfrac{1}{2}(u^2 + w^2)\rho \,\delta x \,\delta z. \tag{4.121}$$

The average kinetic energy per unit of surface area is then given by

$$KE = \frac{\rho}{2LT} \int_t^{t+T} \int_x^{x+L} \int_{-h}^{\eta=0} (u^2 + w^2)\mathrm{d}x\,\mathrm{d}z\,\mathrm{d}t. \tag{4.122}$$

Using the velocity components compatible with the progressive wave $\eta = A\sin(kx - \sigma t)$, we obtain

$$KE = \frac{\rho g A^2}{4}. \tag{4.123}$$

Therefore, the total energy is given by

$$E = PE + KE = \frac{\rho g A^2}{2}. \tag{4.124}$$

4.11 Energy of a complex wave form

It has been shown above that the calculations of both components of wave energy involve the square of the amplitude of the free-surface elevation. It is well known that any periodic function $\eta(t)$ of period $2p$ can be expressed as a Fourier series, that is

$$\eta(t) = \frac{a_0}{2} + \sum_{n=1}^{\infty} \left(a_n \cos \frac{n\pi t}{p} + b_n \sin \frac{n\pi t}{p} \right), \tag{4.125}$$

provided that $\int_{-p}^{p} |\eta(t)|\mathrm{d}t$ is finite, where a_n and b_n are the amplitudes of the nth harmonics of the cosine and sine wave respectively.

It can be easily demonstrated that (4.125) may be written in complex form as (see Rahman (1991)) $\eta(t) = \sum_{n=-\infty}^{\infty} C_n e^{n\pi it/p}$ where $C_0 = a_0/2$, $C_n = (a_n - ib_n)/2$, $C_{-n} = (a_n + ib_n)/2$.

The complex Fourier coefficient C_n may be obtained from

$$C_n = \frac{1}{2p} \int_{-p}^{p} \eta(t) e^{-n\pi it/p} \, dt, \qquad n = 0, \pm 1, \pm 2, \ldots .$$

Then the well-known Parseval theorem for periodic functions gives

$$\frac{\rho g}{2p} \int_{-p}^{p} \eta^2(t) dt = \frac{\rho g}{2p} \int_{-p}^{p} \eta(t) \cdot \eta(t) dt$$

$$= \frac{\rho g}{2p} \int_{-p}^{p} \eta(t) \left(\sum_{n=-\infty}^{\infty} C_n e^{\frac{n\pi it}{p}} \right) dt$$

$$= \rho g \sum_{n=-\infty}^{\infty} C_n \left(\frac{1}{2p} \int_{-p}^{p} \eta(t) e^{\frac{n\pi it}{p}} \, dt \right)$$

$$= \rho g \sum_{n=-\infty}^{\infty} C_n C_{-n}$$

$$= \rho g \sum_{n=-\infty}^{\infty} |C_n|^2 = \rho g \left(\frac{a_0^2}{4} + \frac{1}{2} \sum_{n=1}^{\infty} (a_n^2 + b_n^2) \right).$$

Thus the potential energy is

$$PE = \rho g \left[\frac{a_0^2}{4} + \frac{1}{2} \sum_{n=1}^{\infty} (a_n^2 + b_n^2) \right].$$

The first term of this result clearly matches (4.120) for a monochromatic progressive wave of amplitude $a_0(=A)$.

4.12 Wave shoaling

So far we have been concerned with the behaviour of regular waves propagating over a smooth horizontal sea-bed in the absence of variation in water depth, the presence of current; or obstacles in the flow. However, in practice, as a wave train propagates into shallow water, we observe the change in wave height and wavelength. This process is usually described as *wave shoaling*. The solution to the complete boundary value problem in which the boundary condition at the sea-bed takes into consideration the variation of depth is difficult to solve analytically. However, there are a host of numerical techniques available to solve this type of problem. The shoaling effect may be estimated for a particular wave theory under the assumptions that the motion is two-dimensional, the wave period is constant and the average rate of energy transfer in the direction of wave propagation is constant. These assumptions, however, require that the sea-bed has a gentle slope such that wave reflection can be neglected, and that the wave is neither supplied by the wind nor dissipated by bottom friction.

On the basis of the linear theory, we denote the dispersion relations (4.65) and (4.66) for deep water waves as

$$C_0 = gT/2\pi, L_0 = gT^2/2\pi, k_0 = 2\pi/(gT^2) \tag{4.126}$$

where the subscript 0 is used to denote the deep water case.

The dispersion relation (4.64) can then be written as

$$gk \tanh kh = \sigma^2 = gk_0 = \text{constant.} \tag{4.127}$$

This relation is true because the wave period is assumed constant such that (irrespective of the depth of the ocean)

$$Ck = C_0 k_0 = \sigma = \text{constant.} \tag{4.128}$$

Thus from eqns (4.127) and (4.128) we must have

$$C/C_0 = k_0/k = L/L_0 = \tanh kh. \tag{4.129}$$

The dispersion relation is given as $k \tanh kh = k_0$, or

$$kh \tanh kh = hk_0 = 2\pi h/L_0 = 4\pi^2 h/(gT^2) \tag{4.130}$$

which indicates that kh is a unique function of $h/(gT^2)$. It is now clear that the ratios in eqn (4.129) are uniquely determined at any given depth, $h/(gT^2)$.

In addition, the rate of energy transfer P is independent of depth. We then have

$$P = \tfrac{1}{2} \rho g A^2 C_g = \tfrac{1}{2} \rho g A_0^2 C_{g0} = \text{constant} \tag{4.131}$$

so that

$$\frac{A}{A_0} = \left(\frac{C_{g0}}{C_g}\right)^{\frac{1}{2}} = \left(\frac{2\cosh^2 kh}{2kh + \sinh 2kh}\right)^{\frac{1}{2}}. \tag{4.132}$$

Hence A/A_0 is also uniquely related to the depth, $h/(gT^2)$.

For shallow water waves, the usual approximation results in the following simplified relationships:

$$C/C_0 = L/L_0 = 2\pi(h/gT^2)^{\frac{1}{2}} = (2\pi h/L_0)^{\frac{1}{2}} \tag{4.133}$$

$$A/A_0 = (16\pi^2 h/gT^2)^{-\frac{1}{4}} = (8\pi h/L_0)^{-\frac{1}{4}}. \tag{4.134}$$

These relations are valid for the linear wave theory only.

4.13 Wave refraction

It is observed in the ocean that when waves approach a bottom slope obliquely, the speed of the wave front in shallower water is much smaller than that in deeper water in accordance with the dispersion relation, $C^2 = (g/k)\tanh kh$ (i.e. $C^2 = gh$ for shallow and $C^2 = g/k$ for deep), and as a result the line of the wave crest is bent so as to become more closely aligned with the bottom contours. This wave phenomenon is commonly known as *wave refraction*. It is depicted in Fig. 4.7 for a small time interval δt, occuring across a contour on either side of which the depths are taken as constant and differ by a small amount.

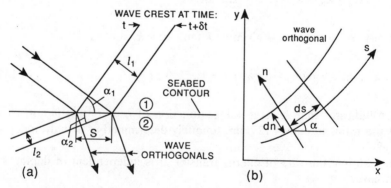

Fig. 4.7. Refraction of wave crests and orthogonals over short intervals (a) relative to a sea-bed contour, (b) relative to a prescribed coordinate system (X, Y).

The wave crest travels a distance l so that the speeds in regions 1 and 2 are given by

$$C_1 = \frac{l_1}{\delta t} = \frac{s \sin \alpha_1}{\delta t} \qquad (4.135)$$

$$C_2 = \frac{l_2}{\delta t} = \frac{s \sin \alpha_2}{\delta t}. \qquad (4.136)$$

Thus we have

$$C_1/C_2 = \sin \alpha_1/\sin \alpha_2 \qquad (4.137)$$

which is exactly Snell's law.

Here α is the angle the wave crest makes with the bottom contour; subscripts denote appropriate regions. Equation (4.137) may be applied to deep and deeper contours so that eventually the deep water conditions may be used as a reference, and for any depth in general

$$C/C_0 = \sin \alpha/\sin \alpha_0. \qquad (4.138)$$

This is the basis for the development of various numerical schemes to trace the paths of wave orthogonals which are the lines orthogonal to the wave crest from deep water to shoaling water in accordance with the contours describing a particular region. Many numerical methods are available to compute the wave refraction, including those of Jen (1969), Keulegan and Harrison (1970) and Skovgaard, Jonsson and Bertelsen (1975). With respect to the differential lengths ds and dn as indicated in Fig. 4.7(b), a differential form of Snell's law, eqn (4.138), may be obtained as (Sarpkaya and Isaacson (1981))

$$\frac{d\alpha}{ds} = -\frac{1}{C}\frac{dC}{dn} \qquad (4.139)$$

which may be expressed as

$$\frac{d\alpha}{ds} = -\frac{1}{C}\left(\frac{dC}{dx}\frac{dx}{dn} + \frac{dC}{dy}\frac{dy}{dn}\right) \qquad (4.140)$$

where

$$dx/dn = -\sin \alpha \qquad (4.141)$$

$$dy/dn = \cos \alpha. \qquad (4.142)$$

Using the relationships in (4.140), we obtain

$$\frac{d\alpha}{ds} = \frac{1}{C}\left(\sin\alpha\frac{dC}{dx} - \cos\alpha\frac{dC}{dy}\right). \tag{4.143}$$

We also have

$$dx/ds = \cos\alpha \tag{4.144}$$

$$dy/ds = \sin\alpha. \tag{4.145}$$

Equations (4.143), (4.144) and (4.145) are usually known as the ray equations and may be solved numerically to determine the variation of α and hence the paths of the orthogonals.

The variations of wave height for refracted waves may be estimated by considering the energy transfer. This energy transfer is assumed not be supplied or dissipated. Consider the distance between two neighbouring orthogonals (see Fig. 4.8). The energy transfer given by (4.131) may be modified to yield

$$\tfrac{1}{2}\rho g A^2 C_g b = \tfrac{1}{2}\rho g A_0^2 C_{g0} b_0 = \text{constant}. \tag{4.146}$$

This can be written as

$$A/A_0 = (b_0/b)^{\frac{1}{2}}(C_{g0}/C_g)^{\frac{1}{2}} = K_r K_s \tag{4.147}$$

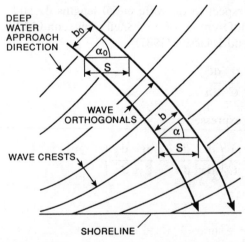

Fig. 4.8. Refraction of waves obliquely approaching a straight shoreline with uniform bed slope.

where $K_r = (b_0/b)^{\frac{1}{2}}$ is the refraction coefficient, and $K_s = (C_{g0}/C_g)^{\frac{1}{2}}$ the shoaling coefficient.

To understand this process, a fundamental example is represented by a wave train obliquely approaching a straight shoreline with a uniform sea-bed slope (see Fig. 4.8). The deep water approach angle which is between wave crest and bottom contour is given as α_0. Using the relations (4.129) and (4.138), we have

$$C/C_0 = L/L_0 = \sin\alpha/\sin\alpha_0 = \tanh kh \qquad (4.148)$$

$$kh\,\mathrm{tahn}\,kh = 4\pi^2 h/(gT^2). \qquad (4.149)$$

From Fig. 4.8, it is clear that the distance s is independent of location and correspondingly

$$s\cos\alpha_0 = b_0, \qquad s\cos\alpha = b$$

$$\text{or} \quad b_0/\cos\alpha_0 = b/\cos\alpha = s = \text{constant}. \qquad (4.150)$$

The variation of wave height is therefore given by

$$A/A_0 = (b_0/b)^{\frac{1}{2}}(C_{g0}/C_g)^{\frac{1}{2}} = (\cos\alpha_0/\cos\alpha)^{\frac{1}{2}}\left(\frac{2\cosh^2 kh}{2kh + \sinh 2kh}\right)^{\frac{1}{2}}$$

$$= \left(\frac{1 - \sin^2\alpha_0\tanh^2(kh)}{\cos^2\alpha_0}\right)^{-\frac{1}{4}}\left(\frac{2\cosh^2(kh)}{2kh + \sinh 2kh}\right)^{\frac{1}{2}}. \qquad (4.151)$$

In shallow water, relations (4.148), (4.149) and (4.151) can be simplified further to yield

$$C/C_0 = L/L_0 = 2\pi\left(\frac{h}{gT^2}\right)^{\frac{1}{2}} \qquad (4.152)$$

$$A/A_0 = \left(\frac{1 - \sin^2\alpha_0 4\pi^2\left(\dfrac{h}{gT^2}\right)}{\cos^2\alpha_0}\right)^{-\frac{1}{4}} (16\pi^2 h/(gT^2))^{-\frac{1}{4}}. \qquad (4.153)$$

These relations are valid for linear wave theory only.

4.14 Wave diffraction

When a wave train encounters a large vertical obstacle it has been observed that the wave motion penetrates into the region of geometric shadow. This phenomenon is usually known as the process of *diffraction*. Thus in establishing the wave action behind breakwaters or around large offshore structures, wave diffraction plays a very important role.

Many authors including Stoker (1957) and Mei (1983) have discussed the methods of solution in general. However, we shall give here a brief outline. First of all we assume that the fluid is incompressible and the flow is irrotational so that the velocity potential $\Phi = \text{Re}(\phi e^{-i\sigma t})$ satisfies Laplace's equation. We shall restrict our discussion to the linear theory only such that the wave height is assumed sufficiently small. Then the total velocity potential ϕ (a complex quantity) can be written as the sum of incident wave potential and scattered or diffracted wave potential as

$$\phi = \phi_I + \phi_S \tag{4.154}$$

where $\phi_I =$ incident wave potential and $\phi_S =$ scattered wave potential.

The free-surface and bottom boundary conditions can be written as

$$\frac{\partial^2 \Phi}{\partial t^2} + g \frac{\partial \Phi}{\partial z} = 0 \qquad \text{at} \qquad z = 0 \tag{4.155}$$

$$\eta = -\frac{1}{g}\left(\frac{\partial \Phi}{\partial t}\right)_{z=0} \tag{4.156}$$

$$\frac{\partial \Phi}{\partial z} = 0 \qquad \text{at} \qquad z = -h. \tag{4.157}$$

Equations (4.155) to (4.157) can subsequently be written as

$$\frac{\partial \phi}{\partial z} - \frac{\sigma^2}{g} \phi = 0 \qquad \text{at} \qquad z = 0 \tag{4.158}$$

$$\eta = \frac{\sigma}{g} \text{Re}(i\phi e^{-i\sigma t}) \tag{4.159}$$

$$\frac{\partial \phi}{\partial z} = 0 \qquad \text{at} \qquad z = -h. \tag{4.160}$$

The boundary condition at the body surface implies that the normal velocity component at the body surface must be zero:

$$\frac{\partial \phi}{\partial n} = \frac{\partial \phi_I}{\partial n} + \frac{\partial \phi_S}{\partial n} \tag{4.161}$$

at the body surface, where n is the distance normal to the body surface.

In order to ensure that a physically acceptable and unique solution is obtained, one more boundary condition is imposed on the scattered potential to correspond to the outgoing waves only. This condition is known as the radiation condition and has been discussed extensively by Sommerfeld (1949) in the context of generalized wave theory. This radiation condition may thus be expressed as

$$\lim_{r \to \infty} \sqrt{r} \left(\frac{\partial \phi_S}{\partial n} \pm i k \phi_S \right) = 0 \tag{4.162}$$

where k is the wave number. This condition is usually satisfied when ϕ_S takes an asymptotic form proportional to $\exp(\pm ikr)/r^{\frac{1}{2}}$. It has been found that a Hankel function of the first kind, $H_n^{(1)}(kr)$, satisfies condition (4.162) with a negative sign whereas the Hankel function of a second kind, $H_n^{(2)}(kr)$, satisfies it with a positive sign.

In certain situations, however, it may be necessary to include the potential due to reflection, ϕ_R. In that situation, we have

$$\phi = \phi_I + \phi_R + \phi_S \tag{4.163}$$

where ϕ_I and ϕ_S will be known functions.

In Chapter 8, we discuss in detail the role of the scattered potential. In this section, we shall present very briefly a problem of harbour resonance (see Rahman (1988)).

Harbour oscillation occurs due to waves arriving from the open sea into the harbour. These waves are partly reflected by the boundaries of the harbour and partly trapped inside the harbour. The waves produce resonance when the frequencies of the various incident and reflected waves happen to coincide with one or more of the free oscillatory modes of the harbour. Thus it is the concern of engineers to find some means of predicting the response of a particular harbour to incident waves.

The incident wave potential may be written in complex form as

$$\phi_I = -\frac{gA}{\sigma} \frac{\cosh k(z+h)}{\cosh kh} \exp(ikx) \tag{4.164}$$

which corresponds to the incident wave

$$\eta_I = A \sin(kx - \sigma t). \tag{4.165}$$

136 Theory of surface waves

The total wave potential may also be expressed in a similar form

$$\phi = -\frac{gA}{\sigma}\frac{\cosh k(z+h)}{\cosh kh}f(x,y) \tag{4.166}$$

which corresponds to the complex form of the total wave field,

$$\hat{\eta} = -\frac{1}{g}\left(\frac{\partial}{\partial t}(\phi e^{-i\sigma t})\right)_{z=0} = Aif(x,y)e^{-i\sigma t} \tag{4.167}$$

which satisfies Laplace's equation and consequently $f(x,y)$ must satisfy the Helmholtz equation

$$\frac{\partial^2 f}{\partial x^2} + \frac{\partial^2 f}{\partial y^2} + k^2 f = 0. \tag{4.168}$$

This function f is called the wave function and it must satisfy the radiation condition and the harbour boundary conditions as follows:

$$\text{(a)} \quad \lim_{r \to \infty} \sqrt{r}\left(\frac{\partial f}{\partial r} - ikf\right) = 0 \tag{4.169}$$

$$\text{(b)} \quad \frac{\partial f}{\partial n} = 0 \tag{4.170}$$

along the solid boundaries including the harbour boundary and coastline. Once the function $f(x,y)$ is determined, the wave height inside the harbour can be given as

$$|\hat{\eta}| = A|f(x,y)|$$

or

$$\frac{|\hat{\eta}|}{A} = |f(x,y)| = \text{amplification factor} = K_d \tag{4.171}$$

where K_d is defined as a *diffraction coefficient* or amplification factor. Rahman (1988) solved this problem and interested readers are referred to his work.

4.15 Exercises

1. (a) Show the linearized surface boundary conditions at $z = \eta$, in terms of the velocity potential function ϕ, may be written as $w = \eta_t = -\phi_z$, and $\phi_{tt} - g\eta_t = 0$

$$\left(\phi = \frac{Ag}{\sigma} \frac{\cosh k(z+h)}{\cosh kh} \cos(kx - \sigma t) \right).$$

(b) Then show that the dispersion relation is $\sigma^2 = gk \tanh kh$, where h is the depth of ocean, k is the wave number $(= 2\pi/L)$, L is the wavelength, σ is the frequency $(= 2\pi/T)$, T is the wave period and g is the acceleration due to gravity.

(c) Compute the wavelength (L) from the dispersion relation, given that $T = 5, 10, 15$ seconds corresponding to ocean depths $h = 5, 10, 20$ metres. Plot your results. $(g = 9.81 \ m/s^2/)$

2. (a) Starting with the velocity potential

$$\phi = \frac{Ag}{\sigma} \frac{\cosh k(z+h)}{\cosh kh} \cos(kx - \sigma t),$$

show that, for the intermediate depth case, individual water particles underneath the progressive wave describe closed elliptic orbits given by $x^2/\alpha^2 + z^2/\beta^2 = 1$, in which

$$\alpha = \frac{A \cosh k(z+h)}{\sinh kh} \quad \text{and} \quad \beta = \frac{A \sinh k(z+h)}{\sinh kh}.$$

(b) Determine the orbits of the water particles for the shallow water case and the deep water case.

(c) Support your claims with the help of schematic diagrams.

3. For two progressive waves in opposite directions in an ocean of uniform depth, the amplitude η_T of the composite wave may be written as $\eta_T = A \sin(kx - \sigma t) + A \sin(kx + \sigma t + \delta)$, where the phase angle $\delta = (2n + 1)\pi - 2kb$, and b is the position of the barrier site. Show that on substitution of δ, η_T may be written as $\eta_T = 2A \sin(kb - \sigma t)\cos(kx - kb)$. Find the nodal position when $\eta_T = 0$ for all time.

4. For standing waves in finite depth, obtain the velocity potential which reduces to $\phi = -(2gA/\sigma)\exp(kz)\cos kx \sin \sigma t$ as kh goes to infinity. Show that this corresponds to the solution of a problem where incident waves are totally reflected by a vertical wall. Using the linearized Bernoulli equation, find the horizontal wave force on the wall.

5. Derive expressions for the kinetic and potential energies of a standing wave in finite depth, and show that the total energy per wavelength is constant. Since the standing wave is equivalent to two identical plane progressive waves propagating in opposite directions, show that the energy of each plane wave is divided equally into kinetic and potential portions.

6. Compute the maximum electric power that can be generated by a device that converts all of the energy from plane progressive waves, of height 1 m and wavelength 100 m, over a width of 1 km parallel to the wave crests.

7. Prove that $W = B \cos k(z + ih - ct)$ is the complex potential for the propagation of simple harmonic surface waves of small height on water of depth h, the origin being in the undisturbed free surface. Express B in terms of the amplitude A of the small oscillations. Prove that $c^2 = (g/k)\tanh kh$, and deduce that every value of c less than $(gh)^{\frac{1}{2}}$ is the velocity of some wave. Prove that each particle describes an elliptical path about its equilibrium position. Obtain the corresponding result when the water is infinitely deep (note: $z = x + iy$).

8. Calculate the kinetic and potential energies associated with a single train of progressive waves on deep water, and, from the condition that these energies are equal, obtain the formula $c^2 = (gL/2\pi)$. Show how this result is modified when the wavelength is so small that the potential energy due to surface tension is not negligible.

9. Show that the wavelength L of stationary waves on a river of depth h, flowing with velocity v, satisfies the equation $v^2 = (gL/2\pi) \times \tanh(2\pi h/L)$. Deduce that, if the velocity of the stream exceeds $(hg)^{\frac{1}{2}}$, such stationary waves cannot exist.

10. In a train of waves on deep water given by

$$\phi = \tfrac{1}{2}Vh \exp[-(2\pi z/l)]\cos\left(\frac{2\pi}{l}(x - Vt)\right),$$

show that, if $(h/l)^{\frac{1}{2}}$ is negligible, the fluid particles describe circular paths with uniform speed.

11. Compute and graph the shoaling curves showing the variations of $C/C_0 = L/L_0$, A/A_0, $(A/L)/(A/L)_0$ and C_g/C with $h/(gT^2)$ as given by the linear theory for the values $0.001 \le h/(gT^2) \le 0.2$ or $0.01 \le h/L_0 \le 1.0$.

12. Establish the results in eqns (4.152) and (4.153).

13. Compute and graph the variation of dimensionless wave height A/A_0 for oblique wave refraction (Fig. 4.8) in terms of $h/(gT^2)$ for various deep water approach angles α_0 such as $\alpha_0 = 0, 10°, 20°, \ldots, 80°$ and $0.001 \le h/(gT^2) \le 0.2$.

References

Jen, Y. (1969). Wave refraction near San Pedro Bay, California., *J Waterw. Harbors, Coastal Eng. Div.*, ASCE, Vol. 95, No. WW3, pp. 379–393.

Keulegan, G. H. and Harrison, J. (1970). Tsunami refraction diagrams by digital computer. *J. Waterw. Harbors, Coastal Eng. Div.*, ASCE, Vol. 96, No. WW2, pp. 219–233.

Mei, C. C. (1983). *The Applied Dynamics of Ocean Surface Waves*, Wiley Interscience, New York.

Rahman, M. (1988). *The Hydrodynamics of Waves and Tides, with Applications*, Computational Mechanics Publications, Southampton.

Rahman, M. (1991). *Applied Differential Equations for Scientists and Engineers, Vol. 1, Ordinary Differential Equations*, Computational Mechanics Publications, Southampton, UK and Boston, USA.

Sarpkaya, T. and Isaacson, M. (1981). *Mechanics of Wave Forces on Offshore Structures*, Van Nostrand Reinhold, New York.

Skovgaard, O., Jonsson, I. and Bertelsen, J. (1975). Computation of wave heights due to refraction and friction. *J. Waterw. Harbours, Coastal Eng. Div.*, ASCE, Vol. 101, No. WW1, pp. 15–32.

Sommerfeld, A. (1949), *Partial Differential Equations in Physics*, Academic Press, New York.

Stoker, J. J. (1957), *Water Waves*, Interscience, New York.

5
Finite-amplitude waves

5.1 Introduction

In the previous chapter the problems of small-amplitude wave theory were discussed. There the free-surface boundary conditions were linearized by assuming that the contribution of second- and higher-order terms was negligible. However, in many engineering applications, the experimental evidence indicates that waves are very nonlinear. For example, in the design of offshore structures, the wave heights of interest are such that the nonlinearities must be considered, requiring that the wave force calculations be based on finite-amplitude wave theory. Therefore, in this chapter we will discuss the wave motion when the wave amplitude is large compared with the wavelength (see Russell (1844) and Reynolds (1883)).

5.2 Mathematical formulation

The mathematical formulation of finite-amplitude wave theory is basically the same as that for small-amplitude wave theory. However, in the former case, higher-order terms in the free-surface boundary conditions are considered important and are retained. It is generally assumed for nonlinear waves that the wave front travels without change of shape and with a constant celerity C. It is then possible to select a reference system moving with constant celerity C, thereby rendering the motion steady with respect to the moving reference frame. The component of the horizontal water particle velocity in the stationary wave system is $u - C$. The problem can be formulated in terms of either a velocity potential ϕ, or a stream function ψ. It is convenient to choose the origin of the horizontal coordinate ($x = 0$) at the wave crest, as shown in Fig. 5.1.

The fluid is assumed to be incompressible and the flow irrotational. Therefore, the velocity potential ϕ or the stream function ψ must satisfy

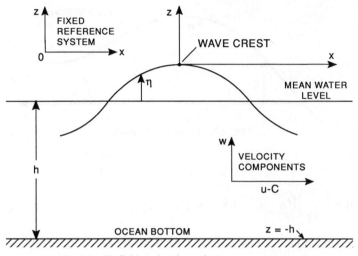

Fig. 5.1. Definition sketch, stationary wave system.

Laplace's equation

$$\frac{\partial^2 \phi}{\partial x^2} + \frac{\partial^2 \phi}{\partial z^2} = 0 \tag{5.1}$$

$$\frac{\partial^2 \psi}{\partial x^2} + \frac{\partial^2 \psi}{\partial z^2}. \tag{5.2}$$

The boundary condition at the bottom of the ocean, $z = -h$, is that no flow occurs across this surface, that is

$$w = -\frac{\partial \phi}{\partial z} = \frac{\partial \psi}{\partial x} = 0 \quad \text{on} \quad z = -h. \tag{5.3}$$

At the free surface, $z = \eta$ (now stationary), there are two boundary conditions which must be satisfied by the velocity potential ϕ and the water elevation $\eta(x)$.

The first of these, the dynamic boundary condition, can be obtained by evaluating the total energy along the free surface which must be a constant Q. This condition is the steady-state form of Bernoulli's equation, namely

$$g\eta + \frac{1}{2}[(u - C)^2 + w^2] = \text{constant}$$

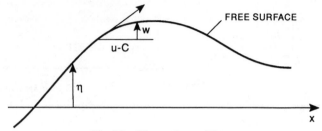

Fig. 5.2.   Kinematic condition.

or

$$\eta + \left(\frac{1}{2g}\right)[(u-C)^2 + w^2] = Q. \tag{5.4}$$

Note: the nonlinear terms u^2 and w^2 were absent in the small-amplitude wave theory of the previous chapter.

The final boundary condition which must be satisfied on the free surface is the kinematic condition which requires that no fluid be transported across the free surface. For a stationary free surface this can be formulated by requiring that the resultant velocity vector at the free surface be everywhere tangent to the free surface (see Fig. 5.2).

The kinematic free-surface boundary condition is then

$$\frac{\partial \eta}{\partial x} = \frac{w}{u-C}. \tag{5.5}$$

Thus, the mathematical problem of finite-amplitude wave theory can be restated as follows: find the solution of Laplace's equation subject to the boundary conditions (5.3), (5.4) and (5.5). The mathematical problem is nonlinear owing to the presence of the free-surface boundary conditions, namely the *dynamic* and *kinematic* boundary conditions, (5.4) and (5.5), respectively.

5.3 Perturbation method of solution

The method is applied to (5.1), (5.3), (5.4) and (5.5) by assuming that the solution can be represented in terms of a power series expansion of some small parameter ε.

Consider the following solutions:

$$\phi = \sum_{n=1}^{\infty} \varepsilon^n \phi_n \quad \text{and} \quad \eta = \sum_{n=1}^{\infty} \varepsilon^n \eta_n. \tag{5.6}$$

The sum of terms up to index n represents the nth-order theory for any particular quantity. The parameter ε is related to the ratio of the wave height to the wavelength, or to water depth, and is usually assumed to be small.

We will seek solutions applicable for a periodic wave system propagating with uniform wave celerity. The wave system will therefore be regarded as stationary, if we choose a coordinate system moving with the wave. In this analysis we assume that the wavelength is a known quantity whereas the velocity potential ϕ and the water wave elevation η are unknown variables. The wave period is considered to be unknown and consequently the wave celerity C, and the total energy Q, are also unknown.

Thus, the unknowns ϕ, η, C and Q may be expressed as perturbation series in the following manner:

$$\phi(x, z) = \sum_{n=1}^{\infty} \varepsilon^n \phi_n(x, z) \tag{5.7}$$

$$\eta(x) = \sum_{n=1}^{\infty} \varepsilon^n \eta_n(x) \tag{5.8}$$

$$C = \sum_{n=0}^{\infty} \varepsilon^n C_n \tag{5.9}$$

$$Q = \sum_{n=0}^{\infty} \varepsilon^n Q_n, \tag{5.10}$$

where ε is a small parameter to be determined later. Here C_0 is the lower-order celerity and Q_0 is the lower-order constant energy.

The difficulties in surface wave problems arise from the free-surface conditions. Because the wave surface is not known apriori, it will be advantageous to use the following Taylor expansion for the velocity potential function ϕ about $z = 0$. Thus

$$\phi[x, \eta(x)] = \phi(x, 0) + \eta \left(\frac{\partial \phi}{\partial z} \right)_{z=0} + \dots$$

$$= (\varepsilon \phi_1 + \varepsilon^2 \phi_2 + \dots) + (\varepsilon \eta_1 + \varepsilon^2 \eta_2 + \dots)$$

$$\times \left(\varepsilon \frac{\partial \phi_1}{\partial z} + \varepsilon^2 \frac{\partial \phi_2}{\partial z} + \dots \right)$$

$$= \varepsilon \phi_1 + \varepsilon^2 \left(\phi_2 + \eta_1 \frac{\partial \phi_1}{\partial z} \right) + O(\varepsilon^3) \qquad \text{at} \qquad z = 0.$$

$$\tag{5.11}$$

Similarly, expanding by Taylor's theorem, we obtain

$$\frac{\partial \phi}{\partial z} = \varepsilon \frac{\partial \phi_1}{\partial z} + \varepsilon^2 \left[\frac{\partial \phi_2}{\partial z} + \eta_1 \frac{\partial}{\partial z}\left(\frac{\partial \phi_1}{\partial z} \right) \right] + \dots \tag{5.12}$$

$$\frac{\partial \phi}{\partial x} = \varepsilon \frac{\partial \phi_1}{\partial x} + \varepsilon^2 \left[\frac{\partial \phi_2}{\partial x} + \eta_1 \frac{\partial}{\partial z}\left(\frac{\partial \phi_1}{\partial x} \right) \right] + \dots \tag{5.13}$$

$$\nabla \phi = \varepsilon \nabla \phi_1 + \varepsilon^2 \left(\nabla \phi_2 + \eta_1 \frac{\partial}{\partial z}(\nabla \phi_1) \right) + \dots . \tag{5.14}$$

These expressions are to be evaluated at $z = 0$.

Each of the velocity potentials given in (5.7) must satisfy Laplace's equation

$$\nabla^2 \phi_n = \frac{\partial^2 \phi_n}{\partial x^2} + \frac{\partial^2 \phi_n}{\partial z^2} = 0, \tag{5.15}$$

where ϕ_n is the nth-order potential.

The bottom boundary condition is

$$\frac{\partial \phi_n}{\partial z} = 0 \qquad \text{at} \qquad z = -h. \tag{5.16}$$

The free-surface boundary conditions (5.4) and (5.5) can be rewritten as follows:

Dynamic free-surface condition:

$$\eta - Q + \frac{1}{2g}\left[\left(\frac{\partial \phi}{\partial x} \right)^2 + 2C\left(\frac{\partial \phi}{\partial x} \right) + C^2 + \left(\frac{\partial \phi}{\partial z} \right)^2 \right] = 0. \tag{5.17}$$

Kinematic free-surface condition:

$$\left(\frac{\partial \phi}{\partial x} \right)\frac{\partial \eta}{\partial x} + C \frac{\partial \eta}{\partial x} = \frac{\partial \phi}{\partial z}. \tag{5.18}$$

By substituting the series forms of the unknowns into (5.17), the dynamic boundary condition can be written as

$$(\varepsilon\eta_1 + \varepsilon^2\eta_2) - (Q_0 + \varepsilon Q_1 + \varepsilon^2 Q_2)$$

$$+ \frac{1}{2g}\left\{\left[\varepsilon\left(\frac{\partial\phi_1}{\partial x}\right) + \varepsilon^2\left(\frac{\partial\phi_2}{\partial x} + \eta_1\frac{\partial^2\phi_1}{\partial z\partial x}\right)\right]^2\right.$$

$$+ 2(C_0 + \varepsilon C_1 + \varepsilon^2 C_2)\left[\varepsilon\frac{\partial\phi_1}{\partial x} + \varepsilon^2\left(\frac{\partial\phi_2}{\partial x} + \eta_1\frac{\partial^2\phi_1}{\partial z\partial x}\right)\right]$$

$$+ (C_0 + \varepsilon C_1 + \varepsilon^2 C_2)^2$$

$$\left. + \left[\varepsilon\frac{\partial\phi_1}{\partial z} + \varepsilon^2\left(\frac{\partial\phi_2}{\partial z} + \eta_1\frac{\partial^2\phi_1}{\partial z^2}\right)\right]^2\right\} + O(\varepsilon^3) = 0. \qquad (5.19)$$

Further simplication of (5.19) yields

$$\varepsilon^0\left(\frac{C_0^2}{2g} - Q_0\right) + \varepsilon\left(\eta_1 - Q_1 + \frac{1}{g}C_0\frac{\partial\phi_1}{\partial x} + \frac{C_0 C_1}{g}\right)$$

$$+ \varepsilon^2\left\{\eta_2 - Q_2 + \frac{C_0}{g}\frac{\partial\phi_2}{\partial x} + \frac{1}{2g}\left[\left(\frac{\partial\phi_1}{\partial x}\right)^2 + \left(\frac{\partial\phi_1}{\partial z}\right)^2\right]\right.$$

$$\left. + 2C_1\frac{\partial\phi_1}{\partial x} + 2C_0\eta_1\frac{\partial^2\phi_1}{\partial z\partial x} + C_1^2 + 2C_0 C_2\right\} + O(\varepsilon^3) = 0.$$

$$(5.20)$$

Similarly, the kinematic boundary condition (5.18) can now be written in terms of the series

$$\left[\varepsilon\frac{\partial\phi_1}{\partial x} + \varepsilon^2\left(\frac{\partial\phi_2}{\partial x} + \eta_1\frac{\partial^2\phi_1}{\partial z\partial x}\right)\right] \times \left(\varepsilon\frac{\partial\eta_1}{\partial x} + \varepsilon^2\frac{\partial\eta_2}{\partial x}\right)$$

$$+ (C_0 + \varepsilon C_1 + \varepsilon^2 C_2)\left(\varepsilon\frac{\partial\eta_1}{\partial x} + \varepsilon^2\frac{\partial\eta_2}{\partial x}\right)$$

$$= \varepsilon\frac{\partial\phi_1}{\partial z} + \varepsilon^2\left(\frac{\partial\phi_2}{\partial z} + \eta_1\frac{\partial^2\phi_1}{\partial z^2}\right) + O(\varepsilon^3). \qquad (5.21)$$

Simplifying (5.21), we obtain

$$\varepsilon \left(C_0 \frac{\partial \eta_1}{\partial x} - \frac{\partial \phi_1}{\partial z} \right) + \varepsilon^2 \left(\frac{\partial \phi_1}{\partial x} \frac{\partial \eta_1}{\partial x} + C_0 \frac{\partial \eta_2}{\partial x} \right.$$

$$\left. + C_1 \frac{\partial \eta_1}{\partial x} - \frac{\partial \phi_2}{\partial z} - \eta_1 \frac{\partial^2 \phi_1}{\partial z^2} \right) + O(\varepsilon^3) = 0. \tag{5.22}$$

From the zeroth-order term (ε^0) we have

$$Q_0 = \frac{C_0^2}{2g}. \tag{5.23}$$

Thus, equating the coefficients of ε and ε^2, and first- and second-order wave theory may be stated as follows:

First-order wave theory:

$$\frac{\partial^2 \phi_1}{\partial x^2} + \frac{\partial^2 \phi_1}{\partial z^2} = 0 \tag{5.24}$$

$$\frac{\partial \phi_1}{\partial z} = 0 \quad \text{on} \quad z = -h \tag{5.25}$$

$$\eta_1 - Q_1 + \frac{1}{g} C_0 \frac{\partial \phi_1}{\partial x} + \frac{1}{g} C_0 C_1 = 0 \quad \text{on} \quad z = 0 \tag{5.26}$$

$$C_0 \frac{\partial \eta_1}{\partial x} - \frac{\partial \phi_1}{\partial z} = 0 \quad \text{on} \quad z = 0. \tag{5.27}$$

Second-order wave theory:

$$\frac{\partial^2 \phi_2}{\partial x^2} + \frac{\partial^2 \phi_2}{\partial z^2} = 0 \tag{5.28}$$

$$\frac{\partial \phi_2}{\partial z} = 0 \quad \text{on} \quad z = -h \tag{5.29}$$

$$\eta_2 - Q_2 + \frac{1}{g} C_0 \frac{\partial \phi_2}{\partial x} + \frac{1}{2g}\left[\left(\frac{\partial \phi_1}{\partial x}\right)^2 + \left(\frac{\partial \phi_1}{\partial z}\right)^2 + 2C_1 \frac{\partial \phi_1}{\partial x}\right.$$

$$\left. + 2C_0\eta_1 \frac{\partial^2 \phi_1}{\partial z \partial x} + C_1^2 + 2C_0 C_2 \right] = 0 \qquad \text{on} \qquad z = 0 \qquad (5.30)$$

$$\frac{\partial \phi_1}{\partial x}\frac{\partial \eta_1}{\partial x} + C_0 \frac{\partial \eta_2}{\partial x} + C_1 \frac{\partial \eta_1}{\partial x} - \frac{\partial \phi_2}{\partial z} - \eta_1 \frac{\partial^2 \phi_1}{\partial z^2} = 0, \qquad z = 0.$$
$$(5.31)$$

A solution which satisfies Laplace's equation (5.24), subject to the boundary condition (5.25), and which is periodic in x with period L, is given by

$$\phi_1 = -B_1 \cosh k(z+h)\sin kx, \qquad (5.32)$$

where $k = 2\pi/L$. Choosing

$$Q_1 = \frac{1}{g} C_0 C_1, \qquad (5.33)$$

then (5.26) reduces to

$$\eta_1 + \frac{1}{g} C_0 \frac{\partial \phi_1}{\partial x} = 0. \qquad (5.34)$$

Combining (5.27) and (5.34), we obtain

$$C_0^2 \frac{\partial^2 \phi_1}{\partial x^2} + g \frac{\partial \phi_1}{\partial z} = 0 \qquad \text{on} \qquad z = 0. \qquad (5.35)$$

Substituting ϕ_1 from (5.32) into (5.35), we obtain the dispersion relation given by

$$C_0^2 = \frac{g}{k} \tanh kh. \qquad (5.36)$$

Thus, once C_0 is known, the value of Q_0 is determined from (5.23). The expression for η_1 can be obtained from (5.34), and is

$$\eta_1 = -\frac{C_0}{g}\frac{\partial \phi_1}{\partial x}$$

$$= \left(\frac{B_1 k C_0}{g}\right)\cosh k(z+h)\cos kx, \tag{5.37}$$

in which the wave crest is located at $x=0$. This first-order theory is usually referred to as Airy wave theory. The value of ε will be determined from the solution of the second-order theory.

A solution ϕ_2, which satisfies Laplace's equation (5.28), subject to the bottom boundary condition (5.29), and which is periodic in x, may be written as

$$\phi_2 = -B_2 \cosh 2k(z+h)\sin 2kx. \tag{5.38}$$

Here the wave number of the second-order velocity potential is assumed to be twice that of the first-order potential. Now differentiating (5.30) with respect to x, we obtain

$$\frac{\partial \eta_2}{\partial x} + \frac{C_0}{g}\frac{\partial^2 \phi_2}{\partial x^2} + \frac{1}{g}\left(\frac{\partial \phi_1}{\partial x}\frac{\partial^2 \phi_1}{\partial x^2} + \frac{\partial \phi_1}{\partial z}\frac{\partial^2 \phi_1}{\partial z\partial x} + C_1\frac{\partial^2 \phi_1}{\partial x^2}\right.$$

$$\left. +C_0\frac{\partial \eta_1}{\partial x}\frac{\partial^2 \phi_1}{\partial z\partial x} + C_0\eta_1\frac{\partial^3 \phi_1}{\partial z\partial x^2}\right) = 0 \qquad \text{on} \qquad z=0. \tag{5.39}$$

Eliminating η_2 between (5.31) and (5.39) yields

$$\frac{1}{C_0}\frac{\partial \phi_2}{\partial z} + \frac{C_0}{g}\frac{\partial^2 \phi_2}{\partial x^2} = -\frac{1}{g}\left[\frac{\partial \phi_1}{\partial x}\frac{\partial^2 \phi_1}{\partial x^2} + \frac{\partial \phi_1}{\partial z}\frac{\partial^2 \phi_1}{\partial z\partial x}\right.$$

$$\left. +C_1\frac{\partial^2 \phi_1}{\partial x^2} + C_0\frac{\partial \eta_1}{\partial x}\frac{\partial^2 \phi_1}{\partial z\partial x} + C_0\eta_1\frac{\partial^3 \phi_1}{\partial z\partial x^2}\right]$$

$$+\frac{1}{C_0}\frac{\partial \eta_1}{\partial x}\frac{\partial \phi_1}{\partial x} + \frac{C_1}{C_0}\frac{\partial \eta_1}{\partial x}$$

$$-\frac{1}{C_0}\eta_1\frac{\partial^2 \phi_1}{\partial z^2} \qquad \text{on} \qquad z=0. \tag{5.40}$$

Substituting the forms of ϕ_1, ϕ_2 and η_1 into (5.40) and simplifying the result by the use of (5.36), we obtain

$$B_2 \sinh^2 kh \sin 2kx = \frac{3}{8} \frac{B_1^2 k}{C_0} \sin 2kx - \frac{1}{2} \frac{C_1}{C_0} B_1 \cosh kh \sin kx.$$

(5.41)

Equating the coefficients of $\sin 2kx$ and $\sin kx$, respectively, yields

$$B_2 = \frac{3}{8} \frac{B_1^2 k}{C_0} \frac{1}{\sinh^2 kh}$$

(5.42)

$$C_1 = 0.$$

(5.43)

Consequently from (5.33), we have $Q_1 = 0$.

From (5.43), we see that the first-order correction to the zeroth-order wave celerity is zero. The second-order velocity potential ϕ_2 can be obtained in a similar manner, and is given by

$$\phi_2 = -\frac{3}{8} \frac{B_1^2 k}{C_0 \sinh^2 kh} \cosh 2k(z+h) \sin 2kx.$$

(5.44)

Similarly, the second-order wave elevation η_2 is

$$\eta_2 = Q_2 - \frac{1}{g} C_0 \frac{\partial \phi_2}{\partial x} - \frac{1}{2g} \left[\left(\frac{\partial \phi_1}{\partial x} \right)^2 + \left(\frac{\partial \phi_1}{\partial z} \right)^2 \right.$$

$$\left. + 2C_0 \eta_1 \frac{\partial^2 \phi_1}{\partial z \partial x} + 2C_0 C_2 \right] \quad \text{on} \quad z = 0$$

(5.45)

from which, after simplification,

$$\eta_2 = Q_2 - \frac{1}{4} \frac{B_1^2 k_1^2}{g} - \frac{C_0 C_2}{g} + \frac{B_1^2 k_1^2}{4g} \frac{\cosh^2 kh(2 + \cosh 2kh)}{\sinh^2 kh} \cos 2kx.$$

(5.46)

Therefore

$$Q_2 = \frac{1}{4} \frac{B_1^2 k_1^2}{g} + \frac{C_0 C_2}{g}.$$

(5.47)

The perturbation parameter ε can be specified by setting the maximum value of the first-order wave profile equal to A.

Since $\eta = A \cos kx = \varepsilon\eta_1 = \varepsilon((kB_1C_0/g)\cosh kh)\cos kx$, it follows that $A = \varepsilon(kB_1C_0/g)\cosh kh$. Consequently the value of ε is

$$\varepsilon = \frac{Ag}{kB_1C_0 \cosh kh} . \tag{5.48}$$

If we define $\varepsilon = kA$, then

$$B_1 = \frac{g}{C_0 k^2 \cosh kh} .$$

With ε defined above, and noting that $H = 2A$, the solutions for velocity potential ϕ, wave elevation η and celerity C, correct to second order (and applicable to a moving wave system), are

$$\phi = \varepsilon\phi_1 + \varepsilon^2\phi_2 = -\frac{H}{2} C \frac{\cosh k(z+h)}{\sinh kh} \sin(kx - \sigma t)$$

$$-\frac{3}{16} \frac{H^2\pi}{T} \frac{\cosh 2k(z+h)}{\sinh^4 kh} \sin 2(kx - \sigma t) \tag{5.49}$$

$$\eta = \varepsilon\eta_1 + \varepsilon^2\eta_2 = \frac{H}{2} \cos(kx - \sigma t)$$

$$+ \frac{\pi}{8} \frac{H^2}{L} \frac{\cosh kh}{\sinh^3 kh} (2 + \cosh 2kh)\cos 2(kx - \sigma t) \tag{5.50}$$

and

$$C = C_0 + \varepsilon C_1 = \sqrt{\frac{g}{k} \tanh kh} . \tag{5.51}$$

The expressions (5.49) to (5.51) comprise the perturbation development of the second-order theory.

These solutions are mainly due to Stokes who developed his theory in 1847. The convergence of the series solution was confirmed by Levi-Civita in 1925. However, in the following we give a brief discussion about the convergence of this series.

We know that for the power series for ϕ in terms of ε to converge, the ratio of the $(n + 1)$th term divided by the nth term must be less than unity as n goes to infinity. Therefore, for the series for ϕ in eqn (5.49) to converge we must have

$$r = \left| \frac{\varepsilon\phi_2}{\phi_1} \right| = \frac{3}{8} \frac{kA \cosh 2kh}{\cosh kh \sinh^3 kh} \ll 1.$$

For deep water waves, when $kh > \pi$, the asymptotic values of the hyperbolic functions can be substituted to obtain r as $r = 3e^{-2kh}(kA)$. Thus for this case, r is very small because kA is a small parameter. It has been found experimentally (Wiegel (1964)) that the highest value in deep water would occur for $kh = \pi, kA = \pi/7$ (for the wave of maximum steepness) such that $r = (3\pi/7)e^{-2\pi} = 0.0025$. Considering this fact, Levi-Civita confirmed that the Stokes perturbation solution is valid up to the second-order term.

In shallow water, $kh < \pi/10$, the asymptotic values of the hyperbolic function can yield the value of r to be $r = 3/8((kA/(kh)^3)) < 1$.

The term within the brackets that is $kA/(kh)^3$, is defined as the Ursell number. It can be easily seen that in shallow water waves, we have constraints like $kA < 8/3(kh)^3$ or $A/h < 8/3(kh)^2$, where kh is small.

The maximum that the ratio A/h can obtain is $A/h = (8/300)\pi^2$ for $kh = \pi/10$, or the maximum wave amplitude is about 26% the water depth. But for the wave amplitude of a breaking wave in shallow water it is almost 40% the water depth. Therefore, for high waves in shallow water, the Stokes expansion is not a good approximation. Thus to overcome this difficulty, we take up this matter again in Chapter 9.

It has been observed in (5.50) that the wave crest appears at $kx - \sigma t = 0$ and the wave trough appears at $kx - \sigma t = \pi$. Using these observations, we obtain

$$\frac{A_c}{H} = \frac{1}{2} + \frac{\pi}{8} \frac{H}{L} \frac{\cosh kh(2 + \cosh 2kh)}{(\sinh kh)^3} \tag{5.52}$$

$$\frac{A_t}{H} = \frac{1}{2} - \frac{\pi}{8} \frac{H}{L} \frac{\cosh kh(2 + \cosh 2kh)}{(\sinh kh)^3}. \tag{5.53}$$

For the deep water case, where $h/L > 1/2$,

$$\frac{A_c}{H} = \frac{1}{2}\left(1 + 1.57\frac{H}{L}\right), \qquad \frac{A_t}{H} = \frac{1}{2}\left(1 - 1.57\frac{H}{L}\right). \tag{5.54}$$

For fairly shallow water, where $h/L < 1/20$,

$$\frac{A_c}{H} = \frac{1}{2}\left(1 + 12.1\frac{H}{L}\right), \qquad \frac{A_t}{H} = \frac{1}{2}\left(1 - 12.1\frac{H}{L}\right) \tag{5.55}$$

152   Finite-amplitude waves

where A_c is the crest amplitude and A_t is the trough amplitude.

It can be easily seen that A_c/H is much greater in shallow water than in deep water. The limiting shallow water wave is, in fact, the solitary wave, whose entire crest is above the mean water level. Because of the physical importance of solitary waves, Chapter 9 is devoted entirely to the study of nonlinear long waves in shallow water.

The wave phase velocity C, to a second-order approximation, is given by $C^2 = (g/k)\tanh kh$ which is the same as the small-amplitude wave result. By third-order theory, the celerity can be determined as (Stoker (1957))

$$C^2 = \frac{g}{k} \tanh kh \left[1 + \left(\frac{\pi H}{L} \right)^2 \frac{5 + 2\cosh 2kh + 2\cosh^2 2kh}{8(\sinh kh)^4} \right]. \tag{5.56}$$

Now the water particle velocities in the x- and z-directions for the second-order theory can be obtained, respectively, as

$$u = -\frac{\partial \phi}{\partial x} = \frac{\pi H}{L} C \frac{\cosh k(z+h)}{\sinh kh} \cos(kx - \sigma t)$$

$$+ \frac{3}{4} \left(\frac{\pi H}{L} \right)^2 C \frac{\cosh 2k(z+h)}{(\sinh kh)^4} \cos 2(kx - \sigma t) \tag{5.57}$$

$$w = -\frac{\partial \phi}{\partial z} = \frac{\pi H}{L} C \frac{\sinh k(z+h)}{\sinh kh} \sin(kx - \sigma t)$$

$$+ \frac{3}{4} \left(\frac{\pi H}{L} \right)^2 C \frac{\sinh 2k(z+h)}{(\sinh kh)^4} \sin 2(kx - \sigma t). \tag{5.58}$$

The water particle orbits beneath a finite-amplitude wave can be obtained (see Wiegel (1964)) as follows:

Motion in the horizontal direction:

$$\xi = -\frac{H}{2} \frac{\cosh k(z+h)}{\sinh kh} \sin(kx - \sigma t)$$

$$- \frac{\pi H^2}{4L} \frac{1}{(\sinh kh)^2} \left(-\frac{1}{2} + \frac{3}{4} \frac{\cosh 2k(z+h)}{(\sinh kh)^2} \right) \sin 2(kx - \sigma t)$$

$$+ \frac{\pi H^2}{L} \frac{C}{2} \frac{\cosh 2k(z+h)}{(\sinh kh)^2} t. \tag{5.59}$$

Motion in the vertical direction:

$$\zeta = \frac{H}{2} \frac{\sinh k(z+h)}{\sinh kh} \cos(kx - \sigma t)$$

$$+ \frac{3}{16} \frac{\pi H^2}{L} \frac{\sinh 2k(z+h)}{(\sinh kh)^4} \cos 2(kx - \sigma t). \tag{5.60}$$

It has been found experimentally that Stokes' theory is in better agreement with the measurements than the small-amplitude wave theory.

5.4 Stokes' theory of breaking waves

In the linear wave theory (infinitely small motion) investigations were carried out with the assumption that the factor (kA) which is essentially the ratio of the maximum elevation to the wavelength must be small. The determination of the wave profiles which satisfy the conditions of uniform propagation without change of shape, when this condition is relaxed, forms the subject of classical research by Stokes and of many subsequent investigations. Stokes treated this problem very conveniently by assuming the steady motion in a paper entitled 'On the theory of oscillatory waves' published in *Cambridge Transactions* viii (1847) [papers, *i*, 197]. It was pointed out by Rayleigh that if we neglect small quantities of order $(kA)^3$, the solution in the case of infinite depth can be obtained from the following velocity potential ϕ and stream function ψ:

$$\frac{\phi}{C} = -x + \beta e^{kz} \sin kx \tag{5.61}$$

$$\frac{\psi}{C} = -z + \beta e^{kz} \cos kx \tag{5.62}$$

where β is a suitable constant to be specified later. If the upper surface coincides with the streamlines $\psi = 0$, whose equation is $z = \beta \cos kx$, the equation of the wave profile is found by successive approximation to

$$z = \beta e^{kz} \cos kx = \beta\left(1 + kz + \tfrac{1}{2}k^2z^2 + \dots\right)\cos kx,$$

which subsequently can be expressed as

$$z = \tfrac{1}{2}k\beta^2 + \beta\left(1 + \tfrac{9}{8}k^2\beta^2\right)\cos kx$$

$$+ \tfrac{1}{2}k\beta^2 \cos 2kx + \tfrac{3}{8}k^2\beta^3 \cos 3kx + \dots \tag{5.63}$$

Let us define the coefficient of cos kx on the right-hand side of eqn (5.63), $\beta(1 + (9/8)k^2\beta^2) = A$. Then the perturbation series (5.63) can be written as

$$z - \tfrac{1}{2}kA^2 = A\cos kx + \tfrac{1}{2}kA^2\cos 2kx + \tfrac{3}{8}k^2A^3\cos 3kx + \dots \tag{5.64}$$

This development coincides with the equation of a trochoid, in which the circumference of the rolling circle is $2\pi/k$ or L, and the length of the arm of the tracing point is A.

The condition of uniform pressure along this streamline ($\psi = 0$) can be satisfied by a suitably chosen value of C. From eqn (5.61), we obtain, without approximation (using Bernoulli's formula), the pressure as

$$\frac{P}{\rho} = \text{constant} - gz - \frac{1}{2}C^2(1 - 2k\beta e^{kz}\cos kx + k^2\beta^2 e^{2kz}) \tag{5.65}$$

and therefore, at points of the line $z = \beta e^{kz}\cos kx$, the pressure becomes

$$\frac{P}{\rho} = \text{constant} + (kC^2 - g)z - \frac{1}{2}C^2k^2\beta^2 e^{2kz}$$

$$= \text{constant} + (kC^2 - g - k^3C^2\beta^2)z + \dots . \tag{5.66}$$

Hence a condition of a free surface is satisfied, to the present order of approximation, provided the coefficient of z is zero, i.e.

$$C^2 = \frac{g}{k} + k^2C^2\beta^2 = \frac{g}{k}(1 - k^2\beta^2)^{-1} \simeq \frac{g}{k}(1 + k^2A^2). \tag{5.67}$$

This determines the velocity of progressive waves of permanent type, and shows that it increases with the amplitude A. A sketch of the wave profile as given by eqn (5.64) is shown in Fig. 5.3 in the case of $kA = 1/2$.

The approximately trochoidal form of the wave profile gives an outline which is sharper near the crests, and flatter in the troughs, than in the case of simple harmonic waves of infinitely small amplitude investigated in

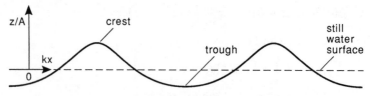

Fig. 5.3.   An approximately trochoidal wave profile.

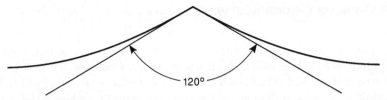

Fig. 5.4.   Breaking wave profile as predicted by Stokes.

Chapter 4, and these features become more pronounced as the amplitude is increased. If the trochoidal form were exact, instead of nearly approximate, the limiting form would have cusps at the crests (see Fig. 5.4) as in the case of Gerstner's waves to be considered in the next section. In the actual problem, which is one of irrotational motion, the extreme has been shown by Stokes, in a very simple manner, to have a sharp angle of 120°. In practice, experiment confirms that when the depth to wavelength ratio (h/L) is greater than $1/10$, the approximately trochoidal profile of the free surface in the nonlinear wave theory predicts the shape of a breaking wave; the angle of the breaking crest obtained is 120° (see Fig. 5.4).

The Stokes theory also predicts a net horizontal convection of the fluid particles per wave cycle for the deep water case, the convection velocity being given by

$$U = \frac{H^2}{4} k^2 C e^{2kz_0}. \tag{5.68}$$

The convection is, therefore, maximum on the free surface and diminishes exponentially with depth z, because z is negative from the free surface to the bottom of the sea.

The path of the particles about any mean depth z_0 is similar to that shown schematically in Fig. 5.5. For a further account of Stoke's breaking wave, readers are referred to Lamb (1945) and McCormick (1973).

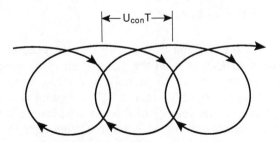

Fig. 5.5.   Convection path beneath a finite wave.

5.5 Gerstner's trochoidal wave theory

Gerstner, in 1802, was the first to produce a theory related to finite-amplitude waves. It is called the trochoidal theory because the profiles of the free surface and other constant pressure surfaces are predicted to be trochoidal. These profiles can be generated by points on a wheel travelling on a straight line located above the x-axis. The radius of the wheel must be equal to $1/k$ since one revolution of the wheel occurs over a distance equal to one wavelength. The wave height, H, is then determined by the radial location of the generating point. If the generating point is on the circumference (as in Fig. 5.6), the curve generated is that of a cycloid with a cusp at the crest, the angle at the crest being zero. The shape is thus that of a breaking wave of height $2/k = L/\pi$. As the generating point approaches the centre of the wheel, the shape of the constant pressure surface approaches that of linear theory.

It is noted that Gerstner's solutions are in closed form in contrast to Stokes' results, which are in the form of an infinite series. At the same time, Gerstner's solution is an 'exact' solution of a particular wave motion problem in that no mathematical approximations are made.

A system of exact equations, expressing a possible form of wave motion when the depth of the fluid is infinite, was given by Gerstner in 1802. The circumstance, however, that the motions in these waves are not irrotational detracts somewhat from the physical interest of the results. A precise account about the singular nature of vorticity for the maximum wave is given below.

The equations of trochoidal wave theory can be derived from those of linear (small-amplitude) wave theory presented in the previous chapter. Gerstner assumed that the position of a particle is given as a function of time by

$$x = x_0 + \frac{1}{k} e^{kz_0} \sin k(x_0 - Ct) \qquad (5.69)$$

$$z = z_0 - \frac{1}{k} e^{kz_0} \cos k(x_0 - Ct) \qquad (5.70)$$

where (x_0, z_0) is the position of the centre of the generating wheel (see Fig. 5.6). The specification is on the Lagrangian plane such that (x_0, z_0) are two parameters serving to identify a particle, and x and z are coordinates of this particle at time t. The constant k determines the wavelength, and C is the velocity of the waves which are travelling in the direction of the positive x-axis. It is obvious from eqns (5.69) and (5.70) that the path of any particle (x_0, z_0) is a circle of radius $k^{-1}e^{kz_0}$. It can be

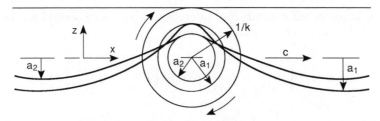

Fig. 5.6. Generation of the trochoidal wave profile.

easily verified that the Lagrangian equation of continuity (see Perrie and Rahman (1991)) is satisfied for deep water waves, i.e.

$$\frac{\partial(x, z)}{\partial(x_0, z_0)} = 1 - e^{2kz_0}. \tag{5.71}$$

By the Lagrangian form of the equations of motion, we find that

$$\frac{\partial}{\partial x_0}\left(\frac{P}{\rho} + gz\right) = kC^2 e^{kz_0} \sin k(x_0 - Ct) \tag{5.72}$$

$$\frac{\partial}{\partial z_0}\left(\frac{P}{\rho} + gz\right) = -kC^2 e^{kz_0} \cos k(x_0 - Ct) + kC^2 e^{2kz_0}. \tag{5.73}$$

So by performing the integration, we obtain

$$\frac{P}{\rho} = \text{constant} - g\left(z_0 - \frac{1}{k}e^{kz_0} \cos k(x_0 - Ct)\right)$$

$$- C^2 e^{kz_0} \cos k(x_0 - Ct) + \tfrac{1}{2}C^2 e^{2kz_0}. \tag{5.74}$$

For a particle on the free surface the pressure must be constant. This requires

$$C^2 = \frac{g}{k}, \tag{5.75}$$

and consequently the pressure is given by

$$\frac{P}{\rho} = \text{constant} - gz_0 + \frac{1}{2}C^2 e^{2kz_0}. \tag{5.76}$$

It has been stated already that the motion of the fluid in Gerstner's waves is rotational. To prove this we remark that the vorticity $\omega = u_z - w_x$ should not be zero; in other words the differential $u\delta x + w\delta z$ should not be an exact differential. This is exactly so, since

$$u\delta x + w\delta z = \left(\dot{x}\frac{\partial x}{\partial x_0} + \dot{z}\frac{\partial z}{\partial x_0} \right)\delta x_0 + \left(\dot{x}\frac{\partial x}{\partial z_0} + \dot{z}\frac{\partial z}{\partial z_0} \right)\delta z_0$$

$$= -\frac{C}{k}\delta\left[e^{kz_0}\sin k(x_0 - Ct) \right] - Ce^{2kz_0}\delta x_0 \qquad (5.77)$$

which is not an exact differential.

Next the amount of vorticity of the element (x_0, z_0) is precisely given by the circulation in the boundary of the parallelogram whose vertices coincide with the particles

$$(x_0, z_0), \quad (x_0 + \delta x_0, z_0), \quad (x_0, z_0 + \delta z_0), \quad (x_0 + \delta x_0, z_0 + \delta z_0),$$

divided by the area of the circuit. Thus the circulation and the area of the circuit are obtained, respectively, as

$$\frac{\partial}{\partial z_0}(Ce^{2kz_0}\delta x_0)\delta z_0 \quad \text{and} \quad \frac{\partial(x,z)}{\partial(x_0, z_0)}\delta x_0\,\delta z_0 = (1 - e^{2kz_0})\delta x_0\,\delta z_0.$$

Hence the vorticity ω of the element (x_0, z_0) is

$$\omega = \frac{2kCe^{2kz_0}}{1 - e^{2kz_0}}. \qquad (5.78)$$

The vorticity becomes infinite at the surface when $z_0 = 0$, and diminishes rapidly with increasing depth, and becomes zero when z_0 goes to negative infinity. This is an unusual behaviour which persists in the vorticity distribution of Gerstner's rotational waves.

The velocity components of the fluid particles in Gerstner's waves are

$$u = \frac{dx}{dt} = -Ce^{kz_0}\cos k(x_0 - Ct) \qquad (5.79)$$

and

$$w = \frac{dz}{dt} = -Ce^{kz_0}\sin k(x_0 - Ct). \qquad (5.80)$$

5.6 Exercises

1. Determine the dimensionless wave profile, η/H, for the finite-amplitude wave theory using the results of Stokes' wave theory in eqn (5.50) and plot it graphically in terms of x/L when $t = 0$. Develop a FORTRAN program.

2. Using the results of exercise 1, determine the profile of a deep water wave 200 m long and 10 m high. Compare this profile with that obtained from the linear theory. Draw the graphs for the linear and nonlinear wave profiles to see the difference.

3. Derive the third-order theory of Stokes and confirm the wave celerity given in eqn (5.56).

4. Using Bernoulli's equation, i.e. $(P/\rho) + gz + (\partial\phi/\partial t) + 1/2(\nabla\phi)^2 = 0$, determine an expression for the pressure P.

5. Determine the horizontal and vertical particle accelerations of the second-order wave theory of Stokes.

6. Using eqns (5.57) and (5.58), find an expression for the vorticity for the surface particles of finite-amplitude waves.

7. Show that the trochoidal wave profile satisfies the continuity equation, but does not satisfy the condition of irrotationality. Determine the value of the vorticity. (*Hint:* see section 5.5.)

References

Gerstner, F. J. (1809). Theorie der Wellen *Ann. Phys*, 32, pp. 412–440.

Lamb, H. (1945). *Hydrodynamics*, 6th ed, Dover, New York, Cambridge University Press, Cambridge.

Levi-Civita, T. (1925). Determination regoureuse des ondes permanentes d'ampleur finie. *Math. Ann.* 93, 264–314.

McCormick, M. E., (1973). *Ocean Engineering Wave Mechanics*, John Wiley, New York.

Perrie, W. and Rahman, M. (1991). Coupling of gravity waves to surface currents and ice motion. Canadian Technical Report of Hydrography and Ocean Sciences, No. 134, 69 pages.

Reynolds, O. (1883). An experimental investigation of the circumstances which determine whether the motion of water shall be direct or sinuous and the law of resistance in parallel channels. *Philos. Trans.* A, 174, 935–982. Also, *Sci. Pap.* 2, 51–105.

Russell, S. (1844). Report on waves. *Br. Assoc. Adv. Sci. Rep.*

Stoker, J. J. (1957). *Water Waves*, Interscience, New York.

Stokes, G. G. (1847). On the theory of oscillatory waves. *Trans. Cambridge Philos. Soc.*, 8, 441–455. Also *Math Phys. Pap*, Vol. 1, Cambridge University Press, 1880.

Wiegel, R. L. (1964). *Oceanographical Engineering*, Prentice Hall, Englewood Cliffs. New Jersey.

6
One-dimensional tidal dynamics

6.1 Introduction

We saw in Chapter 4 that waves propagating in shallow water when the depth to wavelength ratio is less than $\frac{1}{20}$, i.e. $h/L < \frac{1}{20}$ or $kh < \pi/10$, are often called long waves or shallow water waves. By this definition then tidal waves, tsunamis (erroneously called tidal waves) and other waves with extremely long periods and wavelengths are shallow water waves, even in deep ocean. Engineering problems relating to tidal waters have received considerable attention from ocean engineers concerned with the practical aspects, and scientists attempting to express the observed tidal phenomena in terms of solutions of simplified theoretical models. These solutions enable qualitative predictions to be deduced for many problems of flow in estuaries.

The investigation of long waves is of paramount importance to engineers and scientists in the design of harbours and in obtaining the relevant information about tidal estuaries. Tidal propagation in estuaries is affected greatly by the geometry of the estuary. Long waves propagating into the harbours or the estuary can excite the wave energy into resonance which may cause a great deal of havoc to human beings. It is well known that the tides in the Bay of Fundy are among the highest in the world. Its tidal behaviours is semi-diurnal, with two high tides and two low tides of approximately the same height each day. For the extraction of tidal energy at a suitable dam site of the Bay of Fundy or similar tidal estuaries elsewhere, a knowledge of tidal behaviour is important in terms of construction operations and subsequent build-up of sediment.

One of the principal obstacles in obtaining adequate analytical solutions lies in the complicated geometry of natural estuaries. The use of digital computers largely reduces this difficulty by enabling variations in depth and breadth to be incorporated in the calculations, leading to increased accuracy in the numerical solutions. However, the use of numerical methods does not detract from the value of approximate analytic solutions. This chapter is mainly concerned with obtaining and discussing analytic solutions corresponding to some simplified one-dimensional tidal equations.

Also a brief account of the depth average, two-dimensional, tidal wave equations is given at the end of this chapter.

6.2 Tidal waves in a uniform canal

Consider tidal waves travelling along a straight canal, with a horizontal bed and parallel vertical sides. Let us suppose that the x-axis is parallel to the length of the canal and the z-axis is vertically upwards. The motion is assumed to take place entirely in the two-dimensions x and z. Let the ordinate of the free surface, corresponding to the abscissa x, at times be denoted by $h + \eta$ where h is the uniform depth of the canal.

6.2.1 Equations of motion and continuity

The equation of motion in the z-direction is

$$\frac{Dw}{Dt} = -g - \frac{1}{\rho}\frac{\partial P}{\partial z}.$$

Assuming that the vertical acceleration Dw/Dt is negligible in comparison with the other terms in this equation, we have $\partial P/\partial z = -\rho g$ and integrating we get $P = -\rho gz + \text{constant}$. The pressure at any point (x, z) is equal to the static pressure due to the depth h below the free surface. Then if the atmospheric pressure is P_0, we must have $P = P_0$ when $z = \eta + h$, where $\eta(x, t)$ is the elevation of the water above the still water level $z = 0$. So the equation for the pressure becomes

$$P - P_0 = \rho g(h + \eta - z), \tag{6.1}$$

where P_0 is the uniform atmospheric pressure, h is the uniform depth of the canal, and η is the water elevation.

We can substitute this value of P into the equation of motion

$$\frac{Du}{Dt} = -\frac{1}{\rho}\frac{\partial P}{\partial x}$$

and we obtain

$$\frac{Du}{Dt} = -g\frac{\partial \eta}{\partial x}.$$

The right-hand side of this equation is independent of z, so that the horizontal acceleration Du/Dt is the same for all particles in a plane perpendicular to x. Consequently, all particles which lie in such a plane always do so; in other words, the horizontal velocity is a function of x and t only.

Then differentiating eqn (6.1) partially, with respect to x we have

$$\frac{\partial P}{\partial x} = \rho g \frac{\partial \eta}{\partial x}. \tag{6.2}$$

The one-dimensional form of the equation of horizontal motion may be explicitly stated as

$$\frac{\partial u}{\partial t} + u \frac{\partial u}{\partial x} = -\frac{1}{\rho} \frac{\partial P}{\partial x}. \tag{6.3}$$

For infinitely small motions it is assumed that the term $u(\partial u/\partial x)$ is very small compared with the other terms in (6.3) and can be neglected. Therefore,

$$\frac{\partial u}{\partial t} = -g \frac{\partial \eta}{\partial x}, \tag{6.4}$$

which is the *equation of motion* in one dimension.

Next we consider the *equation of continuity* in two-dimensional form which may be written as (see Fig. 6.1)

$$\frac{\partial u}{\partial x} + \frac{\partial w}{\partial z} = 0. \tag{6.5}$$

Equation (6.5) is integrated over a column of water extending from the bottom of the basin $z = -h$ to the free surface $z = \eta$ with respect to z, i.e.

$$\int_{-h}^{\eta} \frac{\partial u}{\partial x} \, dz + \int_{-h}^{\eta} \frac{\partial w}{\partial z} \, dz = 0. \tag{6.6}$$

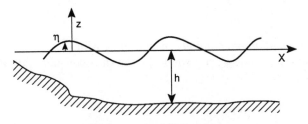

Fig. 6.1. Flow configuration.

Now

$$\frac{\partial}{\partial x}\int_{-h}^{\eta} u\,dz = \int_{-h}^{\eta}\frac{\partial u}{\partial x}\,dz + u(\eta)\frac{\partial \eta}{\partial x} + u(-h)\frac{\partial h}{\partial x},$$

and

$$\int_{-h}^{\eta}\frac{\partial w}{\partial z}\,dz = w(\eta) - w(-h).$$

We know that

$$\frac{d\eta}{dt} = \frac{\partial \eta}{\partial t} + u(\eta)\frac{\partial \eta}{\partial x} = w(\eta).$$

Thus

$$w(\eta) = \frac{\partial \eta}{\partial t} + u(\eta)\frac{\partial \eta}{\partial x}$$

and

$$w(-h) = -u(-h)\frac{\partial h}{\partial x}.$$

Using these results, eqn (6.6) can be written as

$$\frac{\partial}{\partial x}\int_{-h}^{\eta} u\,dz - u(\eta)\frac{\partial \eta}{\partial x} - u(-h)\frac{\partial h}{\partial x} + w(\eta) - w(-h) = 0.$$

The equation may be simplified to yield

$$\frac{\partial}{\partial x}\int_{-h}^{\eta} u\,dz + \frac{\partial \eta}{\partial t} = 0. \tag{6.7}$$

As we have observed that u is not a function of z, eqn (6.7) yields

$$\frac{\partial}{\partial x}u(\eta + h) + \frac{\partial \eta}{\partial t} = 0. \tag{6.8}$$

This is the *equation of continuity* in one dimension.

For small-amplitude waves, when the depth is constant, eqn (6.8) can be written as

$$h\frac{\partial u}{\partial x} + \frac{\partial \eta}{\partial t} = 0 \tag{6.9}$$

which is a simpler form of the *equation of continuity* in one dimension. Eliminating η between (6.4) and (6.9) gives

$$\frac{\partial^2 u}{\partial t^2} = gh \frac{\partial^2 u}{\partial x^2}. \tag{6.10}$$

Similarly, elimination of u gives an equation of the same form, namely

$$\frac{\partial^2 \eta}{\partial t^2} = gh \frac{\partial^2 \eta}{\partial x^2}, \tag{6.11}$$

which can be written as $\eta_{tt} = C^2 \eta_{xx}$, where $C^2 = gh$. This is the familiar wave equation in one dimension, and we deduce that waves travel with speed $C = \sqrt{gh}$. With an unlimited canal there are no boundary conditions involving x, and to our degree of approximation waves with any profile will travel in either direction. However, with a limited canal there will be boundary conditions. Thus if the ends are vertical the velocity u will be zero at each end.

6.2.2 First alternative form of the continuity equation

It is imperative to note that the continuity equation plays a very important role in the investigation of the mechanics of tidal wave. Owing to its importance, we derive its two important alternative forms from basic principles. With reference to Fig. 6.2, an alternative form of the continuity equation is as derived below.

Suppose that the mass present in the elementary volume at time t is $\rho S \Delta x$, the mass present at $t + \Delta t$ is

$$\rho S \Delta x + \frac{\partial}{\partial t}(\rho S) \Delta t \Delta x.$$

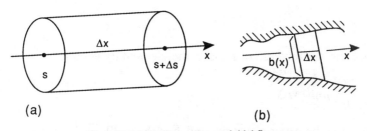

(a)   (b)

Fig. 6.2.   Elementary volume of tidal flow.

Then the net increase of mass in Δt within the element is

$$\frac{\partial}{\partial t}(\rho S)\Delta t\,\Delta x. \tag{6.12}$$

The inflow of mass through the face S is $\rho u S\,\Delta t$. The outflow of mass across the face $S + \Delta S$ is

$$\rho u S\,\Delta t + \frac{\partial}{\partial x}(\rho u S)\Delta x\,\Delta t.$$

Hence the net inflow in the element is

$$-\frac{\partial}{\partial x}(\rho u S)\Delta x\,\Delta t. \tag{6.13}$$

Equating (6.12) and (6.13),

$$\frac{\partial}{\partial t}(\rho S) + \frac{\partial}{\partial x}(\rho u S) = 0$$

is obtained. Then

$$\frac{\partial \rho}{\partial t} + \frac{1}{S}\frac{\partial}{\partial x}(\rho u S) = 0, \tag{6.14}$$

where S is the area, ρ is the density of the fluid and u is the horizontal fluid velocity. This is the first alternative form of the *continuity equation*.

6.2.3 Second alternative form of the continuity equation

Another way of writing the continuity equation (6.14) is as follows. Referring to Fig. 6.2, the mass balance within the elementary volume can be written as: the mass present at time t is $\rho \eta b\,\Delta x$, the mass present at $t + \Delta t$ is

$$\rho \eta b\,\Delta x + \frac{\partial}{\partial t}(\rho \eta b)\Delta t\,\Delta x.$$

Hence the net increase of mass in the elementary volume is

$$\frac{\partial}{\partial t}(\rho \eta b)\Delta t\,\Delta x, \tag{6.15}$$

where η is the water elevation, and b is the width of the estuary.

Then eqn (6.15) with (6.13) gives

$$\frac{\partial}{\partial t}(\rho \eta b) + \frac{\partial}{\partial x}(\rho u S) = 0.$$

For incompressible fluid, assume $\rho = $ constant, and if $b = b(x)$, then

$$\frac{\partial \eta}{\partial t} + \frac{1}{b}\frac{\partial}{\partial x}(uS) = 0. \tag{6.16}$$

Assume $S = bh$, where h is the depth of the estuary and b is the breadth. Then (6.16) can be written as

$$\frac{\partial \eta}{\partial t} + \frac{1}{b}\frac{\partial}{\partial x}(bhu) = 0. \tag{6.17}$$

This is the second alternative form of the *continuity equation*.

6.3 Tidal dynamics in estuaries

6.3.1 Linearized tidal equations (frictionless case)

From the above mathematical formulation, it follows that the motion along the x-axis is

$$\frac{\partial u}{\partial t} + g\frac{\partial \eta}{\partial x} = 0 \tag{6.18}$$

and the continuity condition is

$$\frac{\partial \eta}{\partial t} + \frac{1}{b}\frac{\partial}{\partial x}(bhu) = 0 \tag{6.19}$$

where b is the breadth of the estuary, and h is the depth of the estuary. For a real estuary these are functions of x only. In the following we will examine four important cases of physical interest to obtain some approximate solutions corresponding to four idealized estuaries.

(I): One-dimensional tidal estuary of constant depth and breadth

In this case let us assume that the depth h and breadth b of the estuary are uniformly constant. Then (6.19) reduces to

$$\frac{\partial \eta}{\partial t} + h\frac{\partial u}{\partial x} = 0. \tag{6.20}$$

To eliminate u from (6.18) and (6.20) the cross-differentiation produces

$$\frac{\partial^2 u}{\partial t \partial x} + g \frac{\partial^2 \eta}{\partial x^2} = 0,$$

and

$$\frac{\partial^2 \eta}{\partial t^2} + h \frac{\partial^2 u}{\partial x \partial t} = 0.$$

Eliminating the cross-derivative term yields

$$\frac{\partial^2 \eta}{\partial t^2} = gh \frac{\partial^2 \eta}{\partial x^2}.$$

We define $C = \sqrt{gh}$ as the wave speed or celerity.
   Therefore

$$\frac{\partial^2 \eta}{\partial t^2} = C^2 \frac{\partial^2 \eta}{\partial x^2}, \qquad (6.21)$$

and similarly,

$$\frac{\partial^2 u}{\partial t^2} = C^2 \frac{\partial^2 u}{\partial x^2}. \qquad (6.22)$$

Equations (6.21) and (6.22) are the well-known basic wave equations. These equations are satisfied for the specific case of periodic long waves entering a canal of uniform section and of infinite length by the harmonic function

$$\eta = A \cos(\sigma t - kx) \qquad (6.23)$$

where A is the amplitude of the wave.
   At $x = 0$, $\eta = A \cos \sigma t$ (a wave communicating at the mouth). Substituting from (6.23) into (6.21) gives

$$\frac{\partial^2 \eta}{\partial t^2} = -A\sigma^2 \cos(\sigma t - kx),$$

$$\frac{\partial^2 \eta}{\partial x^2} = -Ak^2 \cos(\sigma t - kx).$$

Therefore $\sigma = Ck$ where $\sigma = 2\pi/T$ is the frequency, and $k = 2\pi/L$ is the wave number. Here T is the time period and L is the wavelength.

Now from (6.18),

$$\frac{\partial u}{\partial t} = -g\frac{\partial \eta}{\partial x} = (-g)(Ak)\sin(\sigma t - kx).$$

Integrating with respect to t, we obtain

$$u = \frac{AC}{h}\cos(\sigma t - kx). \tag{6.24}$$

Therefore the water elevation and velocity are

$$\eta = A\cos(\sigma t - kx) \quad \text{and} \quad u = \frac{AC}{h}\cos(\sigma t - kx). \tag{6.25}$$

The velocity and surface elevation are in phase and are related by

$$u = \left(\frac{\eta}{h}\right)C. \tag{6.26}$$

The velocity is of the same sign as the amplitude; hence the wave is a *progressive wave*. Also it can be easily seen that

$$u_{max} = \left(\frac{A}{h}\right)C. \tag{6.27}$$

(II): One-dimensional tidal estuary of constant depth but variable breadth

In this case we assume the linear variation of breadth of the estuary, which is $b = b_0(x/l)$, and the depth $h = h_0$ where h_0 is a constant (see Fig. 6.3). Combining (6.18) and (6.19), the wave equation can be written as

$$\frac{\partial^2 \eta}{\partial t^2} = \frac{g}{b}\frac{\partial}{\partial x}\left(bh\frac{\partial \eta}{\partial x}\right). \tag{6.28}$$

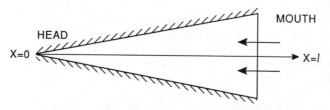

Fig. 6.3.   Schematic diagram of an idealized estuary.

Substituting the values of b and h, we obtain

$$\frac{\partial^2 \eta}{\partial t^2} = \frac{gh_0}{x} \frac{\partial}{\partial x} \left(x \frac{\partial \eta}{\partial x} \right) = \frac{gh_0}{x} \left(x \frac{\partial^2 \eta}{\partial x^2} + \frac{\partial \eta}{\partial x} \right).$$

Therefore we have

$$\frac{\partial^2 \eta}{\partial t^2} = gh_0 \left(\frac{\partial^2 \eta}{\partial x^2} + \frac{1}{x} \frac{\partial \eta}{\partial x} \right). \tag{6.29}$$

The boundary conditions are

at $x = 0$, η must be bounded $\tag{6.30}$

at $x = l$, $\eta = A \cos(\sigma t + \varepsilon)$. $\tag{6.31}$

Considering that tides in the bay are simple harmonics, then

$$\eta \sim \cos(\sigma t + \varepsilon) \qquad \text{or} \qquad \eta = y(x)\cos(\sigma t + \varepsilon). \tag{6.32}$$

Substituting (6.32) into (6.29) gives

$$-\sigma^2 y = gh_0 \left(\frac{d^2 y}{dx^2} + \frac{1}{x} \frac{dy}{dx} \right).$$

Therefore

$$\frac{d^2 y}{dx^2} + \frac{1}{x} \frac{dy}{dx} + \lambda^2 y = 0, \tag{6.33}$$

where $\lambda^2 = \sigma^2/gh_0$.

Equation (6.33) is a Bessel equation, the solution of which exists in the following form:

$$y = A_1 J_0(\lambda x) + B_1 Y_0(\lambda x), \tag{6.34}$$

where J_0 and Y_0 are Bessel functions of the first and second kind of order 0. Therefore

$$\eta = [A_1 J_0(\lambda x) + B_1 Y_0(\lambda x)]\cos(\sigma t + \varepsilon). \tag{6.35}$$

The boundary condition (6.30) is satisfied if $B_1 = 0$, since $Y_0(\lambda x) \to -\infty$ as $x \to 0$. Thus

$$\eta = A_1 J_0(\lambda x)\cos(\sigma t + \varepsilon). \tag{6.36}$$

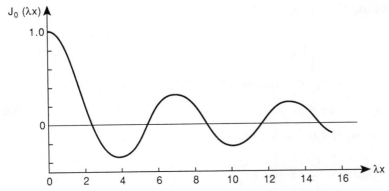

Fig. 6.4.   Standing waves in the estuary of uniform depth and variable breadth.

Now at $x = l$, it is seen from (6.31) and (6.36) that $A_1 J_0(\lambda l) = A$. Hence

$$A_1 = \frac{A}{J_0(\lambda l)}. \tag{6.37}$$

Therefore the water elevation is given by

$$\eta = A \frac{J_0(\lambda x)}{J_0(\lambda l)} \cos(\sigma t + \varepsilon). \tag{6.38}$$

As shown in Fig. 6.4 the zeroth-order Bessel function $J_0(\lambda x)$ calls for a large increase in tidal height as we proceed up the estuary, while a corresponding wavelength is partially constant. If the estuary length l corresponds to a zero of the Bessel function, then the possibility for resonance exists.

(III): One-dimensional tidal estuary of variable depth and constant breadth

In this case we assume the variation is in depth only and this increases uniformly from the head $x = 0$ of the canal to the mouth $x = l$, the remaining circumstances being as before. Thus in this case $h = h_0(x/l)$ and $b = b_0$. Then (6.28) may be written as

$$\frac{\partial^2 \eta}{\partial t^2} = \frac{g}{b_0} \frac{\partial}{\partial x} \left(b_0 h_0 \frac{x}{l} \frac{\partial \eta}{\partial x} \right) = \frac{g h_0}{l} \frac{\partial}{\partial x} \left(x \frac{\partial \eta}{\partial x} \right).$$

Therefore

$$\frac{\partial^2 \eta}{\partial t^2} = \frac{gh_0}{l}\left(x\frac{\partial^2 \eta}{\partial x^2} + \frac{\partial \eta}{\partial x}\right). \tag{6.39}$$

Again considering that the tides in the bay are simple harmonics, then

$$\eta = y(x)\cos(\sigma t + \varepsilon). \tag{6.40}$$

Substituting (6.40) into (6.39), we derive

$$-\sigma^2 y = \frac{gh_0}{l}\left(x\frac{d^2 y}{dx^2} + \frac{dy}{dx}\right).$$

Therefore

$$x\frac{d^2 y}{dx^2} + \frac{dy}{dx} + \lambda^2 y = 0,$$

where $\lambda^2 = \sigma^2 l/(gh_0)$. Thus we obtain

$$\frac{d^2 y}{dx^2} + \frac{1}{x}\frac{dy}{dx} + \frac{\lambda^2}{x}y = 0. \tag{6.41}$$

Equation (6.41) can be recognized as a Bessel equation. On comparing (6.41) with the generalized Bessel differential equation (see McLachlan (1955))

$$\frac{d^2 y}{dx^2} + \left(\frac{2\alpha - 2\beta\nu + 1}{x}\right)\frac{dy}{dx} + \left(\beta^2\gamma^2 x^{2\beta-2} + \frac{\alpha(\alpha - 2\beta\nu)}{x^2}\right)y = 0 \tag{6.42}$$

which has the complete general solution

$$y = x^{\beta\nu-\alpha}\left[A_1 J_\nu(\gamma x^\beta) + B_1 Y_\nu(\gamma x^\beta)\right], \tag{6.43}$$

giving $2\alpha - 2\beta\nu + 1 = 1$, $\alpha(\alpha - 2\beta\nu) = 0$, $2\beta - 2 = -1$ and $\beta^2\gamma^2 = \lambda^2$, which yields $\alpha = 0$, $\nu = 0$, $\beta = \frac{1}{2}$ and $\gamma = 2\lambda$.

Hence the solution is

$$y = A_1 J_0(2\lambda x^{1/2}) + B_1 Y_0(2\lambda x^{1/2}). \tag{6.44}$$

For the solution to be bounded at $x = 0$, $B_1 = 0$, and therefore $y = A_1 J_0(2\lambda x^{1/2})$. Hence the water elevation is given by

$$\eta = A_1 J_0(2\lambda x^{1/2})\cos(\sigma t + \varepsilon).$$

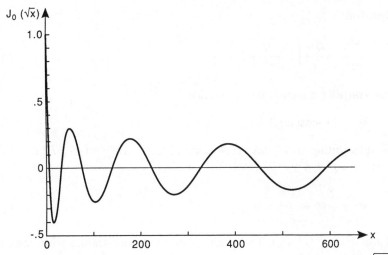

Fig. 6.5. Plot of the Bessel function $J_0(\sqrt{X})$ as against the argument X where $X = \sqrt{4\lambda^2 x}$.

Since, at $x = l$, $\eta = A\cos(\sigma t + \varepsilon)$, then $A_1 = A/[J_0(2\lambda l^{1/2})]$. Hence the water elevation is

$$\eta = \frac{AJ_0(2\lambda x^{1/2})}{J_0(2\lambda l^{1/2})}\cos(\sigma t + \varepsilon). \tag{6.45}$$

From Fig. 6.5 we observe that the zeroth-order Bessel function $J_0(2\lambda\sqrt{x})$ shows how the tidal height continuously increases, and the wavelength diminishes, as we travel up the estuary from the mouth to the head.

(IV): One-dimensional tidal estuary of variable depth and breadth

In this case let us assume that the depth and breadth vary as the distance from the head $x = 0$ of the canal to the mouth $x = l$. Thus in this case we have $h = h_0(x/l)$ and $b = b_0(x/l)$.

Then (6.28) may be written as

$$\frac{\partial^2 \eta}{\partial t^2} = \frac{g}{b}\frac{\partial}{\partial x}\left(bh\frac{\partial \eta}{\partial x}\right) = \frac{g}{b_0\left(\dfrac{x}{l}\right)}\frac{\partial}{\partial x}\left(\frac{h_0 b_0}{l^2}x^2\frac{\partial \eta}{\partial x}\right)$$

$$= \frac{gh_0}{lx}\frac{\partial}{\partial x}\left(x^2\frac{\partial \eta}{\partial x}\right) = \left(\frac{gh_0}{l}\right)\frac{1}{x}\left(x^2\frac{\partial^2 \eta}{\partial x^2} + 2x\frac{\partial \eta}{\partial x}\right).$$

Therefore

$$\frac{\partial^2 \eta}{\partial t^2} = \frac{gh_0}{l}\left(x\frac{\partial^2 \eta}{\partial x^2} + 2\frac{\partial \eta}{\partial x}\right). \tag{6.46}$$

Now consider a simple harmonic tide

$$\eta = y(x)\cos(\sigma t + \varepsilon). \tag{6.47}$$

On substituting (6.47) into (6.46), we obtain $-\sigma^2 y = (gh_0/l)(xy'' + 2y')$, or $xy'' + 2y' + \lambda^2 y = 0$. Thus

$$y'' + \frac{2}{x}y' + \frac{\lambda^2}{x}y = 0, \tag{6.48}$$

where $\lambda^2 = \sigma^2 l/(gh_0)$. The prime denotes differentiation with respect to x.

Comparing (6.48) with (6.42), we obtain $2\alpha - 2\beta\nu + 1 = 2$, $2\beta - 2 = -1$, and $\beta^2\gamma^2 = \lambda^2$, which yield $\alpha = 0$, $\beta = \frac{1}{2}$, $\gamma = 2\lambda$ and $\nu = 1$.

Hence the solution is

$$y = x^{\beta\nu - \alpha}\left[A_1 J_\nu(\gamma x^\beta) + B_1 Y_\nu(\gamma x^\beta)\right]$$
$$= x^{-1/2}\left[A_1 J_1(2\lambda x^{1/2}) + B_1 Y_1(2\lambda x^{1/2})\right]. \tag{6.49}$$

The boundary condition (6.30) at $x = 0$ is satisfied if $B_1 = 0$, and hence $y = x^{-1/2}A_1 J_1(2\lambda x^{1/2})$. The water elevation is $\eta = x^{-1/2}A_1 J_1(2\lambda x^{1/2}) \times \cos(\sigma t + \varepsilon)$. The boundary condition (6.31) at $x = l$ implies that $A_1 = Al^{1/2}/[J_1(2\lambda l^{1/2})]$. Therefore the water elevation is

$$\eta = A\frac{J_1(2\lambda x^{1/2})}{J_1(2\lambda l^{1/2})}\left(\frac{l}{x}\right)^{1/2}\cos(\sigma t + \varepsilon). \tag{6.50}$$

It is worth mentioning here that this solution gives a fair representation of the case of the Bristol Channel. The tidal elevations observed at various stations along the channel are found to be in good agreement with the present formula considered in this case (see Taylor (1921)).

We do not want to pursue any more cases of free oscillations of tidal estuaries in this text. However, for modes of free oscillations in canals of simple geometrical shape of uniform breadth but variable depths of circular, parabolic shapes, etc., we refer to the standard texts including Lamb (1945) and Dean and Dalrymple (1984) for more information. The following section concerns only the two fundamental solutions of uniform channel.

6.4 Tidal oscillations without friction: fundamental solutions

The linearized tidal equations in a channel of slowly varying breadth $b(x)$ and mean depth $h(x)$ are, upon neglecting friction,

$$\frac{\partial^2 \eta}{\partial t^2} = \frac{g}{b} \frac{\partial}{\partial x} \left(hb \frac{\partial \eta}{\partial x} \right) \tag{6.51}$$

$$\frac{\partial u}{\partial t} = -g \frac{\partial \eta}{\partial x}, \tag{6.52}$$

where η is the surface elevation above the mean tidal level and u is the average current across a section. The fundamental solutions of uniform channels are investigated. The fundamental solutions of diverging channels (see Hunt (1964)) are very much involved and therefore this case is left as an exercise.

6.4.1 Uniform channel

The channel is of constant depth, $h = h_0$, and breadth, $b = b_0$. In this case (6.51) and (6.52) become

$$\frac{\partial^2 \eta}{\partial t^2} = gh_0 \frac{\partial^2 \eta}{\partial x^2} \quad \text{and} \quad \frac{\partial u}{\partial t} = -g \frac{\partial \eta}{\partial x}.$$

Putting $C^2 = gh_0$,

$$\frac{\partial^2 \eta}{\partial t^2} = C^2 \frac{\partial^2 \eta}{\partial x^2}$$

is obtained.

By the separation of variables, assume a solution $\eta = X(x)T(t)$, which on substituting into the above equation yields $XT'' = C^2 X'' T$. Hence

$$\frac{1}{C^2} \frac{T''}{T} = \frac{X''}{X} = -k^2.$$

Therefore

$$X'' + k^2 X = 0, \quad T'' + k^2 C^2 T = 0. \tag{6.53}$$

Solving (6.53), gives

$$X = a_1 \cos kx + a_2 \sin kx$$

$$T = b_1 \cos(kCt) + b_2 \sin(kCt)$$

$$\eta = XT = (a_1 \cos kx + a_2 \sin kx)(b_1 \cos Ckt + b_2 \sin Ckt). \qquad (6.54)$$

Putting $\sigma^2 = C^2 k^2$, then

$$\eta = (a_1 \cos kx + a_2 \sin kx)(b_1 \cos \sigma t + b_2 \sin \sigma t)$$

$$= \alpha_1 \cos kx \cos \sigma t + \alpha_2 \sin kx \cos \sigma t + \alpha_3 \cos kx \sin \sigma t$$

$$+ \alpha_4 \sin kx \sin \sigma t. \qquad (6.55)$$

(a) First fundamental solution

Considering the first and third terms on the right-hand side of (6.55), the first fundamental solution is constructed:

$$\eta_1 = A_1 \cos kx \cos(\sigma t + \varepsilon_1). \qquad (6.56)$$

Therefore

$$\frac{\partial \eta_1}{\partial x} = -A_1 k \sin kx \cos(\sigma t + \varepsilon_1)$$

$$\frac{\partial u_1}{\partial t} = -g \frac{\partial \eta_1}{\partial x} = A_1 kg \sin kx \cos(\sigma t + \varepsilon_1).$$

Integrating this last equation with respect to t yields

$$u_1 = \frac{A_1 kg}{\sigma} \sin kx \sin(\sigma t + \varepsilon_1). \qquad (6.57)$$

Considering the second and fourth terms on the right-hand side of (6.55), another first fundamental solution is constructed:

$$\eta_2 = B_1 \sin kx \cos(\sigma t + \varepsilon_2). \qquad (6.58)$$

Therefore

$$\frac{\partial \eta_2}{\partial x} = B_1 k \cos kx \cos(\sigma t + \varepsilon_2)$$

$$\frac{\partial u_2}{\partial t} = -g \frac{\partial \eta_2}{\partial x} = -gB_1 k \cos kx \cos(\sigma t + \varepsilon_2).$$

Integrating this last equation with respect to t yields

$$u_2 = -\frac{B_1 kg}{\sigma} \cos kx \sin(\sigma t + \varepsilon_2),$$ (6.59)

where

$$C^2 = \frac{\sigma^2}{k^2} = gh_0.$$ (6.60)

The first fundamental solutions obtained in this section exhibit *standing wave solutions*.

(b) Second fundamental solution

The second fundamental solution can be obtained from (6.55) by suitable combinations of the terms on the right-hand side. For example, combining the first and fourth terms yields

$$\eta_1 = A_2 \cos k(x - Ct + \varepsilon_3).$$ (6.61)

Hence

$$\frac{\partial \eta_1}{\partial x} = -A_2 k \sin k(x - Ct + \varepsilon_3),$$

and

$$\frac{\partial u_1}{\partial t} = -g \frac{\partial \eta_1}{\partial x} = A_2 kg \sin k(x - Ct + \varepsilon_3).$$

Integrating with respect to t yields

$$u_1 = \frac{A_2 kg}{\sigma} \cos k(x - Ct + \varepsilon_3).$$ (6.62)

Another suitable combination yields

$$\eta_2 = B_2 \cos k(x + Ct + \varepsilon_4).$$ (6.63)

Hence

$$\frac{\partial \eta_2}{\partial x} = -B_2 k \sin k(x + Ct + \varepsilon_4),$$

and

$$\frac{\partial u_2}{\partial t} = -g\frac{\partial \eta_2}{\partial x} = B_2 kg \sin k(x + Ct + \varepsilon_4).$$

Integrating with respect to t yields

$$u_2 = -\frac{B_2 kg}{\sigma} \cos k(x + Ct + \varepsilon_4). \tag{6.64}$$

Now from the first fundamental solutions, we obtain

$$\eta_1 = A_1 \cos kx \cos(\sigma t + \varepsilon_1); \qquad u_1 = \frac{A_1 kg}{\sigma} \sin ks \sin(\sigma t + \varepsilon_1)$$

$$\eta_2 = B_1 \sin kx \cos(\sigma t + \varepsilon_2); \qquad u_2 = -\frac{B_1 kg}{\sigma} \cos kx \sin(\sigma t + \varepsilon_2).$$

The standing wave solutions exhibit a characteristic phase difference of a quarter period between *current* and *surface elevation*, and simultaneous high-water times within a half wavelength. The simultaneous low water over the remaining half wavelength may be disregarded in the present context since we do not encounter estuaries of sufficient length to include a *nodal line* or *amphidromic region*.

From the second fundamental solutions, we obtain

$$\eta_1 = A_2 \cos k(x - Ct + \varepsilon_3); \qquad u_1 = \frac{A_2 kg}{\sigma} \cos k(x - Ct + \varepsilon_3)$$

$$\eta_2 = B_2 \cos k(x + Ct + \varepsilon_4); \qquad u_2 = -\frac{B_2 kg}{\sigma} \cos k(x + Ct + \varepsilon_4).$$

The *progressive wave solutions* show the current and elevation to be in phase everywhere, while the high-water time in each component, indicated by the wave crest, progresses with speed C.

The solution permitting the introduction of a tidal barrier, say at $x = 0$, where $u = 0$, is only given by

$$\eta_1 = A_1 \cos kx \cos(\sigma t + \varepsilon_1)$$

$$u_1 = \frac{A_1 kg}{\sigma} \sin kx \sin(\sigma t + \varepsilon_1).$$

The only combination of progressive wave solutions to satisfy this condition is

$$\eta = \frac{A}{2}[\cos k(x - Ct + \varepsilon_3) + \cos k(x + Ct + \varepsilon_3)] \tag{6.65}$$

$$-g\frac{\partial \eta}{\partial x} = \frac{gAk}{2}[\sin k(x - Ct + \varepsilon_3) + \sin k(x + Ct + \varepsilon_3)]$$

$$\frac{\partial u}{\partial t} = \frac{gAk}{2}[\sin k(x - Ct + \varepsilon_3) + \sin k(x + Ct + \varepsilon_3)].$$

Integration yields

$$u = \frac{gAk}{2\sigma}[\cos k(x - Ct + \varepsilon_3) - \cos k(x + Ct + \varepsilon_3)], \tag{6.66}$$

where both incoming and 'reflected' outgoing waves show the wave identity between η and u, although the total flow expressed by (6.56) and (6.57) shows a quarter-period phase difference. The second fundamental solutions obtained in this section exhibit *progressive wave solutions*.

6.5 Tidal oscillations with friction

It is assumed that tidal propagation in estuaries can be described by the one-dimensional shallow water wave equations expressed in a linear form and including a linearized friction term. It is further assumed that estuarine geometry may be approximated by a breadth variation of the form $B_L(x/\lambda)^n$ and a depth variation $H_L(x/\lambda)^m$, where X is the distance from the head of the estuary. Here the head denotes the location of the upstream boundary condition where the channel is assumed to be closed. By rewriting the relevant equations in dimensionless form, analytical solutions are obtained applicable to any estuary within the limits of the assumptions stated.

Tidal propagation in a narrow channel of rectangular cross-section may be described by the following linearized equations (Dorrestein (1961)): Motion along the X-axis:

$$\frac{\partial U}{\partial T} + g\frac{\partial Z}{\partial X} + FU = 0. \tag{6.67}$$

Continuity:

$$\frac{\partial Z}{\partial T} + \frac{1}{B}\frac{\partial}{\partial X}(BHU) = 0. \tag{6.68}$$

Symbols are defined as follows: U = velocity along the X-direction, Z = elevation above a horizontal datum, g = gravitational constant, F = friction coefficient, B = channel breadth, H = channel depth, T = time.

Assuming a breadth variation of the form

$$B(X) = B_L(X/\lambda)^n \tag{6.69}$$

and a depth variation

$$H(X) = H_L(X/\lambda)^m, \tag{6.70}$$

with X measured from the head of the estuary, solutions are found for U and Z. Since we are interested in the general application of these equations to estuaries of widely varying dimensions it is useful to rewrite the equations in dimensionless form. For this purpose tidal period τ is introduced as the unit of time, λ as the unit of horizontal dimension (may be considered as wavelength) and H_L as the unit of vertical dimension. The following relationship is assumed between these parameters:

$$\lambda = (gH_L)^{1/2}\tau. \tag{6.71}$$

Thus, the transformed dimensionless variables are $x = X/\lambda$, $t = T/\tau$, $z = Z/H_L$, $b = B/\lambda$, $b_L = B_L/\lambda$, $h = H/H_L$, $u = U(\tau/\lambda)$, $f = F\tau$. Equations (6.67) to (6.71) may then be rewritten as

$$\frac{\partial u}{\partial t} + \frac{\partial z}{\partial x} + fu = 0, \tag{6.72}$$

$$\frac{\partial z}{\partial t} + \frac{1}{b}\frac{\partial}{\partial x}(bhu) = 0, \tag{6.73}$$

$$b(x) = b_L x^n, \tag{6.74}$$

$$h(x) = x^m. \tag{6.75}$$

Equations (6.73), (6.74) and (6.75) give

$$\frac{\partial z}{\partial t} + x^{m-1}\left((n+m)u + x\frac{\partial u}{\partial x}\right) = 0. \tag{6.76}$$

Eliminating u from (6.72) and (6.76) yields

$$\frac{\partial^2 z}{\partial t^2} + f\frac{\partial z}{\partial t} = x^{m-1}\left(x\frac{\partial^2 z}{\partial x^2} + (n+m)\frac{\partial z}{\partial x}\right). \tag{6.77}$$

We are interested in solutions of the form

$$z = \phi(x)e^{i\sigma t}, \tag{6.78}$$

$$u = \theta(x)e^{i\sigma t}. \tag{6.79}$$

Substituting (6.78) into (6.77) we obtain

$$x^m\frac{d^2\phi}{dx^2} + (n+m)x^{m-1}\frac{d\phi}{dx} + \sigma^2\left(1 - \frac{if}{\sigma}\right)\phi = 0. \tag{6.80}$$

Introducing the parameters

$$k^2 = \left(1 - \frac{if}{\sigma}\right), \tag{6.81}$$

$$y = \sigma k\left(\frac{2}{2-m}\right)x^{(2-m)/2}. \tag{6.82}$$

Equation (6.80) may be rewritten as

$$\frac{d^2\phi}{dy^2} + \frac{1}{y}\left(\frac{2n+m}{2-m}\right)\frac{d\phi}{dy} + \phi = 0. \tag{6.83}$$

By using the parameter $\alpha = -(n+m-1)/(2-m)$, eqn (6.83) becomes

$$\frac{d^2\phi}{dy^2} + \frac{(1-2\alpha)}{y}\frac{d\phi}{dy} + \phi = 0. \tag{6.84}$$

Then writing $\phi = \hat{\phi}y^\alpha$ transforms (6.84) into

$$\frac{d^2\hat{\phi}}{dy^2} + \frac{1}{y}\frac{d\hat{\phi}}{dy} + \left(1 - \frac{\alpha^2}{y^2}\right)\hat{\phi} = 0. \tag{6.85}$$

The general solution of this equation is (see McLachlan (1955))

$$\hat{\phi} = A_1 J_{-\alpha}(y) + B_1 Y_{-\alpha}(y), \tag{6.86}$$

where $J_{-\alpha}$ and $Y_{-\alpha}$ are Bessel functions of the first and second kind, respectively, and of order α.

Introducing the parameter $\nu = 1 - \alpha = (n + 1)/(2 - m)$ from eqn (6.84) and eqn (6.78) the solution for z is

$$z = y^{1-\nu}[A_1 J_{\nu-1}(y) + B_1 Y_{\nu-1}(y)]e^{i\sigma t}. \tag{6.87}$$

At the downstream boundary, $y = y_0$, we specify a tidal amplitude A_0; hence

$$A_0 = y_0^{1-\nu}[A_1 J_{\nu-1}(y_0) + B_1 Y_{\nu-1}(y_0)]. \tag{6.88}$$

For an estuary without a barrier, for z to remain finite at the head $x = 0$ ($y = 0$), it follows that $B_1 = 0$ and so

$$z = \frac{A_0 y^{1-\nu} J_{\nu-1}(y)}{y_0^{1-\nu} J_{\nu-1}(y_0)} e^{i\sigma t}. \tag{6.89}$$

For an estuary with a barrier at $X = b$, $u = 0$ and, therefore, $\partial z/\partial x = 0$ at $y = y_b$. Then making use of the relationships

$$L_{\nu-1}(y) + L_{\nu+1}(y) = \frac{2\nu}{y} L_\nu(y) \tag{6.90}$$

$$L'_\nu(y) = L_{\nu-1}(y) - \frac{\nu}{y} L_\nu(y), \tag{6.91}$$

where L denotes a Bessel function of the first or second kind and the prime denotes differentation with respect to y, and differentiating (6.87) with respect to x gives

$$A_1 J_\nu(y_b) + B_1 Y_\nu(y_b) = 0. \tag{6.92}$$

Hence (6.86) and (6.88) give

$$z = A_0 \left(\frac{y}{y_0}\right)^{1-\nu} \left(\frac{J_\nu(y_b)Y_{\nu-1}(y) - J_{\nu-1}(y)Y_\nu(y_b)}{J_\nu(y_b)Y_{\nu-1}(y_0) - J_{\nu-1}(y_0)Y_\nu(y_b)}\right) e^{i\sigma t}. \tag{6.93}$$

The equivalent solutions for u are obtained by substituting (6.89) and (6.93), respectively, into (6.72). Thus, putting (6.79) in (6.72) yields

$$u = -\frac{\partial z}{\partial x} \bigg/ (ik^2\sigma). \tag{6.94}$$

Then for an estuary without barriers

$$u = \frac{-A_0}{ik^2 y_0^{1-\nu} J_{\nu-1}(y_0)} kx^{-m/2} \left[-y^{1-\nu} J_\nu(y) \right] e^{i\sigma t}$$

or

$$u = -\frac{i}{k} A_0 x^{-m/2} \left(\frac{y}{y_0} \right)^{1-\nu} \frac{J_\nu(y)}{J_{\nu-1}(y_0)} e^{i\sigma t}. \tag{6.95}$$

While from (6.93) for an estuary with barriers

$$u = -\frac{i}{k} A_0 x^{-m/2} \left(\frac{y}{y_0} \right)^{1-\nu}$$

$$\times \left(\frac{J_\nu(y_b)Y_\nu(y) - J_\nu(y)Y_\nu(y_b)}{J_\nu(y_b)Y_{\nu-1}(y_0) - J_{\nu-1}(y_0)Y_\nu(y_b)} \right) e^{i\sigma t}. \tag{6.96}$$

Note that it is convenient to replace the complex parameter y by ky. In addition, the angular velocity $\sigma = 2\pi$ for the dimensionless time variable t. The preceding analysis of tidal oscillations with friction is mainly due to Prandle and Rahman (1980).

6.6 Depth-averaged 2-D tidal waves

In this section we briefly describe the depth-averaged two-dimensional tidal equation (see Rahman (1988)). The tidal equations, which describe the movement of the tidal wave in terrestrial waters, are derived from the hydrodynamic equations of continuity and motion. Since the flow is turbulent, the time-smoothed velocity and pressure distributions must be considered in the application of these equations. The external forces associated with tidal propagation are the gravity force, the Coriolis force, and the bottom friction together with the Reynolds stresses. Wind forces on the surface of the water must also be included when necessary.

The equation of continuity, which describes the rate of change of density at a fixed point resulting in the change of the mass velocity vector $\rho\mathbf{V}$, is

$$\frac{\partial \rho}{\partial t} = -\left(\frac{\partial \rho u}{\partial x} + \frac{\partial \rho v}{\partial y} + \frac{\partial \rho w}{\partial z} \right) \tag{6.97}$$

in which u, v and w are the velocity components in the three-dimensional Cartesian coordinates x, y and z. They vary with time t. Here x and y are horizontal coordinates positive to the east and to the north respectively; z is the vertical coordinate positive upwards from the mean water level. The customary Boussinesq approximation is introduced so that the variations of density are allowed to affect only the gravitational acceleration. The main simplification resulting from this approximation is that the water is effectively incompressible and thus the continuity equation reads

$$\frac{\partial u}{\partial x} + \frac{\partial v}{\partial y} + \frac{\partial w}{\partial z} = 0. \tag{6.98}$$

If eqn (6.98) is integrated over a column of water extending from the bottom of the basin to the free surface, it clearly establishes a relationship between the change of surface level and the volume of water leaving or entering the column. Thus if the depth of the lake or estuary is denoted by h and the free surface by η, then integrating (6.98) from $z = -h(x, y)$ to $z = \eta(x, y, t)$ with respect to z gives

$$\int_{-h}^{\eta} \frac{\partial u}{\partial x}\, dz + \int_{-h}^{\eta} \frac{\partial v}{\partial y}\, dz + \int_{-h}^{\eta} \frac{\partial w}{\partial z}\, dz = 0. \tag{6.99}$$

Now,

$$\frac{\partial}{\partial x} \int_{-h}^{\eta} u\, dz = \int_{-h}^{\eta} \frac{\partial u}{\partial x}\, dz + u(\eta)\frac{\partial \eta}{\partial x} + u(-h)\frac{\partial h}{\partial x}$$

$$\frac{\partial}{\partial y} \int_{-h}^{\eta} v\, dz = \int_{-h}^{\eta} \frac{\partial v}{\partial y}\, dz + v(\eta)\frac{\partial \eta}{\partial y} + v(-h)\frac{\partial h}{\partial y}$$

$$\int_{-h}^{\eta} \frac{\partial w}{\partial z}\, dz = w(\eta) - w(-h) \tag{6.100}$$

also define

$$\frac{d\eta}{dt} = \frac{\partial \eta}{\partial t} + u(\eta)\frac{\partial \eta}{\partial x} + v(\eta)\frac{\partial \eta}{\partial y} = w(\eta).$$

Thus

$$w(\eta) = \frac{\partial \eta}{\partial t} + u\frac{\partial \eta}{\partial x} + v\frac{\partial \eta}{\partial y}.$$

Application of (6.100) to (6.99) reduces to

$$\frac{\partial}{\partial x} \int_{-h}^{\eta} u\, dz + \frac{\partial}{\partial y} \int_{-h}^{\eta} v\, dz + \frac{\partial \eta}{\partial t} = 0. \tag{6.101}$$

Now define the following average values of the velocity components:

$$U = \frac{1}{\eta + h} \int_{-h}^{\eta} u \, dz$$

$$V = \frac{1}{\eta + h} \int_{-h}^{\eta} v \, dz. \tag{6.102}$$

The continuity eqn (6.101) now becomes

$$\frac{\partial \eta}{\partial t} + \frac{\partial}{\partial x}[(\eta + h)U] + \frac{\partial}{\partial y}[(\eta + h)V] = 0. \tag{6.103}$$

The equation of motion may be derived by writing a momentum balance for the volume element by considering the forces which act on fluid particles. It is well known that the Reynolds stresses are associated with the turbulent velocity fluctuations. It can be assumed that the Newtonian stresses, which are associated with the viscosity, may be neglected. In the classical theory of long waves, the vertical accelerations of the fluid particles are neglected because these accelerations are very small with respect to the acceleration of the gravity field. Hence, the velocities of the water particles in the z-direction may be neglected in dealing with long waves.

The change of the horizontal velocity components from the bottom to the surface, defined by $\partial u / \partial z$ and $\partial v / \partial z$, are usually considerably smaller than the change in the velocity components in the x- and y-directions. Consequently, the shear component derivatives $\partial \tau_x / \partial z$ and $\partial \tau_y / \partial z$ are usually much greater than $\partial \tau_x / \partial x$, etc. Thus taking these assumptions into account, the tidal equations of motion can be written as

$$\rho \left(\frac{\partial u}{\partial t} + u \frac{\partial u}{\partial x} + v \frac{\partial u}{\partial y} - \Omega v \right) + \frac{\partial P}{\partial x} - \frac{\partial \tau_x}{\partial z} = 0 \tag{6.104}$$

$$\rho \left(\frac{\partial v}{\partial t} + u \frac{\partial v}{\partial x} + v \frac{\partial v}{\partial y} + \Omega u \right) + \frac{\partial P}{\partial y} - \frac{\partial \tau_y}{\partial z} = 0 \tag{6.105}$$

$$\frac{\partial P}{\partial z} + \rho g = 0 \tag{6.106}$$

in which P denotes the pressure, which is hydrostatic according to (6.106); g is the acceleration due to gravity (m/s^2), τ_x is the shear stress component in the x-direction (N/m^2), τ_y is the shear stress component in the

y-direction (N/m^2), Ωv, Ωu are components of the Coriolis forces where $\Omega = 2\omega \sin \phi$, in which ω is the angular velocity of the earth around its axis (rad/s), and ϕ is the latitude.

It follows from eqn (6.106), after integration with respect to z from $z = z$ to $z = \eta(x, y, t)$, that in the case of variable density,

$$\int_z^\eta \frac{\partial P}{\partial z}\,dz + \int_z^\eta g\rho\,dz = 0, \qquad P(z) = P(\eta) + g\int_z^\eta \rho\,dz$$

$$P(z) = P_0 + g\int_z^\eta \rho\,dz \tag{6.107}$$

where P_0 is atmospheric pressure. Thus

$$\frac{\partial P}{\partial x} = g\int_z^\eta \frac{\partial \rho}{\partial x}\,dz + g\rho(\eta)\frac{\partial \eta}{\partial x} \tag{6.108}$$

$$\frac{\partial P}{\partial y} = g\int_z^\eta \frac{\partial \rho}{\partial y}\,dz + g\rho(\eta)\frac{\partial \eta}{\partial y}. \tag{6.109}$$

Now define $\sigma = (\rho - \rho_0)g$ which is a measure of the density anomaly defined by Simons (1973). Then eqn (6.107) reduces to

$$P(z) = P_0 + \rho_0 g(\eta - z) + \int_z^\eta \sigma\,dz. \tag{6.110}$$

Next, the equation of motion will be considered in the form of momentum balance.

The first momentum is given by

$$\frac{\partial u}{\partial t} + u\frac{\partial u}{\partial x} + v\frac{\partial u}{\partial y} + w\frac{\partial u}{\partial z} = \Omega v - \frac{1}{\rho}\frac{\partial P}{\partial x} + \frac{1}{\rho}\frac{\partial \tau_x}{\partial z}$$

which can be written as

$$\frac{\partial u}{\partial t} + u\frac{\partial u}{\partial x} + v\frac{\partial u}{\partial y} + w\frac{\partial u}{\partial z} + u\left(\frac{\partial u}{\partial x} + \frac{\partial v}{\partial y} + \frac{\partial w}{\partial z}\right)$$

$$= \Omega v - \frac{1}{\rho}\frac{\partial P}{\partial x} + \frac{1}{\rho}\frac{\partial \tau_x}{\partial z}$$

or

$$\frac{\partial u}{\partial t} + \frac{\partial}{\partial x}(uu) + \frac{\partial}{\partial y}(uv) + \frac{\partial}{\partial z}(uw) = \Omega v - \frac{1}{\rho}\frac{\partial P}{\partial x} + \frac{1}{\rho}\frac{\partial \tau_x}{\partial z}.$$

(6.111)

Now integrating eqn (6.111) with respect to z from $z = -h(x,y)$ to $z = \eta(x,y,t)$

$$\int_{-h}^{\eta}\left(\frac{\partial u}{\partial t} + \frac{\partial}{\partial x}(uu) + \frac{\partial}{\partial y}(uv) + \frac{\partial}{\partial z}(uw)\right)dz$$

$$= -\int_{-h}^{\eta}\frac{1}{\rho}\frac{\partial P}{\partial x}dz + \Omega\int_{-h}^{\eta}v\,dz + \int_{-h}^{\eta}\frac{1}{\rho}\frac{\partial \tau_x}{\partial z}dz \quad (6.112)$$

where

$$\int_{-h}^{\eta}\frac{\partial u}{\partial t}dz = \frac{\partial}{\partial t}\int_{-h}^{\eta}u\,dz - u(\eta)\frac{\partial \eta}{\partial t} - u(-h)\frac{\partial h}{\partial t}$$

$$= \frac{\partial}{\partial t}[(\eta + h)U] - u(\eta)\frac{\partial \eta}{\partial t}.$$

Therefore,

$$\frac{\overline{\partial u}}{\partial t} = \frac{\partial U}{\partial t} + \frac{U}{\eta + h}\frac{\partial \eta}{\partial t} - \frac{u(\eta)}{\eta + h}\frac{\partial \eta}{\partial t}$$

(6.113)

where

$$\frac{1}{\eta + h}\int_{-h}^{\eta}\frac{\partial u}{\partial t}dz = \frac{\overline{\partial u}}{\partial t}.$$

Also,

$$\int_{-h}^{\eta}\frac{\partial}{\partial x}(uu)dz = \frac{\partial}{\partial x}\int_{-h}^{\eta}(uu)dz - (uu)_{\eta}\frac{\partial \eta}{\partial x} - (uu)_{-h}\frac{\partial h}{\partial x}. \quad (6.114)$$

Define

$$\frac{1}{\eta + h}\int_{-h}^{\eta}(uu)dz = (\overline{uu})$$

and

$$\frac{1}{\eta+h} \int_{-h}^{\eta} \frac{\partial}{\partial x}(uu)\mathrm{d}z = \overline{\frac{\partial}{\partial x}(uu)}.$$

Thus,

$$\overline{\frac{\partial}{\partial x}(uu)} = \frac{\partial}{\partial x}(\overline{uu}) + \frac{\overline{uu}}{\eta+h} \frac{\partial}{\partial x}(\eta+h)$$

$$- \frac{1}{\eta+h}\left((uu)_\eta \frac{\partial \eta}{\partial x} + (uu)_{-h}\frac{\partial h}{\partial x} \right). \tag{6.115}$$

Similarly we have

$$\overline{\frac{\partial}{\partial y}(uv)} = \frac{\partial y}{\partial y}(\overline{uv}) + \frac{\overline{uv}}{\eta+h} \frac{\partial}{\partial y}(\eta+h)$$

$$- \frac{1}{\eta+h}\left((uv)_\eta \frac{\partial \eta}{\partial y} + (uv)_{-h}\frac{\partial h}{\partial y} \right) \tag{6.116}$$

and

$$\overline{\frac{\partial}{\partial z}(uv)} = \frac{1}{\eta+h}[(uw)_\eta - (uw)_{-h}]. \tag{6.117}$$

The magnitude $|V|$ of the velocity components u and v can be empirically represented by the parabolic formula $|V| = V_1(z+h)^{1/q}$ where V_1 is a characteristic velocity. Also, $u = |V|\cos \alpha$, $v = |V|\sin \alpha$. Hence,

$$U^2 = (\bar{u})^2 = \frac{V_1^2 \cos^2 \alpha}{(\eta+h)^2}\left(\int_{-h}^{\eta} (z+h)^{1/q}\, \mathrm{d}z \right)^2 \tag{6.118}$$

and

$$\overline{uu} = \frac{v_1^2 \cos^2 \alpha}{\eta+h} \int_{-h}^{\eta} (z+h)^{2/q}\, \mathrm{d}z.$$

Thus after some calculations,

$$\overline{uu} = \left(1 + \frac{1}{q^2+2q}\right)(\bar{u})^2 \simeq \left(1 + \frac{1}{q^2+2q}\right)U^2, \quad \bar{u} = U. \tag{6.119}$$

In practice, $5 < q < 7$, in which case $\overline{uu} \simeq U^2$. Similarly it can be shown that

$$\overline{uu} \simeq UV, \qquad \overline{v} = V$$

$$\frac{\partial}{\partial x}\left(\overline{uu}\right) \simeq \frac{\partial}{\partial x}U^2 \tag{6.120}$$

$$\frac{\partial}{\partial y}\left(\overline{uv}\right) \simeq \frac{\partial}{\partial y}(UV).$$

The continuity equation (6.103) can be written as

$$\frac{\partial \eta}{\partial t} + \frac{\partial}{\partial x}[(h+\eta)U] + \frac{\partial}{\partial y}[(\eta+h)V] = 0 \tag{6.121}$$

or, we have

$$\frac{\partial \eta}{\partial t} + (\eta+h)\left(\frac{\partial U}{\partial x} + \frac{\partial V}{\partial y}\right) + U\frac{\partial}{\partial x}(\eta+h) + V\frac{\partial}{\partial y}(\eta+h) = 0 \tag{6.122}$$

or

$$\frac{U}{\eta+h}\frac{\partial \eta}{\partial t} + \frac{U^2}{\eta+h}\frac{\partial}{\partial x}(\eta+h) + \frac{UV}{\eta+h}\frac{\partial}{\partial y}(\eta+h)$$

$$= -U\left(\frac{\partial U}{\partial x} + \frac{\partial V}{\partial y}\right). \tag{6.123}$$

Thus from eqn (6.112), the mean value of the sum of terms on the left-hand side can be simply obtained by using eqns (6.113), (6.115), (6.116), (6.117) and (6.123) as follows:

$$\frac{\partial \overline{u}}{\partial t} + \frac{\partial}{\partial x}\left(\overline{uu}\right) + \frac{\partial}{\partial y}\left(\overline{uv}\right) - \overline{u}\left(\frac{\partial \overline{u}}{\partial x} + \frac{\partial \overline{v}}{\partial y}\right) \tag{6.124}$$

and it is found that expression (6.124) can be replaced by

$$\frac{\partial U}{\partial t} + U\frac{\partial U}{\partial x} + V\frac{\partial U}{\partial y}. \tag{6.125}$$

The various approximations applied in the derivative of expression (6.125) follow immediately from the preceding discussion.

The mean value of $\partial P/\partial x$ is given by

$$\frac{1}{\eta+h}\int_{-h}^{\eta}\left(\frac{\partial P}{\partial x}\right)dz = \frac{g}{\eta+h}\int_{-h}^{\eta}\int_{z}^{\eta}\left(\frac{\partial \rho}{\partial x}\right)dz^2$$

$$+\frac{g}{\eta+h}\int_{-h}^{\eta}\rho(\eta)\left(\frac{\partial \eta}{\partial x}\right)dz$$

i.e.

$$\left(\overline{\frac{\partial P}{\partial x}}\right) = \frac{g}{\eta+h}\int_{-h}^{\eta}\int_{z}^{\eta}\left(\frac{\partial \rho}{\partial x}\right)dz^2 + g\,\rho(\eta)\frac{\partial \eta}{\partial x}. \tag{6.126}$$

Similarly, the mean value of $(\partial P/\partial y)$ is

$$\left(\overline{\frac{\partial P}{\partial y}}\right) = \frac{g}{\eta+h}\int_{-h}^{\eta}\int_{z}^{\eta}\left(\frac{\partial \rho}{\partial y}\right)dz^2 + g\,\rho(\eta)\frac{\partial \eta}{\partial y}. \tag{6.127}$$

In case density variation does not occur, it is found that

$$\left(\overline{\frac{\partial P}{\partial x}}\right) = g\,\rho\frac{\partial \eta}{\partial x} \tag{6.128}$$

$$\left(\overline{\frac{\partial P}{\partial g}}\right) = g\,\rho\frac{\partial \eta}{\partial y}. \tag{6.129}$$

It must be noted that x and y are taken in the horizontal plane and η is defined with respect to the mean water level. If the mean water plane is not horizontal, the above pressure equations must be modified by the following equations:

$$\left(\overline{\frac{\partial P}{\partial x}}\right) = g\frac{\partial(a_0 + \eta)}{\partial x} \tag{6.130}$$

$$\left(\overline{\frac{\partial P}{\partial y}}\right) = g\frac{\partial(a_0 + \eta)}{\partial y} \tag{6.131}$$

where $a_0(x, y)$ is the mean water level with respect to the datum level.

The mean values of the shear stresses are

$$\frac{1}{\eta+h}\int_{-h}^{\eta}\frac{\partial \tau_x}{\partial z}\,dz = \frac{1}{\eta+h}[\tau_x(\eta)-\tau_x(-h)] \tag{6.132}$$

$$\frac{1}{\eta+h}\int_{-h}^{\eta}\frac{\partial \tau_y}{\partial z}\,dz = \frac{1}{\eta+h}\left[\tau_y(\eta)-\tau_y(-h)\right]. \tag{6.133}$$

The shear stress at the water surface is mainly caused by the wind effects and at the bottom by friction.

Let W be the velocity of the wind, and w_x and w_y its components (m/s). It can be assumed that

$$\tau_x(\eta) = \frac{C_1}{\eta+h}|W|w_x \tag{6.134}$$

$$\tau_y(\eta) = \frac{C_1}{\eta+h}|W|w_y \tag{6.135}$$

$$|W| = \left(w_x^2 + w_y^2\right)^{1/2}$$

where $C_1 =$ an empirical constant $= 0.35 \times 10^{-6}$ kg/m^2. The empirical formula for the *frictional forces* at the bottom of an estuary are

$$\tau_x(-h) = \frac{g\,\rho U\sqrt{U^2+V^2}}{C_2^2} \tag{6.136}$$

$$\tau_y(-h) = \frac{g\,\rho V\sqrt{U^2+V^2}}{C_2^2} \tag{6.137}$$

in which C_2 is Chezy's coefficient (m$^{1/2}$/s), $C_1 \neq C_2$. When Manning's coefficient is used, the relation between C_2 and n is

$$C_2 = \frac{1.49}{n}(\eta+h)^{1/6}. \tag{6.138}$$

Thus the depth-averaged tidal equations for constant density are the following:

$$\frac{\partial \eta}{\partial t} + \frac{\partial}{\partial x}[(\eta + h)U] + \frac{\partial}{\partial y}[(\eta + h)V] = 0 \tag{6.139}$$

$$\rho \left(\frac{\partial U}{\partial t} + U\frac{\partial U}{\partial x} + V\frac{\partial U}{\partial y} - \Omega V \right) = -\rho g \left(\frac{\partial \eta}{\partial x} + \frac{\partial a_0}{\partial x} \right)$$

$$-\frac{\rho g U\sqrt{U^2 + V^2}}{C_2^2(\eta + h)} + \frac{C_1|\mathbf{W}|w_x}{(\eta + h)^2} \tag{6.140}$$

$$\rho \left(\frac{\partial V}{\partial t} + U\frac{\partial V}{\partial x} + V\frac{\partial V}{\partial y} + \Omega U \right) = -\rho g \left(\frac{\partial \eta}{\partial y} + \frac{\partial a_0}{\partial y} \right)$$

$$-\frac{g \rho V\sqrt{U^2 + V^2}}{C_2^2(\eta + h)} + \frac{C_1|\mathbf{W}|w_y}{(\eta + h)^2}. \tag{6.141}$$

The governing equations, namely the continuity equation (6.139) and the equations of motion (6.140) and (6.141), are nonlinear. To obtain analytical solutions of this set of equations, we assume that U, V and η are all small. Therefore their products are also very small. Thus in the absence of wind shear stresses, Coriolis forces and bottom friction, the linearized two-dimensional tidal equations are as follows:

The continuity equation:

$$\frac{\partial \eta}{\partial t} + \frac{\partial}{\partial x}(hU) + \frac{\partial}{\partial y}(hV) = 0. \tag{6.142}$$

The equations of motion:

$$\frac{\partial U}{\partial t} + g\frac{\partial \eta}{\partial x} = 0 \tag{6.143}$$

$$\frac{\partial V}{\partial t} + g\frac{\partial \eta}{\partial y} = 0. \tag{6.144}$$

If the bottom is horizontal, in which case the depth h is a constant, the equations can be cross-differentiated to eliminate U and V, yielding

$$\frac{\partial^2 \eta}{\partial t^2} = C^2 \left(\frac{\partial^2 \eta}{\partial x^2} + \frac{\partial^2 \eta}{\partial y^2} \right) \tag{6.145}$$

where $C = \sqrt{gh}$. This of course is known as the *wave equation*, the methods associated with its solution being discussed in Chapter 3. For the one-dimensional wave motions eqn (6.145) reduces to $\eta_{tt} = C^2 \eta_{xx}$, the solution of which can be obtained exactly as in eqn (6.25) for a uniform estuary. The dispersion relation for the long wave profile is $C^2 = gh$ as derived before in the case of shallow water waves.

6.7 Tidal waves of finite amplitude

In this section we discuss an important aspect of tidal oscillations in estuaries. When the tidal elevation η is not small compared with the mean water depth h, the characteristics of the waves will be different from those of small-amplitude waves. This problem was first investigated by Airy (1845), by the method of successive approximation. This investigation assumes that the vertical acceleration may be neglected. It follows that the horizontal velocity component V may be taken to be uniform over any section of the canal. As before, neglecting the wind stresses, bottom friction and Coriolis forces, the equations of continuity and motion, respectively, can be written as

$$\eta_t + ((h + \eta)U)_x = 0 \tag{6.146}$$

$$U_t + UU_x + g\eta_x = 0. \tag{6.147}$$

Equation (6.146) may be written as

$$\eta_t + U\eta_x = -(h + \eta)U_x. \tag{6.148}$$

The set of equations is nonlinear. To obtain analytic solutions we follow the method of successive approximation as illustrated by Airy (1845). Let us consider a canal communicating at the mouth of the estuary, $x = 0$ with an open ocean, where the water elevation is given by

$$\eta = A \cos \sigma t. \tag{6.149}$$

For a first approximation we have

$$U_t = -g\eta_x \tag{6.150}$$

$$\eta_t = -hU_x. \tag{6.151}$$

The solution which is consistent with (6.149) is

$$\eta = A\cos(\sigma t - kx) \tag{6.152}$$

$$U = \frac{AC}{h}\cos(\sigma t - kx). \tag{6.153}$$

For a second approximation we substitute these values of η and U in the nonlinear terms of eqns (6.147) and (6.148), and obtain

$$U_t = -g\eta_x - \frac{\sigma g A^2}{2hC}\sin 2(\sigma t - kx) \tag{6.154}$$

$$\eta_t = -hU_x - \frac{\sigma A^2}{h}\sin 2(\sigma t - kx). \tag{6.155}$$

Eliminating U from eqns (6.154) and (6.155) we obtain

$$\eta_{tt} = gh\eta_{xx} - \gamma\cos 2(\sigma t - kx) \tag{6.156}$$

where

$$\gamma = \frac{\sigma g k A^2}{C} + \frac{2\sigma^2 A^2}{h}.$$

The solutions for η and U can be obtained as (see Lamb (1945))

$$\eta = A\cos(\sigma t - kx) - \frac{3}{4}\frac{\sigma A^2 x}{hC}\sin 2(\sigma t - kx) \tag{6.157}$$

$$U = \frac{AC}{h}\cos(\sigma t - kx) - \frac{1}{8}\frac{gA^2}{hC}\cos 2(\sigma t - kx)$$

$$-\frac{3}{4}\frac{\sigma A^2 x}{h^2}\sin 2(\sigma t - kx). \tag{6.158}$$

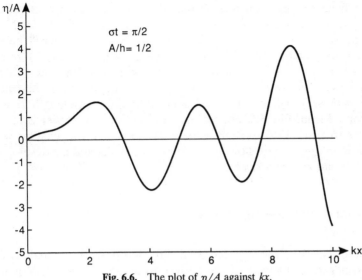

Fig. 6.6. The plot of η/A against kx.

Figure 6.6 shows the plot of η/A against kx for $A/h = 0.5$ and $\sigma t = \pi/2$, while Fig. 6.7 shows the plot against σt for $kx = 0.5$. It is to be noted from these results that at a particular position of the canal, the rise and fall of the water do not take place symmetrically, and in fact the fall occupies a longer time than the rise (Fig. 6.7). Another interesting result can be

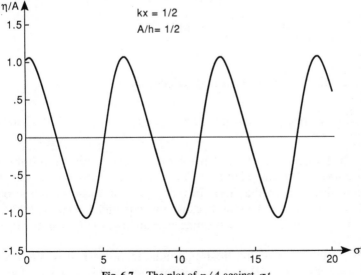

Fig. 6.7. The plot of η/A against σt.

derived from the solution (6.157) that due to the presence of factor x outside the trigonometric term in eqn (6.157), there is a limit beyond which the approximation breaks down. This approximate solution will be valid provided the factor $\sigma Ax/hC$ or equivalently $2\pi(A/h)(x/L)$ remains small, where L is a wavelength. However, it should be noted that no matter how small the factor A/h (amplitude to depth ratio), the fraction ceases to be small when x is a sufficient multiple of wavelength L as can be observed from Fig. 6.6. For further information, the reader is referred to Airy's treatise (1845) on this subject. We conclude this chapter with a brief description in the following section of the mathematical development of tidal equations for a river.

6.8 Tidal equations for a river

Assuming that the flow is in the axial direction only, the equations for a tidal propagation in a river can be derived from the general tidal equations (6.140) and (6.141) by setting $V = 0$.

The equations then reduce to

$$\rho\left(\frac{\partial U}{\partial t} + U\frac{\partial U}{\partial x}\right) = -\rho g\left(\frac{\partial \eta}{\partial x} + \frac{\partial a_0}{\partial x}\right)$$

$$-\rho g\left(\frac{U|U|}{C_2^2(\eta+h)}\right) + \frac{C_1|\mathbf{W}|w_x}{(\eta+h)^2}. \tag{6.159}$$

$$\Omega U = -\rho g\left(\frac{\partial \eta}{\partial y} + \frac{\partial a_0}{\partial y}\right) + \frac{C_1|\mathbf{W}|w_y}{(\eta+h)^2} \tag{6.160}$$

and the continuity equation (6.139) reduces to

$$\frac{\partial \eta}{\partial t} + \frac{\partial}{\partial x}[(\eta+h)U] = 0. \tag{6.161}$$

It is worth mentioning here that for wide rivers, it may be possible that the Coriolis force or the wind force has some effect on water levels in the transverse direction of the river. In that case, eqn (6.141) applies; otherwise, this equation is redundant because two equations, namely (6.139) and (6.140), will be sufficient to determine two unkowns U and η as functions of x and t.

In the formulation of the tidal equations for a river, it is often convenient to use discharge Q as a physical variable instead of axial velocity U.

The variable Q denotes the total amount of water flowing through a cross-section for unit time. Then

$$Q = SU = (\eta + h)b_s U \qquad (6.162)$$

where $S(x, h, t) = $ cross-sectional area of the stream bed of water, $b_s(x, h, t) = $ width of the stream bed, and $h(x, t) = $ depth of the river.

Introducing Q into eqns (6.159) and (6.161), the following equations are derived:

$$\frac{\partial Q}{\partial t} - \frac{Q}{\eta + h}\frac{\partial \eta}{\partial t} - \frac{Q}{b_s}\frac{\partial b_s}{\partial h}\frac{\partial h}{\partial t}$$

$$+ \alpha\frac{Q}{S}\left(\frac{\partial Q}{\partial x} - \frac{Q}{\eta + h}\frac{\partial(\eta + h)}{\partial x} - \frac{Q}{b_s}\frac{\partial b_s}{\partial h}\frac{\partial h}{\partial x}\right)$$

$$= -gS\left(\frac{\partial \eta}{\partial x} + \frac{\partial a_0}{\partial x}\right) - \frac{g|Q|Q}{C_2^2 S(\eta + h)} + \frac{C_1 b_s}{\rho}\frac{|W|w_x}{(\eta + h)} \qquad (6.163)$$

$$\frac{\partial \eta}{\partial t} + \frac{\partial}{\partial x}\left(\frac{Q}{b_s}\right) = 0.$$

The second equation of (6.163) can be subsequently written as

$$\frac{\partial Q}{\partial x} + b_s\frac{\partial \eta}{\partial t} - \frac{Q}{b_s}\frac{\partial b_s}{\partial h}\frac{\partial h}{\partial x} = 0. \qquad (6.164)$$

A coefficient α in eqn (6.163) due to Dronkers (1964) has been introduced to account for the effect of the distribution of the velocities in the vertical and in the transverse and lengthwise directions of the river.

Equation (6.164) can be written as

$$\frac{\partial Q}{\partial x} + b\frac{\partial \eta}{\partial t} + q = 0 \qquad (6.165)$$

where $b(h, t) = $ the storage width, and $q(h, t) = $ the supplementary discharges per unit length, e.g. by overflow, etc. This mathematical model has been applied in the study of tidal propagation in the St Lawrence river by Kamphuis (1968).

6.9 Exercises

1. For long waves in a canal, prove that

$$\frac{\partial P}{\partial x} = g\rho\frac{\partial \eta}{\partial x}.$$

Use the fact that the right-hand side is independent of z to infer that the particles in a vertical plane perpendicular to the direction of propagations remain in that plane.

2. If (u,v) are small components of velocity in a long wave, use the equation of motion and the results of exercise 1 to show that

$$\frac{\partial u}{\partial t} = -\frac{1}{\rho}\frac{\partial P}{\partial x} = -g\frac{\partial \eta}{\partial x}$$

and that

$$\frac{\partial^2 \xi}{\partial t^2} = -g\frac{\partial \eta}{\partial x},$$

where $\xi = \int_0^t u\,dt$. Obtain the equation of continuity in the form

$$\eta = -h\frac{\partial \xi}{\partial x},$$

where h is the mean depth.

3. Using the results of exercise 2, show that

$$\frac{\partial^2 \eta}{\partial t^2} = C^2\frac{\partial^2 \eta}{\partial x^2},$$

$C^2 = gh$, and hence prove that $\eta = f_1(x + Ct) + f_2(x - Ct)$.

4. Establish the equation

$$b\frac{\partial^2 \eta}{\partial t^2} = g\frac{\partial}{\partial x}\left(S\frac{\partial \eta}{\partial x}\right)$$

for the elevation η of the surface in the tidal wave motion in a canal of variable section, where b denotes the breadth at the surface, and S is the area of the section.

Prove from this equation that the amplitude of a progressive wave is nearly proportional to $b^{-\frac{1}{2}}h^{-\frac{1}{4}}$, where h is the mean depth across a section. If b and h, and their rates of change along the canal, vary only by small fractions of themselves in distances of the order of a wavelength, verify that this corresponds to assuming continuous propagation of energy without reflection.

5. Obtain the theory of long waves in a canal of uniform depth h due to a disturbing potential, $W = H \exp[i(\sigma t - Kx)]$. If the bottom yields to the disturbing force so that its elevation is $\eta_0 = \alpha \exp[i(\sigma t - Kx)]$, prove that the relative height of the waves is the same as if the potential had been diminished in the ratio $1 - \mu$, where μ denotes the ratio of α to the 'equilibrium height', $-H/g$, due to the disturbance. Prove that this conclusion is not confined to simple harmonic waves.

6. Obtain the theory of 'long waves' in a canal uniform width and of depth h, proving that the velocity of free waves is $(gh)^{\frac{1}{2}}$. An earthquake wave, $\eta_0 = A \cos k(Ct - x)$, travels along the bottom. Prove that the consequent wave on the free surface is

$$\eta = \frac{AC^2}{C^2 - gh} \cos k(Ct - x).$$

7. Tidal waves are due to a body moving in the plane of the equator at a constant angular rate relative to the point P at which a great circle canal crosses the equator at inclination α. Show that this results in a permanent change of level proportional to $\sin^2 \alpha \cos 2x/a$, and the addition of two semi-diurnal tides of standing waves, having amplitudes proportional to $(1 + \cos^2 \alpha)\cos 2x/a$ and $2\cos \alpha \sin 2x/a$, respectively, x being the distance along the canal measured from P and a the radius of the earth.

8. Two-dimensional long waves are travelling parallel to the x-axis in water of variable depth h. Prove that, if η is the height of the free surface above the equilibrium level, then η satisfies the equation

$$\frac{\partial^2 \eta}{\partial t^2} = \frac{\partial}{\partial x}\left(gh \frac{\partial \eta}{\partial x} \right).$$

If $h = x^2/2b$, prove that

$$\eta = ax^{-\frac{1}{2}} \cos\left[\sigma\left(\sqrt{\frac{2b}{g} - \frac{1}{4\sigma^2}} \log x + t\right) + \alpha\right]$$

is a typical solution of period $2\pi/\sigma$. Use this result to illustrate the variations of amplitude and wavelength to be expected in the case of waves moving in from deep water up a gradually shelving beach.

9. The bottom of a canal is slightly corrugated, so that the depth is given by $h + c \sin Kx$, c and Kh being small. Prove that, if a stream of velocity U flows along the canal, there will be standing waves in the latter, of height η, which are given by

$$\eta = c \sin(Kx)\bigg/\left(\frac{gh}{U^2} - 1\right).$$

Do the corrugations affect the velocity of progressive waves along the canal?

10. Derive the fundamental solutions of diverging channels for tidal oscillations without friction.

11. Verify the nonlinear solutions for the water elevation η and the current U given in eqns (6.157) and (6.158).

References

Airy, G. B. (1845). Tides and waves. *Encycl. Metrop.*, Arts 192, 198 and 308.

Dean, R. G. and Dalrymple, R. A. (1984). *Water wave mechanics for engineers and scientists*. Prentice-Hall, Inc., Englewood Cliffs, New Jersey.

Dorrestein, R. (1961). Amplification of long waves in bays. *Eng. Prog. Univ. Florida, Gainsville*, 15(12).

Dronkers, J. J. (1964). *Tidal Computations in Rivers and Coastal Waters*, North-Holland, Amsterdam.

Hunt, J. N. (1964). Tidal oscillations in estuaries. *Geophys. J. R. Astron. Soc.*, 8, 440–455.

Kamphuis, J. W. (1968). Mathematical model study of the propagation of tides in the St. Lawrence river and estuary, NRC, DME Mech. Eng. Report, MH-105, August, National Research Council of Canada, Ottawa.

Lamb, H. (1945). *Hydrodynamics*, 6th edn, Cambridge University Press, Cambridge.

McLachlan, N. W. (1955). *Bessel Functions for Engineers*, Clarendon Press, Oxford.

Prandle, D. and Rahman, M. (1980). Tidal response in estuaries. *J. Phys. Oceanogr.*, 10, 1552–1573.

Rahman, M. (1988). *The Hydrodynamics of Waves and Tides, with Applications*, Computational Mechanics Publications, Southampton.

Simons, T. J. (1973). Development of three dimensional numerical models of the Great Lakes. Scientific Series No. 12, Inland Waters Directorate, Canada Centre for Inland Waters, Burlington, Ontario.

Taylor, G. I. (1921). Tides in Bristol Channel. *Proc. Cambridge Philos. Soc.*, 20, 320–325.

7
Wave statistics and spectral method

7.1 Introduction

In the previous chapters we have discussed the waves that are regular and remain invariant in their properties from one cycle to the next. However, ocean waves are in general not regular but random in the sense that the ocean surface is composed of waves moving in different directions and with different amplitudes, frequencies and phases. In this situation ocean waves should be described by their statistical properties. In the following section, therefore, we shall define the statistical parameters from a mathematical viewpoint. These statistical definitions are then applied to the random wave field.

7.2 Some statistical definitions

Consider a random signal $x(t)$ (Fig. 7.1) which is a function of time t alone. Then the mean of the process is defined as the expected value of $x(t)$ and in symbolic form can be written as

$$\mu_x = E[x(t)] \tag{7.1}$$

where E denotes the expected value of $x(t)$.

The variance of a random signal $x(t)$ about the mean is defined as

$$\sigma_x^2 = E\left[(x(t) - \mu_x)^2\right] \tag{7.2}$$

and the root-mean-square value of the process, which is called the standard deviation, is simply σ_x.

The covariance of two random signals $x(t)$ and $y(t)$ with means μ_x and μ_y, respectively, is defined as

$$\sigma_{xy} = E[xy] - \mu_x \mu_y. \tag{7.3}$$

Fig. 7.1.   A random signal.

The autocorrelation function is obtained as

$$R(t_1, t_2) = E[x(t_1)x(t_2)].\tag{7.4}$$

The process $x(t)$ is stationary if μ_x and σ_x are constant for all values of t, and R is a function only of $\tau = t_2 - t_1$.

Under some assumptions, these parameters of a random signal are known to follow probability density functions. A probability density function is defined as the function of time that a particular event is expected to occur. Associated with the probability density function is the cumulative probability density function (or probability distribution). A cumulative probability density function $P(x)$, of a random variable x, with probability density function $p(x)$ is defined as

$$P(x) = Prob[x(t) \leq x]\tag{7.5}$$

and for a continuous random variable $x(t)$, eqn (7.5) may be written as

$$P(x) = \int_{-\infty}^{x} p(x)\mathrm{d}x.\tag{7.6}$$

Thus the probability density function $p(x)$ is obtained as the rate of change of $P(x)$ and is given by

$$p(x) = \frac{\mathrm{d}P(x)}{\mathrm{d}x}.\tag{7.7}$$

The area between two values a and b under the probability density function $p(x)$ defines the probability that the results of an event will lie between the values a and b. Thus mathematically,

$$P(a \leq x(t) \leq b) = \int_{a}^{b} p(x)\mathrm{d}x.\tag{7.8}$$

The total probability of the event is unity and in integral form it can be written as

$$P(-\infty \leq x(t) \leq \infty) = \int_{-\infty}^{\infty} p(x)\mathrm{d}x = 1.\tag{7.9}$$

In terms of the probability density function $p(x)$, the mean value of the random variable $x(t)$ defined by eqn (7.1) may be given by

$$\mu_x = E[x(t)] = \int_{-\infty}^{\infty} xp(x)\mathrm{d}x. \tag{7.10}$$

Similarly, the variance of the random variable, $x(t)$, about the mean, μ_x, defined by eqn (7.2) is given by $\sigma_x^2 = E[(x(t) - \mu_x)^2] = E[x^2(t)] - \mu_x^2$ which is the same as

$$\sigma_x^2 = \int_{-\infty}^{x} (x(t) - \mu_x)^2 p(x)\mathrm{d}x. \tag{7.11}$$

Physically, the mean, μ_x, may be considered as the distance to the centroid of the probability density function, whereas the variance may be looked upon as the moment of inertia of the probability density function about the mean.

Two probability distributions are discussed relevant to use in the ocean engineering problem; they are the Gaussian (or normal distribution) and the Rayleigh distribution. Without going into detail, we define them in the following. The cumulative probability of a Gaussian distribution given by

$$P(x) = \frac{1}{\sigma_x\sqrt{2\pi}} \int_{-\infty}^{\infty} \exp\left(-\frac{(x-\mu_x)^2}{2\sigma_x^2}\right)\mathrm{d}x \tag{7.12}$$

and its probability density function is

$$p(x) = \frac{\mathrm{d}P(x)}{\mathrm{d}x} = \frac{1}{\sigma_x\sqrt{2\pi}} \exp\left(-\frac{(x-\mu_x)^2}{2\sigma_x^2}\right). \tag{7.13}$$

Using the standard notation for this distribution, we can write eqn (7.13) as

$$p(x) = N(x; \mu_x, \sigma_x). \tag{7.14}$$

The difficulty of finding the integrals of the normal density function necessitates the tabulation of normal curve areas for ready reference. However, it would be a tremendous task to attempt to set up separate tables for every conceivable value of μ_x and σ_x. This inconvenience can be eased by a suitable transformation in the following form:

$$z(t) = \frac{x(t) - \mu_x}{\sigma_x} \tag{7.15}$$

such that $z(t)$ is a new random variable with zero mean and unity variance. Thus $P(x_1 < x(t) < x_2) = (1/\sqrt{2\pi}\,\sigma_x)\int_{x_1}^{x_2} \exp\{-[(x(t)-\mu_x)^2/2\sigma_x^2]\}\,dx$ can be transformed by the transformation (7.15) as follows:

$$P(x_1 < x(t) < x_2) = \frac{1}{\sqrt{2\pi}} \int_{z_1}^{z_2} \exp\left(-\frac{z^2(t)}{2}\right) dz$$

$$= \int_{z_1}^{z_2} \phi(z(t); 0, 1) dz = P(z_1 < z(t) < z_2) \qquad (7.16)$$

where $z_1 = (x_1 - \mu_x)/\sigma_x$ and $z_2 = (x_2 - \mu_x)/\sigma_x$.

The distribution of a random variable with mean zero and variance unity is called a standard normal distribution. Here $N(z(t); 0, 1)$ is called a standard normal distribution. Thus in variable x, the standard normal density function is

$$p(x) = \frac{1}{\sqrt{2\pi}} \exp\left(-\frac{x^2}{2}\right) \qquad (7.17)$$

while the standard cumulative distribution is given by the well-known error function

$$P(x) = \frac{1}{\sqrt{2\pi}} \int_{-\infty}^{x} \exp\left(-\frac{x^2}{2}\right) dx. \qquad (7.18)$$

The probability density function of the Rayleigh distribution and the corresponding cumulative probability obtained by integration may be written as

$$p(x) = \begin{cases} \dfrac{\pi x}{2\mu_x^2} \exp\left(-\dfrac{\pi}{4}\left(\dfrac{x}{\mu_x}\right)^2\right) & \text{for } x \geq 0 \\ 0 & \text{otherwise.} \end{cases} \qquad (7.19)$$

$$P(x) = \begin{cases} 1 - \exp\left(-\dfrac{\pi}{4}\left(\dfrac{x}{\mu_x}\right)^2\right) & \text{for } x \geq 0 \\ 0 & \text{otherwise.} \end{cases} \qquad (7.20)$$

Figures 7.2 and 7.3 display the probability plots for the Gaussian and Rayleigh distributions. It should be noted that the Gaussian probability density function is symmetric about the mean value and can take on both

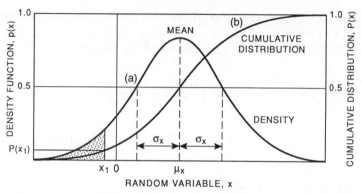

Fig. 7.2. The Gaussian probability distribution: (a) probability density function, (b) cumulative probability distribution.

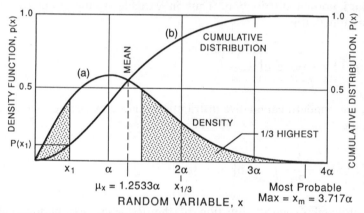

Fig. 7.3. The Rayleigh probability distribution: (a) probability density function, (b) cumulative probability distribution.

positive and negative values. On the other hand, the Rayleigh density function has only positive values, and has a value of zero at $x = 0$, rises to a maximum and approaches zero exponentially at large values of x. The Rayleigh distribution is also known as the chi-square distribution. This distribution is not symmetric.

We have seen in eqn (7.11) that the variance of the random variable $x(t)$ about its mean is given by $\sigma_x^2 = E[(x(t) - \mu_x)^2] = \int_{-\infty}^{\infty}(x(t) - \mu_x)^2 p(x)\mathrm{d}x$. In other words, this is known as the second central moment of $x(t)$.

The nth central moment of $x(t)$ is then defined by

$$\sigma_x^n = E\left[(x(t) - \mu_x)^n\right] = \int_{-\infty}^{\infty} (x(t) - \mu_x)^n p(x)\mathrm{d}x. \tag{7.21}$$

To observe skewness of the Gaussian and Rayleigh distributions we have to compute the third central moment of the random variable $x(t)$ which is $E[(x(t) - \mu_x)^3]$. It can be easily verified that for the Gaussian distribution the skewness is zero, but not zero for the Rayleigh distribution. Now for ocean engineering problems, the quantities μ_x and σ_x^2 defined in eqns (7.10) and (7.11) are obtained from a random signal, $x(t)$, of duration T as

$$\mu_x = E[x(t)] = \langle x(t) \rangle = \lim_{T \to \infty} \left\{ \frac{1}{T} \int_{-T/2}^{T/2} x(t) dt \right\} \tag{7.22}$$

$$\sigma_x^2 = E\left[(x(t) - \mu_x)^2 \right] = \langle (x(t) - \mu_x)^2 \rangle$$

$$= \lim_{T \to \infty} \left\{ \frac{1}{T} \int_{-T/2}^{T/2} (x(t) - \mu_x)^2 \, dt \right\}. \tag{7.23}$$

Physically, if the random variable $x(t)$ represents the recorded wave heights, then μ_x is the mean and σ_x^2 the variance about the mean of the wave heights recorded over $-T/2 \le t \le T/2$.

Another important quantity frequently used in ocean engineering problems is the concept of autocorrelation. The autocorrelation function or autocovariance can be defined as the integral over the record length of the lagged product of the variable. Thus an autocorrelation function, $R_x(\tau)$, is calculated from the given record of length T as

$$R_x(\tau) = \langle x(t)x(t + \tau) \rangle = \lim_{T \to \infty} \left\{ \frac{1}{T} \int_{-T/2}^{T/2} x(t)x(t + \tau) dt \right\} \tag{7.24}$$

where $x(t)$ is the wave height measured at a point of time t and $x(t + \tau)$ is the wave height measured at the same point at time $(t + \tau)$. Here τ is the time lag.

The correlation coefficient $\rho_x(\tau)$ is defined as

$$\rho_x(\tau) = \frac{R_x(\tau)}{\langle x^2(t) \rangle} = \frac{\langle x(t)x(t + \tau) \rangle}{\langle x^2(t) \rangle}$$

$$= \frac{\lim_{T \to \infty} \left\{ \dfrac{1}{T} \int_{-T/2}^{T/2} x(t)x(t + \tau) dt \right\}}{\lim_{T \to \infty} \left\{ \dfrac{1}{T} \int_{-T/2}^{T/2} x^2(t) dt \right\}}. \tag{7.25}$$

The cross-correlation function between two random variables $x(t)$ and $y(t)$ can then be defined as

$$R_{xy}(\tau) = \langle x(t)y(t+\tau)\rangle = \lim_{T\to\infty}\left\{\frac{1}{T}\int_{-T/2}^{T/2}x(t)y(t+\tau)dt\right\}. \quad (7.26)$$

When the random variable $y(t)$ is the same as the variable $x(t)$, then we arrive at the autocorrelation function given in eqn (7.24).

It can be easily shown that $R_x(\tau) = R_x(-\tau)$ and $R_{xy}(\tau) = R_{xy}(-\tau)$, which means that both autocorrelation and cross-correlation functions are symmetric about the origin.

7.3 Distribution of wave heights

In the design of a structure, the engineer must have knowledge of the maximum expected wave heights. In characterizing the ocean state several wave heights can be defined. Among them, $H_{1/3}$, which is the significant wave height, and H_{max}, which is the maximum wave height, are the most popular wave heights. To see the physical meaning of these wave heights, we consider a group of N wave heights measured at a certain point of the ocean. These wave heights are sequentially arranged from the largest to the smallest assigning the number 1 to N. The $H_{1/3}$ wave is defined as the average of the first highest $N/3$ waves. Thus the significant wave height is mathematically defined as

$$H_{1/3} = \frac{3}{N}\sum_{i=1}^{N/3} H_i. \quad (7.27)$$

Similarly H_p is defined as the average of the first highest pN waves where $0 \le p \le 1$. With this definition then, H_1 is the average of the first highest N waves or simply the average wave height. The concept of significant wave height, $H_{1/3}$, was first introduced by Sverdrup and Munk (1947).

If there are N waves of heights $H_1, H_2, \ldots, H_N$, then the root-mean-square (rms) wave height is calculated by the formula

$$H_{rms} = \left(\frac{1}{N}\sum_{i=1}^{N} H_i^2\right)^{1/2}. \quad (7.28)$$

This value of H_{rms} is always larger than H_1 in real ocean. Thus it is clear that for a single sinusoidal wave, i.e. $\eta(t) = (H_0/2)\cos\sigma t$, the waves are all of same height, which leads to the conclusion that $H_p = H_0$ for any p and $H_p = H_{rms}$. The maximum wave height is obtained from the largest of all wave heights in the record. Longuet-Higgins (1952) derived a relationship between the most probable maximum and the root-mean-square wave height for a given number of waves and this relationship is given by

$$H_{max} = \left(\sqrt{\ln N} + \frac{0.2886}{\sqrt{\ln N}} \right) H_{rms}. \tag{7.29}$$

7.3.1 Wave groups

We know that in real ocean surface there are a large number of waves different varieties of which move in different directions with different frequencies, phases and amplitudes. Consider, for simplicity, two wave trains of different frequencies but having the same amplitudes and then superimposed to form a single wave as follows:

$$\eta(t) = \frac{H_0}{2} \cos(\sigma - \varepsilon)t + \frac{H_0}{2} \cos(\sigma + \varepsilon)t \tag{7.30}$$

where $\varepsilon = \Delta\sigma/2$ is a small quantity.

After simplifying, eqn (7.30) can be written as

$$\eta(t) = H_0 \cos \sigma t \cos \varepsilon t = \frac{A(t)}{2} \cos \sigma t \tag{7.31}$$

where $A(t) = 2H_0 \cos \varepsilon t$.

In eqn (7.31), $\cos \sigma t$ is known as the carrier wave with frequency σ and $H_0 \cos(\Delta\sigma/2)t$ is called the slowly modulated wave. A plot of eqn (7.31) leads to Fig. 7.4.

Because ε is a small quantity, the period of the factor $\cos \varepsilon t$ is large. $\eta(t)$ is the product of the function $\cos \sigma t$ and the factor $H_0 \cos \varepsilon t$ which serves to make the amplitude of $\cos \sigma t$, i.e., $A(t)/2$, vary slowly between 0 and $A(t)/2$. A wave form of variable amplitude as shown in Fig. 7.4 is said to be amplitude modulated, and the two enclosed curves to which the

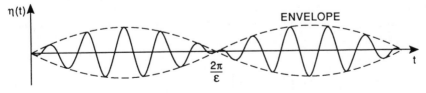

Fig. 7.4. Amplitude modulated wave.

actual wave periodically rises and falls are called the envelope. This is one of the simplest examples of the phenomenon of beats which occur whenever two slightly different frequencies are impressed upon a system. Therefore, to examine the wave height distribution to this wave system, we need only look at the envelope from the antinode to the first node, i.e. from $t = 0$ to π/ε.

Thus, to determine H_p, we arrange the wave height envelope for $t = 0$ to $p\pi/\varepsilon$, such that

$$H_p = \frac{1}{p\pi/\varepsilon} \int_0^{p\pi/\varepsilon} 2H_0 \cos \varepsilon t \, dt = \frac{4H_0}{p\pi} \sin \frac{p\pi}{2}. \tag{7.32}$$

The root-mean-square wave height can be obtained as

$$H_{rms} = \left(\frac{1}{\pi/\varepsilon} \int_0^{\pi/\varepsilon} 4H_0^2 \cos^2 \varepsilon t \, dt \right)^{1/2} = \sqrt{2} \, H_0. \tag{7.33}$$

Thus the relationship between H_p and H_{rms} is given by

$$H_p = \frac{2\sqrt{2} \, H_{rms}}{p\pi} \sin \frac{p\pi}{2}. \tag{7.34}$$

We know from $2H_0 \cos \varepsilon t$ that $H_{max} = 2H_0$. Therefore,

$$H_{max} = \sqrt{2} \, H_{rms}. \tag{7.35}$$

7.3.2 Probability distribution of η

We assume that the probability distribution of the ocean surface is Gaussian. The Gaussian model implicitly assumes symmetry about the still water level. The density function is then given by

$$p(\eta) = \frac{1}{\sqrt{2\pi} \, \sigma_\eta} \exp\left(-\frac{(\eta - \mu_\eta)^2}{2\sigma_\eta^2} \right) \tag{7.36}$$

where μ_η and σ_η are respectively the mean and standard deviation of the water surface elevation.

When the mean of the Gaussian distribution is zero, i.e. $\mu_\eta = 0$, then (7.36) becomes

$$p(\eta) = \frac{1}{\sqrt{2\pi} \, \sigma_\eta} \exp\left(-\frac{\eta^2}{2\sigma_\eta^2} \right). \tag{7.37}$$

It has been found from the experimental data that this distribution for wave elevations is appropriate.

7.3.3 Probability distribution of H

We consider that the ocean surface is composed of a large number of sinusoidal waves with different frequencies, amplitudes and phases. When these frequencies are in a narrow frequency band about σ then we say that it is a narrow-banded ocean. Then superposing all these waves we have

$$\eta(t) = \sum_{k=1}^{K} \frac{H_k}{2} \cos(\sigma_k t - \varepsilon_k) \tag{7.38}$$

where H_k is the amplitude of the kth harmonic, σ_k is the frequency of the kth harmonic, ε_k is the phase of the kth harmonic, and K is the number of harmonic waves.

Equation (7.38) can be subsequently written as

$$\eta(t) = \mathrm{Re}\left\{ \sum_{k=1}^{K} \frac{H_k}{2} \exp[i(\sigma_k t - \varepsilon_k)] \right\} \tag{7.39}$$

where Re stands for real part of $\{\cdot\}$. Then using the concept of carrier wave $\exp(i\sigma t)$, eqn (7.39) can be arranged as

$$\eta(t) = \mathrm{Re}\{\exp(i\sigma t)c(t)\} \tag{7.40}$$

where

$$c(t) = \sum_{k=1}^{K} \frac{H_k}{2} \exp[i(\sigma_k t - \sigma t - \varepsilon_k)] \tag{7.41}$$

is the slowly varying envelope from which we can examine the probability of the wave height distribution.

It has been shown by Longuet-Higgins (1952) that a narrow-band Gaussian ocean wave whose components are in random phase follows the Rayleigh distribution. If a large value of K is used, then the probability of a wave height greater than or equal to an arbitrary wave height H_0 is given by

$$P(H \geq H_0) = \exp\left[-\left(\frac{H_0}{H_{rms}} \right)^2 \right] \tag{7.42}$$

which is called the Rayleigh distribution.

However, if n is the number of waves higher than H_0, then the probability that the wave height is greater than or equal to this arbitrary wave height is

$$P(H \geq H_0) = \frac{n}{N} \qquad (7.43)$$

where N is the total number of waves.

Equating these two relations we have

$$\frac{n}{N} = \exp\left(-\frac{H_0^2}{H_{rms}^2} \right). \qquad (7.44)$$

From eqn (7.44), we can predict the height H which is exceeded by n waves in one group of N. Hence $N/n = \exp(H_0^2/H_{rms}^2)$ and taking the natural logarithm and simplifying we obtain

$$H_0 = H_{rms}\left[\ln\left(\frac{N}{n} \right) \right]^{1/2}. \qquad (7.45)$$

Thus the height that is exceeded by the pN of the waves is

$$H_p = H_{rms}\left[\ln\left(\frac{1}{p} \right) \right]^{1/2}. \qquad (7.46)$$

Let us now examine the Rayleigh distribution given by (7.42). The Rayleigh probability distribution function can be written as

$$P(H < H_0) = 1 - \exp\left(-\frac{H_0^2}{H_{rms}^2} \right). \qquad (7.47)$$

The Rayleigh probability density function is then

$$p(H_0) = \frac{\mathrm{d}}{\mathrm{d}H_0} P(H < H_0)$$

$$= \frac{\mathrm{d}}{\mathrm{d}H_0}\left(1 - \exp\left(-\frac{H_0^2}{H_{rms}^2} \right) \right) = \frac{2H_0 \exp\left(-\dfrac{H_0^2}{H_{rms}^2} \right)}{H_{rms}^2}.$$

Thus in general we have

$$p(H) = \frac{2H\exp\left(-\dfrac{H^2}{H_{rms}^2}\right)}{H_{rms}^2}. \tag{7.48}$$

To find the maximum of this function, we differentiate with respect to H and equate to zero, which yields

$$\frac{dp(H)}{dH} = \frac{2}{H_{rms}^2}\left[1 + H\left(-\frac{2H}{H_{rms}}\right)\frac{1}{H_{rms}}\right]e^{-(H^2/H_{rms}^2)} = 0.$$

Simplifying we get $H/H_{rms} = 1/\sqrt{2}$, i.e. $H = 0.707 H_{rms}$, which signifies the most frequent waves. The plot of eqn (7.48) is shown in Fig. 7.5. Using the Rayleigh density function, we can derive some important statistical properties as follows. The mean wave height can be obtained as

$$\langle H \rangle = \frac{\displaystyle\int_0^\infty H p(H)\,dH}{\displaystyle\int_0^\infty p(H)\,dH} = \frac{\displaystyle\int_0^\infty H\left(\frac{2H}{H_{rms}^2}\right)\exp\left(-\frac{H^2}{H_{rms}^2}\right)dH}{\displaystyle\int_0^\infty \frac{2H}{H_{rms}^2}\exp\left(-\frac{H^2}{H_{rms}^2}\right)dH}.$$

Put $x = H/H_{rms}$ such that $dH = H_{rms}\,dx$ and therefore

$$\langle H \rangle = (H_{rms})\frac{\displaystyle\int_0^\infty x^2 e^{-x^2}\,dx}{\displaystyle\int_0^\infty x e^{-x^2}\,dx}.$$

The denominator. $\int_0^\infty x e^{-x^2}\,dx = \frac{1}{2}\int_0^\infty e^{-x^2}\,dx^2 = \frac{1}{2}[-e^{-x^2}]_0^\infty = \frac{1}{2}.$

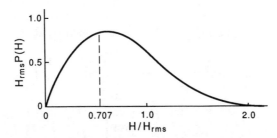

Fig. 7.5. The Rayleigh probability density function.

To evaluate the numerator, $\int_0^\infty x^2 e^{-x^2} dx$, put

$$x^2 = t, \quad x = \sqrt{t}, \quad dx = \frac{1}{2}\frac{dt}{\sqrt{t}}$$

and hence $N = \int_0^\infty (t/2\sqrt{t})e^{-t} dt = \frac{1}{2}\int_0^\infty \sqrt{t}\, e^{-t} dt = \frac{1}{2}\Gamma(3/2) = \frac{1}{2}\cdot\frac{1}{2}\Gamma(\frac{1}{2}) = \frac{1}{4}\sqrt{\pi}$. Thus we obtain

$$\langle H \rangle = (H_{rms})\frac{\dfrac{\sqrt{\pi}}{4}}{\dfrac{1}{2}} = \frac{\sqrt{\pi}}{2}H_{rms} = 0.866 H_{rms}. \tag{7.49}$$

The average height of the highest pN waves is then given by

$$\langle H_p \rangle = \frac{\displaystyle\int_{H_p}^{\infty} Hp(H)\,dH}{\displaystyle\int_{H_p}^{\infty} p(H)\,dH}. \tag{7.50}$$

Using as similar approach as above, we can write (7.50) as

$$\langle H_p \rangle = (H_{rms})\frac{\displaystyle\int_{H_p/H_{rms}}^{\infty} x^2 e^{-x^2}\,dx}{\displaystyle\int_{H_p/H_{rms}}^{\infty} x e^{-x^2}\,dx}.$$

The denominator

$$\int_{\left(\frac{H_p}{H_{rms}}\right)}^{\infty} x e^{-x^2}\,dx = \tfrac{1}{2}\exp\left(-\left(\frac{H_p^2}{H_{rms}}\right)\right).$$

The numerator can be integrated by parts, which yields

$$\int_{H_p/H_{rms}}^{\infty} \frac{x}{2} e^{-x^2}\,dx^2 = \frac{1}{2}[x(-e^{-x^2})]_{H_p/H_{rms}}^{\infty} + \frac{1}{2}\int_{H_p/H_{rms}}^{\infty} e^{-x^2}\,dx$$

$$= \frac{1}{2}\left(\frac{H_p}{H_{rms}}\right)\exp\left(-\frac{H_p^2}{H_{rms}^2}\right) + \frac{1}{2}\int_{H_p/H_{rms}}^{\infty} e^{-x^2}\,dx.$$

Thus we get

$$\frac{\langle H_p \rangle}{H_{rms}} = \left(\frac{H_p}{H_{rms}} \right) + \frac{\int_{H_p/H_{rms}}^{\infty} e^{-x^2} dx}{\exp\left(-\frac{H_p^2}{H_{rms}^2} \right)}.$$

We know the error function is defined as $\mathrm{erf}(x) = (2/\sqrt{\pi})\int_0^x e^{-z^2} dz$ and the error complementary function is $\mathrm{erfc}(x) = 1 - \mathrm{erf}(x) = (2/\sqrt{\pi})\int_x^{\infty} e^{-z^2} dz$. Therefore we obtain

$$\frac{\langle H_p \rangle}{H_{rms}} = \frac{H_p}{H_{rms}} + \exp\left(\frac{H_p^2}{H_{rms}^2} \right)\left(\frac{\sqrt{\pi}}{2} \right)\mathrm{erfc}\left(\frac{H_p}{H_{rms}} \right). \tag{7.51}$$

Using the relation (7.46), we obtain

$$\frac{\langle H_p \rangle}{H_{rms}} = \left(\ln \frac{1}{p} \right)^{1/2} + \frac{\sqrt{\pi}}{2p} \, \mathrm{erfc}\left[\left(\ln \frac{1}{p} \right)^{1/2} \right]. \tag{7.52}$$

Similarly, using the Rayleigh distribution, we can easily obtain the following relations of H_p and H_{rms}:

$$H_1 = 0.886 H_{rms}, \quad H_{1/3} = 1.416 H_{rms} \quad \text{and} \quad H_{1/10} = 1.80 H_{rms}. \tag{7.53}$$

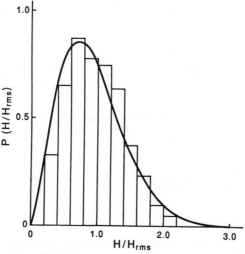

Fig. 7.6. Normalized wave record compared with the Rayleigh density.

The probability density of North Atlantic storm data of September 1961 is compared with the Rayleigh density in Fig. 7.6. The histogram is an average of many wave records over a period of storm conditions. The histogram of the normalized wave heights from ocean wave records is compared with the probability density function as shown by Chakrabarti (1987).

7.4 Wave spectral method

Currently two methods are available in choosing the design wave environment. The first method is a single wave method where the design wave is represented by a regular sinusoidal wave. In practice ocean waves are not regular; rather they are random waves. The second method is the wave spectrum. With reference to Fig. 7.7, we can distinguish if the given wave is a regular, irregular or random wave. Consider a pure sinusoidal wave as given by

$$\eta = \frac{H}{2}\cos(kx - \sigma t). \tag{7.54}$$

Figure 7.7(a) is the plot of η as against t when $x = 0$, whereas Fig. 7.7(b) is the plot of η as against x when $t = 0$. These two graphs demonstrate the behavior of regular waves. The waves recorded at a certain position, say at $x = 0$, can be found to be composed of many sinusoidal waves with many

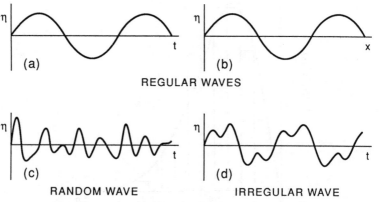

Fig. 7.7. Representation of various types of wave profiles.

frequencies σ_n, amplitudes A_n and phases ε_n. Superimposing all these waves, we can form a single wave which is given by

$$\eta = \sum_{n=0}^{\infty} A_n \cos(\sigma_n t - \varepsilon_n). \tag{7.55}$$

This wave is definitely a random wave and the plot of η against t is given by Fig. 7.7(c). It is random because it does not have any definite fundamental period. However, if the wave in (7.55) is written in the form of a Fourier series composed of the sum of sines and cosines having a fundamental period T such that

$$\eta = \sum_{n=0}^{\infty} (a_n \cos n\sigma t + b_n \sin n\sigma t) \tag{7.56}$$

where $\sigma = 2\pi/T$, then is called the irregular wave. Figure 7.7(d) is the plot of η against t (eqn (7.56)) which clearly shows the period T and this can be regarded as the irregular wave.

7.4.1 Definition of spectrum

The wave spectrum can be defined in many ways. In the following, we shall discuss only four of them by graphical plots. Referring to eqn (7.55), if the amplitudes A_n are plotted against σ_n, then an amplitude spectrum results. This spectrum is shown in Fig. 7.8(a). However, if we plot $A_n^2/2$ against σ_n as shown in Fig. 7.8(b), then another spectrum can be obtained. This spectrum is usually known as the energy spectrum. Both of these spectra are line or discrete spectra because the frequency component is discrete. On the other hand, Fig. 7.8(c) shows the plot of $A_n^2/2\Delta\sigma$ against σ, known as the energy density spectrum, which is more popular because the area under this curve is a measure of total energy in the wave field.

For a continuous range of frequencies, eqn (7.55) can be put into complex integral form as follows:

$$\eta(t) = \mathrm{Re}\left[\int_0^{\infty} A(\sigma)\exp(i(\sigma t - \varepsilon(\sigma)))d\sigma\right] \tag{7.57}$$

where $A(\sigma)$ might be called the amplitude density function. A plot of $A(\sigma)$ against σ is shown in Fig. 7.8(d). This spectrum is called the continuous amplitude spectrum.

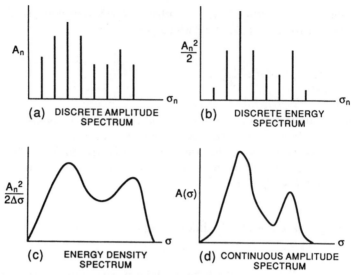

(a)　DISCRETE AMPLITUDE
SPECTRUM

(b)　DISCRETE ENERGY
SPECTRUM

(c)　ENERGY DENSITY
SPECTRUM

(d) CONTINUOUS AMPLITUDE
SPECTRUM

Fig. 7.8.　Plots of different spectra.

7.4.2 Spectral analysis: Fourier series

The Fourier expansion plays a very important role in the study of spectral analysis. Joseph Fourier (1768–1830), a confidant of Napoleon, first undertook the systematic study of such expansions in a memorable monograph, *Theorie analytique de la chaleur*, published in 1822. The theory states that if $\eta(t)$ is a piecewise continuous function defined over $d \le t \le d + 2p$ where $2p$ is the period of the function $\eta(t)$, then $\eta(t)$ can be expressed as a sum of sines and cosines as follows:

$$\eta(t) = \frac{a_0}{2} + \sum_{n=1}^{\infty} \left(a_n \cos \frac{n\pi t}{p} + b_n \sin \frac{n\pi t}{p} \right) \tag{7.58}$$

where

$$a_n = \frac{1}{p} \int_d^{d+2p} \eta(t)\cos \frac{n\pi t}{p} \, dt \qquad n = 0, 1, 2, \ldots \tag{7.59}$$

$$b_n = \frac{1}{p} \int_d^{d+2p} \eta(t)\sin \frac{n\pi t}{p} \, dt \qquad n = 1, 2, 3, \ldots . \tag{7.60}$$

Equation (7.58) is known as the Fourier series, a_n and b_n are called the Fourier coefficients which can be obtained from the formulae (7.59) and (7.60). Here a_0 is the mean of the periodic function η. If the Fourier series

(7.58) contains a finite number of components, then the infinite series can be formulated as

$$\eta(t) = \frac{a_0}{2} + \sum_{n=1}^{N} \left(a_n \cos \frac{n\pi t}{p} + b_n \sin \frac{n\pi t}{p} \right). \tag{7.61}$$

It is to be noted that this representation of an ocean state may be used to approximate a particular wave record. This form is often used in a design where time domain analysis is employed.

In our usual notation we have: fundamental period $= T = 2p$, fundamental frequency $= \sigma = 2\pi/T = \pi/p$, fundamental frequency in cycles $= f = \sigma/2\pi = 1/T = 1/2p$.

If we multiply eqn (7.58) by $\cos(n\pi t/p)$ and $\sin(n\pi t/p)$ in turn and integrate with respect to t from $t = d$ to $t = d + 2p$ we obtain

$$\int_d^{d+2p} \eta(t) \begin{cases} \cos \dfrac{n\pi t}{p} \\[2ex] \sin \dfrac{n\pi t}{p} \end{cases} dt$$

$$= \int_d^{d+2p} \frac{a_0}{2} \begin{cases} \cos \dfrac{n\pi t}{p} \\[2ex] \sin \dfrac{n\pi t}{p} \end{cases} dt$$

$$+ \int_d^{d+2p} \begin{cases} \cos \dfrac{n\pi t}{p} \\[2ex] \sin \dfrac{n\pi t}{p} \end{cases} \sum_{n=1}^{\infty} \left(a_n \cos \frac{n\pi t}{p} + b_n \sin \frac{n\pi t}{p} \right) dt. \tag{7.62}$$

Using the following orthogonality properties of the trigonometric functions:

$$\int_d^{d+2p} \cos \frac{n\pi t}{p} \cos \frac{m\pi t}{p} \, dt = \begin{cases} 2p & m = n = 0 \\ p & m = n \neq 0 \\ 0 & m \neq n \end{cases}$$

$$\int_d^{d+2p} \sin \frac{n\pi t}{p} \sin \frac{m\pi t}{p} \, dt = \begin{cases} p & m = n \neq 0 \\ 0 & m \neq n \end{cases}$$

the formulae (7.59) and (7.60) for a_n and b_n can be easily verified.

It is of interest in practice to know the average wave energy delivered per cycle. This can be found in the usual way by integrating the product $\eta^2(t)(\frac{1}{2}\rho g)$ over one period and then dividing by the length of the period $2p$. Therefore, if $\langle E \rangle$ represents the average wave energy per cycle then

$$\langle E \rangle = \frac{1}{2}\rho g \langle \eta^2(t) \rangle. \tag{7.63}$$

However, $\langle \eta^2(t) \rangle = (1/2p)\int_d^{d+2p} \eta^2(t)dt$ is the mean square value of $\eta(t)$ over a wave cycle.

Thus

$$\langle \eta^2(t) \rangle = \frac{1}{2p} \int_d^{d+2p} \eta(t) \cdot \eta(t)\, dt$$

$$= \frac{1}{2p} \int_d^{d+2p} \eta(t) \left[\frac{a_0}{2} + \sum_{n=1}^{\infty} \left(a_n \cos \frac{n\pi t}{p} + b_n \sin \frac{n\pi t}{p} \right) \right] dt$$

$$= \frac{a_0}{2} \left(\frac{1}{2p} \int_d^{d+2p} \eta(t)dt \right)$$

$$+ \sum_{n=1}^{\infty} a_n \left(\frac{1}{2p} \int_d^{d+2p} \eta(t)\cos \frac{n\pi t}{p}\, dt \right)$$

$$+ \sum_{n=1}^{\infty} b_n \left(\frac{1}{2p} \int_d^{d+2p} \eta(t)\sin \frac{n\pi t}{p}\, dt \right).$$

But

$$a_n = \frac{1}{p} \int_d^{d+2p} \eta(t)\cos \frac{n\pi t}{p}\, dt \qquad n = 0,1,2,3,\ldots$$

$$b_n = \frac{1}{p} \int_d^{d+2p} \eta(t)\sin \frac{n\pi t}{p}\, dt \qquad n = 1,2,3,\ldots.$$

Therefore substituting these values in the above expression we obtain

$$\langle \eta^2(t) \rangle = \frac{a_0^2}{4} + \frac{1}{2} \sum_{n=1}^{\infty} (a_n^2 + b_n^2). \tag{7.64}$$

This relation is usually known as Parseval's identity.

Hence, eqn (7.63) becomes

$$\langle E \rangle = \left(\frac{1}{2}\rho g \right) \left[\frac{a_0^2}{4} + \frac{1}{2} \sum_{n=1}^{\infty} (a_n^2 + b_n^2) \right]. \tag{7.65}$$

For practical use, the upper limit of the summation can be put as N where N is large. The complex exponential form of a Fourier series may be obtained by substituting the exponential equivalents of the cosine and sine terms in eqns (7.58), (7.59) and (7.60):

$$\eta(t) = \frac{a_0}{2} + \sum_{n=1}^{\infty} a_n \left(\frac{e^{in\pi t/p} + e^{-in\pi t/p}}{2} \right) + b_n \left(\frac{e^{in\pi t/p} - e^{-in\pi t/p}}{2i} \right)$$

$$= \frac{a_0}{2} + \sum_{n=1}^{\infty} \left(\frac{a_n - ib_n}{2} e^{in\pi t/p} + \frac{a_n + ib_n}{2} e^{-in\pi t/p} \right).$$

If we now define $c_0 = a_0/2$, $c_n = a_n - ib_n/2$, $c_{-n} = a_n + ib_n/2 = c_n^*$ where c_n^* is the complex conjugate of c_n, the last series can be written as

$$\eta(t) = c_0 + \sum_{n=1}^{\infty} (c_n e^{in\pi t/p} + c_{-n} e^{-in\pi t/p}). \tag{7.66}$$

But $\sum_{n=1}^{\infty} c_{-n} e^{-in\pi t/p} = \sum_{n=-1}^{-\infty} c_n e^{in\pi t/p}$. Hence we have

$$\eta(t) = \sum_{n=-\infty}^{\infty} c_n e^{in\pi t/p}, \tag{7.67}$$

which is known as the complex Fourier series and c_n are the complex Fourier coefficients. Note that c_0 is included in the summation and it corresponds to $n = 0$. To find these coefficients we use the formulae (7.59) and (7.60):

$$c_0 = \frac{1}{2} a_0 = \frac{1}{2p} \int_d^{d+2p} \eta(t) dt$$

$$c_n = \frac{1}{2}(a_n - ib_n) = \frac{1}{2} \left[\frac{1}{p} \int_d^{d+2p} \eta(t) \left(\cos \frac{n\pi t}{p} - i \sin \frac{n\pi t}{p} \right) dt \right]$$

$$= \frac{1}{2p} \int_d^{d+2p} \eta(t) e^{-in\pi t/p} dt$$

and

$$c_{-n} = \frac{1}{2}(a_n + ib_n) = \frac{1}{2} \left[\frac{1}{p} \int_d^{d+2p} \eta(t) \left(\cos \frac{n\pi t}{p} + i \sin \frac{n\pi t}{p} \right) dt \right]$$

$$= \frac{1}{2p} \int_d^{d+2p} \eta(t) e^{in\pi t/p} dt.$$

Clearly, whether the index n is positive, negative or zero, the formula for c_n can be written as

$$c_n = \frac{1}{2p} \int_d^{d+2p} \eta(t) e^{-in\pi t/p} \, dt \qquad n = 0, \pm 1, \pm 2, \ldots . \tag{7.68}$$

In most practical situations, d can be either $-p$ or 0. If $d = -p$, then eqns (7.67) and (7.68) become

$$\eta(t) = \sum_{n=-\infty}^{\infty} c_n e^{in\pi t/p}$$

$$c_n = \frac{1}{2p} \int_{-p}^{p} \eta(t) e^{-in\pi t/p} \, dt \qquad n = 0, \pm 1, \pm 2, \ldots . \tag{7.69}$$

Parseval's identity (7.64) can very easily be deduced using complex Fourier series:

$$\langle \eta^2(t) \rangle = \frac{1}{2p} \int_d^{d+2p} \eta^2(t) dt = \frac{1}{2p} \int_d^{d+2p} \eta(t) \cdot \eta(t) dt$$

$$= \frac{1}{2p} \int_d^{d+2p} \eta(t) \left(\sum_{n=-\infty}^{\infty} c_n e^{in\pi t/p} \right) dt$$

$$= \sum_{n=-\infty}^{\infty} c_n \left(\frac{1}{2p} \int_d^{d+2p} \eta(t) e^{in\pi t/p} \, dt \right)$$

$$= c_0 \left(\frac{1}{2p} \int_d^{d+2p} \eta(t) dt \right)$$

$$\quad + \sum_{n=-\infty, n \neq 0}^{\infty} c_n \left(\frac{1}{2p} \int_d^{d+2p} \eta(t) e^{in\pi t/p} \, dt \right)$$

$$= c_0^2 + \sum_{n=-\infty, n \neq 0}^{\infty} c_n c_{-n} = c_0^2 + 2 \sum_{n=1}^{\infty} |c_n|^2$$

$$= \frac{a_0^2}{4} + 2 \sum_{n=1}^{\infty} \left(\frac{a_n - ib_n}{2} \right) \left(\frac{a_n + ib_n}{2} \right)$$

$$= \frac{a_0^2}{4} + \frac{1}{2} \sum_{n=1}^{\infty} (a_n^2 + b_n^2) \tag{7.70}$$

which is the same as in eqn (7.64). Therefore the average wave energy per wave cycle can be written as

$$\langle E \rangle = \left(\frac{1}{2} \rho g \right) \left(c_0^2 + 2 \sum_{n=1}^{\infty} |c_n|^2 \right) \tag{7.71}$$

where c_0 and c_n are the complex Fourier coefficients.

7.4.3 Continuous spectrum of wave record

For random waves the foregoing analysis must be modified so that there is no finite time period of the wave and in fact the time period of random waves is infinite. Keeping this in mind, we begin an analysis with the complex Fourier series, eqn (7.69):

$$\eta_p(t) = \sum_{n=-\infty}^{\infty} c_n e^{in\pi t/p}$$

$$c_n = \frac{1}{2p} \int_{-p}^{p} \eta_p(t) e^{-in\pi t/p} \, dt = \frac{1}{2p} \int_{-p}^{p} \eta_p(\tau) e^{-in\pi\tau/p} \, d\tau$$

where $\eta_p(t)$ is a periodic function of period $2p$.

Substitution of the second expression for c_n into $\eta_p(t)$ yields

$$\eta_p(t) = \sum_{n=-\infty}^{\infty} \left(\frac{1}{2p} \int_{-p}^{p} \eta_p(\tau) e^{-in\pi\tau/p} \, d\tau \right) e^{in\pi t/p}$$

$$= \sum_{n=-\infty}^{\infty} \left(\frac{1}{2\pi} \int_{-p}^{p} \eta_p(\tau) e^{-in\pi\tau/p} \, d\tau \right) e^{in\pi t/p} \left(\frac{\pi}{p} \right).$$

Define $\sigma_n = n\pi/p$, $\sigma_{n+1} = (n+1)\pi/p$, and $\Delta\sigma = \sigma_{n+1} - \sigma_n = \pi/p$. With these definitions then we have

$$\eta_p(t) = \sum_{n=-\infty}^{\infty} \left[\frac{1}{2\pi} \int_{-p}^{p} \eta_p(\tau) e^{-i\sigma_n\tau} \, d\tau \right] e^{i\sigma_n t}(\Delta\sigma).$$

Let p go to infinity such that the nonperiodic limit of $\eta_p(t)$ becomes $\eta(t)$, and as $p \to \infty$ implies $\Delta\sigma \to 0$, then the above expression goes to the integral for the continuous random wave records or wave elevations

$$\eta(t) = \int_{-\infty}^{\infty} \left(\frac{1}{2\pi} \int_{-\infty}^{\infty} \eta(\tau) e^{-i\sigma\tau} \, d\tau \right) e^{i\sigma t} \, d\sigma. \tag{7.72}$$

Equation (7.72) is actually a valid representation of the nonperiodic limit function $\eta(t)$ provided that (a) in every finite interval $\eta(t)$ satisfies the Dirichlet conditions, and (b) the improper integral $\int_{-\infty}^{\infty}|\eta(t)|dt$ exists. Equation (7.72) is known as the Fourier integral.

Now define

$$g(\sigma) = \int_{-\infty}^{\infty} \eta(t)e^{-i\sigma t}\,dt. \tag{7.73}$$

Then

$$\eta(t) = \frac{1}{2\pi}\int_{-\infty}^{\infty} g(\sigma)e^{i\sigma t}\,d\sigma. \tag{7.74}$$

Here $g(\sigma)$ is called the Fourier transform of $\eta(t)$ where $\eta(t)$ is a continuous nonperiodic random function. This is also known as the continuous spectrum of $\eta(t)$. The expressions (7.73) and (7.74) are jointly known as the Fourier transform pair.

Rewriting eqn (7.72), we have

$$\eta(t) = \frac{1}{2\pi}\int_{-\infty}^{\infty}\int_{-\infty}^{\infty} \eta(\tau)e^{-i\sigma(\tau-t)}\,d\tau\,d\sigma$$

$$= \frac{1}{2\pi}\int_{-\infty}^{\infty}\int_{-\infty}^{\infty} \eta(\tau)\cos\sigma(\tau-t)\,d\tau\,d\sigma$$

This result is due to the fact that $\int_{-\infty}^{\infty}\sin\sigma(\tau-t)d\sigma=0$. Also we know that $\int_{-\infty}^{\infty}\cos\sigma(\tau-t)d\sigma = 2\int_{0}^{\infty}\cos\sigma(\tau-t)dt$ so that

$$\eta(t) = \frac{1}{\pi}\int_{0}^{\infty}\int_{-\infty}^{\infty} \eta(\tau)\cos\sigma(\tau-t)\,d\tau\,d\sigma$$

$$\eta(t) = \frac{1}{\pi}\int_{0}^{\infty}\int_{-\infty}^{\infty} \eta(\tau)(\cos\sigma\tau\cos\sigma t + \sin\sigma\tau\sin\sigma t)\,d\tau\,d\sigma$$

$$= \frac{1}{\pi}\int_{0}^{\infty}\left[\left(\int_{-\infty}^{\infty}\eta(\tau)\cos\sigma\tau\,d\tau\right)\cos\sigma t\right.$$

$$\left.+\left(\int_{-\infty}^{\infty}\eta(\tau)\sin\sigma\tau\,d\tau\right)\sin\sigma t\right]d\sigma.$$

Define

$$a(\sigma) = \int_{-\infty}^{\infty} \eta(t)\cos \sigma t\, dt, \qquad b(\sigma) = \int_{-\infty}^{\infty} \eta(t)\sin \sigma t\, dt$$

$$\eta(t) = \frac{1}{\pi}\int_{0}^{\infty} [a(\sigma)\cos \sigma t + b(\sigma)\sin \sigma t]\, d\sigma. \tag{7.75}$$

Here $a(\sigma)$ and $b(\sigma)$ are called the continuous amplitude spectra.

The total energy of a wave, E, per unit surface area in the wave record between infinite time limits is given by

$$E = \frac{1}{2}\rho g \int_{-\infty}^{\infty} \eta^2(t)\, dt. \tag{7.76}$$

Using the expression for $\eta(t)$ given in eqn (7.75), we have

$$E = \frac{1}{2}\rho g \int_{-\infty}^{\infty} \eta(t)\cdot \eta(t)\, dt$$

$$= \frac{1}{2}\rho g \int_{-\infty}^{\infty} \eta(t)\left(\frac{1}{\pi}\int_{0}^{\infty} \{a(\sigma)\cos \sigma t + b(\sigma)\sin \sigma t\}\, d\sigma\right) dt.$$

Interchanging the integrals

$$E = \left(\frac{\rho g}{2\pi}\right)\int_{0}^{\infty}\left[a(\sigma)\int_{-\infty}^{\infty} \eta(t)\cos \sigma t\, dt + b(\sigma)\int_{-\infty}^{\infty} \eta(t)\sin \sigma t\, dt\right] d\sigma$$

$$= \frac{\rho g}{2\pi}\int_{0}^{\infty} [a^2(\sigma) + b^2(\sigma)]\, d\sigma \tag{7.77}$$

which is written as

$$E = \frac{\rho g}{2\pi}\int_{0}^{\infty} A^2(\sigma)\, d\sigma \tag{7.78}$$

where $A^2(\sigma) = a^2(\sigma) + b^2(\sigma)$. Now comparing eqns (7.76) and (7.77) we see that

$$\int_{-\infty}^{\infty} \eta^2(t)\, dt = \frac{1}{\pi}\int_{0}^{\infty} A^2(\sigma)\, d\sigma \tag{7.79}$$

which is Parseval's identity for a continuous nonperiodic random wave function $\eta(t)$. The relation (7.79) really gives rise to the concept of the wave energy spectrum.

If $\langle \eta^2(t) \rangle$ is the mean square value of $\eta(t)$ over a specified record length, T, then

$$\langle \eta^2(t) \rangle = \lim_{T \to \infty} \left\{ \frac{1}{T} \int_{-T/2}^{T/2} \eta^2(t) dt \right\}. \tag{7.80}$$

With this definition then the mean energy per unit area is

$$\langle E \rangle = \frac{1}{2} \rho g \lim_{T \to \infty} \left\{ \frac{1}{T} \int_{-T/2}^{T/2} \eta^2(t) dt \right\}$$

$$= \lim_{T \to \infty} \left(\frac{1}{2} \rho \frac{g}{T} \right) \int_{-T/2}^{T/2} \eta(t)$$

$$\times \left(\frac{1}{\pi} \int_0^\infty [a(\sigma)\cos \sigma t + b(\sigma)\sin \sigma t] d\sigma \right) dt$$

$$= \lim_{T \to \infty} \left(\frac{\rho g}{2\pi T} \right) \int_0^\infty \left(a(\sigma) \int_{-T/2}^{T/2} \eta(t)\cos \sigma t \, dt \right.$$

$$\left. + b(\sigma) \int_{-T/2}^{T/2} \eta(t)\sin \sigma t \, dt \right) d\sigma.$$

Considering that T is large but finite and that the integrals $\int_{-T/2}^{T/2} \eta(t)\cos \sigma t \, dt$ and $\int_{-T/2}^{T/2} \eta(t)\sin \sigma t \, dt$ are approximated as $a(\sigma)$ and $b(\sigma)$, respectively, then

$$\langle E \rangle = \frac{\rho g}{2\pi T} \int_0^\infty [a^2(\sigma) + b^2(\sigma)] d\sigma = \frac{\rho g}{2} \int_0^\infty \frac{A^2(\sigma)}{\pi T} d\sigma. \tag{7.81}$$

Defining the spectral energy density as

$$s(\sigma) = \frac{A^2(\sigma)}{\pi T} \tag{7.82}$$

the total energy is obtained for the area covered by the energy density curve as a function of frequency

$$\langle E \rangle = \frac{\rho g}{2} \int_0^\infty s(\sigma) d\sigma, \tag{7.83}$$

where $s(\sigma)$ has the units of (length^2) (time). There are two commonly known methods, namely the autocorrelation method and the fast Fourier transform method, of calculating the energy spectrum of an ocean wave record. In the following these two methods will be discussed briefly.

(a) Autocorrelation method

The method of finding the energy density spectrum by the autocorrelation method is as follows.

First compute the autocorrelation function of the wave profile. The Fourier transform of the autocorrelation function gives the energy density spectrum. The autocorrelation function defined in eqn (7.24) can be rewritten as

$$R_\eta(\tau) = \lim_{T \to \infty} \left\{ \frac{1}{T} \int_{-T/2}^{T/2} \eta(t)\eta(t+\tau)\,dt \right\} \tag{7.84}$$

where τ is the lag or time interval between measures of surface elevations $\eta(t)$ and $\eta(t+\tau)$.

If we now substitute the Fourier series representation for $\eta(t+\tau)$ into eqn (7.84), we obtain

$$R_\eta(\tau) = \lim_{T \to \infty} \left\{ \frac{1}{T} \int_{-T/2}^{T/2} \eta(t) \sum_{n=-\infty}^{\infty} c_n e^{in\sigma(t+\tau)} \right\} dt$$

$$= \lim_{T \to \infty} \left\{ \frac{1}{T} \int_{-T/2}^{T/2} \sum_{n=-\infty}^{\infty} (\eta(t)e^{in\sigma t})(c_n e^{in\sigma\tau})\,dt \right\}$$

or

$$R_\eta(\tau) = \sum_{n=-\infty}^{\infty} \left\{ \lim_{T \to \infty} \frac{1}{T} \int_{-T/2}^{T/2} \eta(t)e^{in\sigma t}\,dt \right\} (c_n e^{in\sigma\tau})$$

$$= \sum_{n=-\infty}^{\infty} c_n^* c_n e^{in\sigma\tau} \tag{7.85}$$

where $c_n^* = $ complex conjugate of $c_n = \lim_{T \to \infty} (1/T) \int_{-T/2}^{T/2} \eta(t) e^{in\sigma t}\,dt$.

Taking the Fourier transform of $R_\eta(\tau)$, we obtain

$$s(\sigma) = \lim_{T \to \infty} \left\{ \frac{1}{T} \int_{-T/2}^{T/2} R_\eta(\tau)e^{-in\sigma\tau}\,d\tau \right\}$$

$$= \lim_{T \to \infty} \left\{ \frac{1}{T} \int_{-T/2}^{T/2} \left(\sum_{n=-\infty}^{\infty} c_n^* c_n e^{in\sigma\tau} \right) e^{-in\sigma\tau}\,d\tau \right\}$$

$$= |c_n|^2 \quad \text{for} \quad -\infty < n < \infty \tag{7.86}$$

which is known as the energy density spectrum in the discrete case.

More specifically

$$s_n(\sigma) = |c_n|^2 \qquad \text{for} \qquad -\infty < n < \infty \tag{7.87}$$

which is the two-sided energy spectrum.

For physical consideration, the one-sided energy spectrum is usually used and in that situation

$$s'_n(\sigma) = 2|c_n|^2, \qquad n > 0, \qquad s'_n(\sigma) = |c_0|^2, \qquad n = 0. \tag{7.88}$$

(b) Fast Fourier transform (FFT) method

In a fast Fourier transform technique, the transformation is taken directly from the time domain to the frequency domain and then the result is squared to convert to the energy unit. Thus if $\eta(t)$ is the wave profile as a function of time, then the energy spectrum by this method is obtained as

$$s(\sigma) = \frac{1}{T} \left(\sum_{n=1}^{N} \eta(n\Delta t) e^{i2\pi f(n\Delta t)} \Delta t \right)^2 \tag{7.89}$$

where $2\pi f = \sigma$ and $f = 1/T$.

Chakrabarti (1987) has given a good account of this method. However, most computer library systems have FFT algorithms available. The interested reader may consult these algorithms for further reference.

7.5 Theoretical models of energy spectrum

In this section several mathematical spectrum models will be discussed. For a more comprehensive study of the energy spectrum, the reader is referred to the work of Chakrabarti (1987). The Pierson–Moskowitz (1964) spectrum model, which is based on the significant wave height or wind speed, is the most common single-parameter spectrum. Some spectral models, including Bretschneider (1969), International Ship Structures Congress (ISSC) (1964) and International Towing Tank Congress (ITTC) (1966), are of two parameters. However, the Joint North Sea Wave Project (JONSWAP) spectrum (Hasselmann et al. 1973, 1976) is a five-parameter spectrum, but three of the parameters are usually held constant.

7.5.1 Pierson–Moskowitz spectrum

On the basis of the similarity theory of Kitaigorodskii (1962), a new formula for the energy spectrum distribution of a wind-generated ocean state was proposed by Pierson and Moskowitz (1964). This spectrum is commonly known as the P–M model and written as

$$s(\sigma) = \frac{\alpha g^2}{\sigma^5} \exp\left[-0.74\left(\frac{\sigma U_w}{g}\right)^{-4}\right] \tag{7.90}$$

where $\alpha = 0.0081$ and U_w is the wind speed.

7.5.2 Bretschneider spectrum

When the spectrum is narrow banded and the individual wave height and wave period follow the Rayleigh distribution, the following form of the spectral model due to Bretschneider (1969) may be derived:

$$s(\sigma) = 0.1687H_s^2 \frac{\sigma_s^4}{\sigma^5} \exp\left[-0.675\left(\frac{\sigma_s}{\sigma}\right)^4\right] \tag{7.91}$$

where $\sigma_s = 2\pi/T_s$, H_s = significant wave height, and T_s = significant wave period defined as the average period of the significant waves.

From the Bretschneider spectral model it follows that

$$T_s = 0.94\sigma T_0 \tag{7.92}$$

where T_s is the peak period. This relationship makes the P–M and Bretschneider models equivalent.

7.5.3 ISSC spectrum

In 1964 the International Ship Structures Congress (ISSC) suggested a slight modification in the form of the Bretschneider spectrum,

$$s(\sigma) = 0.1107H_s^2 \frac{\bar{\sigma}^4}{\sigma^5} \exp\left[-0.4427\left(\frac{\bar{\sigma}}{\sigma}\right)^4\right]. \tag{7.93}$$

The relationship between the peak frequency σ_0 and $\bar{\sigma}$ for the ISSC spectrum is

$$\bar{\sigma} = 1.296\sigma_0. \tag{7.94}$$

7.5.4 ITTC spectrum

A modification of the P–M spectrum in terms of the significant wave heights and zero-crossing frequency, σ_z, was proposed by the ITTC (1966). The ITTC spectrum has been written as

$$s(\sigma) = \frac{\alpha g^2}{\sigma^5} \exp\left(-\frac{4\alpha g^2}{H_s^2 \sigma^4}\right) \tag{7.95}$$

where

$$\alpha = \frac{0.0081}{k^4} \quad \text{and} \quad k = \frac{\sqrt{g/\sigma_\eta}}{3.54\sigma_z} \tag{7.96}$$

in which σ_η is the standard deviation of the water surface elevation.

7.5.5 JONSWAP spectrum

In 1973 Hasselmann and his co-workers developed the JONSWAP spectrum during a joint North Sea wave project; hence the name. The JONSWAP spectrum is a modification of the P–M model and can be written as follows:

$$s(\sigma) = \frac{\alpha g^2}{\sigma^5} \exp\left[-1.25\left(\frac{\sigma_0}{\sigma}\right)^4\right] \gamma^{\exp\left[-\frac{(\sigma-\sigma_0)^2}{2\tau^2\sigma_0^2}\right]} \tag{7.97}$$

in which $\gamma =$ peakedness parameter, and $\tau =$ shape parameter.

In the design of an offshore structure, the most widely used spectral model is the P–M model. However, the JONSWAP model spectrum with the appropriate γ values between 1 and 4 can often be regarded as the representative form of a design storm wave.

7.6 The directional wave spectrum

In real situations, an ocean surface can be represented by a great number of waves; this may be true when there is a storm. To account fully for this physical situation, a directional wave spectrum is used consisting of wave frequency and the direction of wave, θ. This directional wave system can be expressed as

$$\eta(x, y, t) = \sum_{n=-\infty}^{\infty} \int_0^{2\pi} c_n(\theta) e^{i(n\sigma t - (k_n \cos \theta)x - (k_n \sin \theta)y)} \, d\theta \tag{7.98}$$

where θ is the angle between the wave orthogonal and the x-axis.

If the waves are measured at the origin $x = 0$, $y = 0$, then (7.98) reduces to

$$\eta(t) = \sum_{n=-\infty}^{\infty} \int_0^{2\pi} c_n(\theta) e^{in\sigma t} \, d\theta. \tag{7.99}$$

The energy density spectrum $s_\eta(\sigma)$ is obtained analytically by first determining the autocorrelation function $R_{\eta(\tau)}$.

$$R_\eta(\tau) = \lim_{T\to\infty} \left\{ \frac{1}{T} \int_{-T/2}^{T/2} \eta(t)\eta(t+\tau) \, dt \right\}.$$

Replacing $\eta(t+\tau)$ by eqn (7.99) gives

$$R_\eta(\tau) = \lim_{T\to\infty} \left\{ \frac{1}{T} \int_{-T/2}^{T/2} \eta(t) \sum_{n=-\infty}^{\infty} \int_0^{2\pi} c_n(\theta) e^{in\sigma(t+\tau)} \, d\theta \, dt \right\}$$

$$= \sum_{n=-\infty}^{\infty} \lim_{T\to\infty} \left\{ \frac{1}{T} \int_{-T/2}^{T/2} \eta(t) e^{in\sigma t} \, dt \right\} \left(\int_0^{2\pi} c_n(\theta) e^{in\sigma\tau} \, d\theta \right)$$

$$= \sum_{n=-\infty}^{\infty} \int_0^{2\pi} c_n^*(\theta') d\theta' \int_0^{2\pi} c_n(\theta) e^{in\sigma\tau} \, d\theta. \tag{7.100}$$

Because

$$\int_0^{2\pi} c_n(\theta) d\theta = \lim_{T\to\infty} \left\{ \frac{1}{T} \int_{-T/2}^{T/2} \eta(t) e^{-in\sigma\tau} \, dt \right\} \tag{7.101}$$

the energy density spectrum of the water surface elevation is then the Fourier transform of $R_\eta(\tau)$, i.e.

$$s_\eta(\sigma) = \lim_{T\to\infty} \left\{ \frac{1}{T} \int_{-T/2}^{T/2} R_\eta(\tau) e^{-in\sigma\tau} \, d\tau \right\}$$

$$= \lim_{T\to\infty} \left\{ \frac{1}{T} \int_{-T/2}^{T/2} \left(\sum_{n=-\infty}^{\infty} \int_0^{2\pi} c_n^*(\theta') d\theta' \right. \right.$$

$$\left. \left. \times \int_0^{2\pi} c_n(\theta) e^{in\sigma\tau} \, d\theta \right) e^{-in\sigma\tau} \, d\tau \right\}. \tag{7.102}$$

By expanding $c_n(\theta)$ in a complex Fourier series, eqn (7.100) can be put in the following form:

$$R_\eta(\tau) = \sum_{n=-\infty}^{\infty} \int_0^{2\pi} |c_n(\theta)|^2 e^{in\sigma\tau} \, d\theta. \tag{7.103}$$

Therefore

$$s_\eta(\sigma) = \lim_{T\to\infty} \left\{ \frac{1}{T} \int_{-T/2}^{T/2} \sum_{n=-\infty}^{\infty} \int_0^{2\pi} |c_n(\theta)|^2 e^{in\sigma\tau} \, d\theta e^{-in\sigma\tau} \, d\tau \right\}$$

$$= \int_0^{2\pi} |c_n(\theta)|^2 \, d\theta \qquad \text{for} \qquad -\infty < n < \infty. \tag{7.104}$$

$s_\eta(\sigma)$ is the energy at each frequency σ_n, and it is seen to be integral over the direction θ. However, the quantity $|c_n(\theta)|^2$ which gives the distribution of energy with direction as well as frequency is called the directional energy density spectrum. A similar analysis can be made with the energy density spectrum of the horizontal velocities. Interested readers are referred to the standard textbooks including Dean and Dalrymple (1984).

7.7 Exercises

1.. A group of two sinusoidal waves of equal height and slightly different periods is generated in a laboratory wave tank and recorded by a fixed wave gauge. Determine the values of $H_{1/10}$, $H_{1/3}$, H_1 and H_{max} in terms of H_{rms}.

2. At a pier in Halifax harbour, Nova Scotia, 200 consecutive wave heights are measured. Assuming that the sea state is narrow banded, determine (a) how many waves are expected to exceed $H = 3H_{rms}$; (b) what height is exceeded by half the waves; (c) what height is exceeded by only one wave. (Note: For narrow-banded waves, the Rayleigh probability distribution is suitable).

3. Using the Rayleigh probability distribution determine the average height of the highest of the pN waves when $p = 1/10$, $1/3$ and 1 in terms of H_{rms}.

4. A square wave of period 4 seconds is described functionally by

$$\eta(t) = \begin{cases} 1 & 0 < t < 1 \\ 0 & 1 < t < 2 \\ -1 & 2 < t < 3 \\ 0 & 3 < t < 4. \end{cases}$$

(a) Find the Fourier expansion of this wave function; (b) determine Parseval's identity.

5. Find the Fourier series of the following periodic functions whose definitions in one period are (a) $\eta(t) = \cos t, 0 < t < \pi$; (b) $\eta(t) = \sin t$, $0 < t < \pi$.

6. Determine the Fourier series of a square wave of period 2 seconds with an amplitude of unity and find the average wave energy per cycle when $\rho = 1$ and $g = 980$ cm/s^2.

7. In a wave train consisting of 400 waves with $H_{rms} = 3$ m, (a) what is the probability that the height of a particular wave will exceed 5 m? (b) What is the probability that the height of at least one of the 400 waves will exceed 5 m? (c) Determine H_{max}.

8. For the following time functions determine: (a) the Fourier coefficients a_n and b_n; (b) the complex Fourier coefficients; (c) the two-sided energy spectra; (d) the cross-spectrum. $x(t) = 1 + \cos \sigma t + 2 \sin \sigma t - 3 \cos 2\sigma t$ and $y(t) = 3 + 2 \cos(\sigma t - \pi/4) + 3 \sin 3\sigma t$.

9. The cross-correlation function $R_{xy}(\tau)$ associated with a pair of time functions $x(t)$ and $y(t)$ is given by $R_{xy}(t) = 3 \sin^2 \sigma t \cos \sigma t$. If $x(t)$ is given as $x(t) = 1 + \frac{1}{2} \cos \sigma t + \frac{1}{4} \sin 2\sigma t + \frac{3}{2} \cos 3\sigma t - 4 \sin 4\sigma t$, determine $y(t)$.

References

Bretschneider, C. L. (1969). *Wave Forecasting, Handbook of Ocean and Underwater Engineering*, Meyers, J. J., *et al.* (Editors), McGraw-Hill, New York, Chapter 11.
Chakrabarti, S. K. (1987). *Hydrodynamics of Offshore Structures*, Computational Mechanics Publications, Southampton.
Dean, R. G. and Dalrymple, R. A. (1984). *Water Wave Mechanics for Engineers and Scientists*, Prentice Hall, Englewood Cliffs, New Jersey.

Hasselmann, K., Barnett, T. P., Bouws, E., Carlson, H., Cartwright, D. E., Enke, K., Ewing, J. A., Gienapp, H., Hasselmann, D. E., Kruseman, P., Meerburg, A., Muller, P., Olbers, D. J., Richter, K., Sell, W. and Walden, H. (1973). Measurements of wind-wave growth and swell decay during the Joint North Sea Wave Project (JONSWAP). *Dtsch. Hydrogr. Z.*, Erganzunscheft, A 8(12), 95 pp.

Hasselmann, K., Ross, D. B., Muller, P. and Sell, W. (1976). A parametric wave prediction model. *J. Phys. Oceanogr.*, 6, 200–228.

ISSC (1964). *Proceedings of the Second International Ship Structures Congress, Delft, The Netherlands.*

ITTC (1966). Recommendations of the 11th International Towing Tank Conference. *Proceedings 11th ITTC, Tokyo, Japan.*

Kitaigorodskii, S. A. (1962). Applications of the theory of similarity to the analysis of wind generated wave motion as a stochastic process. *Izv., Geophys. Ser. Acad. Sci., USSR*, 1, 105–117.

Longuet-Higgins, M. S. (1952). On the statistical distribution of the heights of sea-waves. *J. Mar. Res.*, 11, 245–266.

Pierson, W. J. and Moskowitz, L. (1964). A proposed spectral form for fully developed wind seas based on the similarity theory of S. A. Kitaigorodskii. *J. Geophys. Res.*, December 1964, 69(24), 5181–5203.

Sverdrup, H. U. and Munk, W. H. (1947). Wind, sea and swell. The theory of relations for forecasting. *US Hydrogr. Office, Washington, DC*, Publ. No. 601.

Part III

Advanced water waves

8
Wave forces on offshore structures

8.1 Introduction

The determination of the forces induced by water waves on an offshore structure is one of the central interests in the design of the structure. There is a large variety of offshore structures which can be described as being composed of small tubular members, large gravity platforms, tension-legged platforms, with large vertical, horizontal, diagonal, circular or square cylinders, semi-submersibles and arctic structures. These structures may be fixed and compliant and also may be floating in waves. Whatever the case, the complexity of the interaction of waves with the structures makes the calculation of wave forces more difficult. Furthermore, owing to the inherent nonlinear nature of ocean waves, no suitable nonlinear mathematical theory is available at present to predict the wave forces on an offshore structure, the following three methods are available to calculate wave forces: (a) the Morison equation, (b) Froude–Krylov theory and (c) diffraction theory. The distinction between these methods is described below.

When the size of a structure is small compared with the water wavelength, the Morison equation is suitable to evaluate the wave forces on the structure. The Morison equation is composed of two forces, namely inertia and drag forces, added together. The coefficients of these two forces are the inertia (or mass) and drug coefficients (due to viscosity) which must be determined experimentally.

The Froude–Krylov theory can be applied when the drag force is small in comparison with the inertia force but the size of the structure is still relatively small. In this case, the force is computed using the incident wave pressure and the pressure area method on the surface of the structure. This method has the advantage that for certain symmetric structures the force may be obtained in closed form and the force coefficients can easily be determined.

When the size of the structure is large, i.e. the structure spans a significant fraction of a wavelength, the incoming waves, after striking it, generally undergo significant scattering or diffraction. In this case the diffraction of the waves should be taken into account in the evaluation of the wave forces. This is generally known as the diffraction theory. By this method, analytic solutions in closed form may be possible for a few simple cases; however, the solution generally involves a numerical technique such as the Green function method, finite-element method or boundary element method to solve Laplace's equation with boundary conditions. We shall discuss the numerical methods in section 8.6.

In deciding which theory will be suitable for a particular problem, the three-dimensionless numbers, namely Reynolds number, $Re = U_m D/\nu$, Keulegan–Carpenter parameter, $KC = U_m T/D$, and diffraction parameter $\pi D/L$, play a very important role, where U_m = maximum horizontal water particle velocity, D = diameter of the structure, ν = kinematic viscosity, T = wave period, L = wavelength, and $\pi = 3.14159$.

The KC parameter is a measure of the importance of the drag force effect and is equivalent to the ratio of the horizontal particle orbit velocity to the structure diameter. The diffraction parameter is equivalent to the ratio of the structure diameter to the wavelength. The relative importance of these two parameters will be discussed later. However, the distinction between these three methods can be described briefly as follows.

The Morison equation is applicable if the flow incident on the given structure separates from the structure which forms a wake (low-pressure region) alternatively in front of and behind the structure. If the incident wave experiences scattering upon reaching the surface of the structure in the form of reflected waves of the same order of magnitude as the incident wave, then the diffraction theory is applied to compute the wave forces. Froude–Krylov theory is applicable when the structure is neither too small nor too large compared with the incident wavelength such that neither appreciable separation from the surface of the structure nor large reflection is evident.

8.2 The Morison equation

In 1950, Morison and his co-workers proposed an empirical formula to determine the horizontal wave forces acting on a vertical pile which extends from the bottom to the free surface (see Fig. 8.1). According to their proposition, the total wave force is composed of two components, drag and inertia, i.e.

$$\Delta F = \Delta F_D + \Delta F_I \qquad (8.1)$$

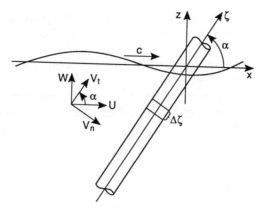

Fig. 8.1.   An inclined circular pile in waves.

where

$$\Delta F_D = \tfrac{1}{2} C_D A \rho U |U|$$

$$\Delta F_I = C_M \rho V \frac{DU}{Dt},\tag{8.2}$$

in which ΔF_D = elementary drag force, ρ = density of the fluid, C_D = drag coefficient, A = the projected area perpendicular to the stream velocity U, ΔF_I = elementary inertia force, $C_M = 1 + \dfrac{M_a}{\rho V}$ = the coefficients of mass, M_a = the added mass, V = the volumetric displacement of the body, DU/Dt = the water particle acceleration. Here $U|U|$ has been introduced in place of U^2 in order to maintain the proper sign for the drag force. The drag force is mainly due to viscosity in a real fluid, which depends upon the Reynolds number, whereas the inertia force is the force exerted on a submerged body in a frictionless, incompressible fluid. By combining the drag and inertia components of force, eqn (8.1) can be rewritten as

$$\Delta F = \tfrac{1}{2} \rho C_D A U |U| + \rho C_M V \frac{DU}{Dt}.\tag{8.3}$$

Equation (8.3) is often referred to as the 'Morison equation'.

8.2.1 Wave forces on a pile

Offshore structures are usually cylindrical piles which may be vertical for load bearing, or inclined due to wave or current forces. In this section, we shall discuss the forces exerted by waves on the piles. The analysis is empirical, and leads to the so-called Morison equation as discussed in the previous section.

Consider the surface-piercing circular pile inclined at an angle α to the horizontal, as shown in Fig. 8.1.

The fluid velocity vector at any given depth has components both normal and tangential to the axis of the pile. These velocity components can be written, respectively, as

$$V_n = U \sin \alpha - W \sin \alpha \tag{8.4}$$

$$V_t = U \cos \alpha + W \sin \alpha. \tag{8.5}$$

Corresponding to these velocity components, there are normal forces due to pressure, viscosity, and fluid inertia, and tangential forces due to viscous stress on the cylindrical surface. The viscous and pressure forces are usually described in terms of a dimensionless quantity called the drag coefficient, which is the ratio of the force in question to the dynamic pressure force of the fluid. Thus, for an elemental length, $\Delta \zeta$ (see Fig. 8.1), the normal and tangential drag coefficients are, respectively,

$$C_n = \frac{\Delta F_n}{\frac{1}{2} \rho V_n^2 D \Delta \zeta} \tag{8.6}$$

$$C_t = \frac{\Delta F_t}{\frac{1}{2} \rho V_t^2 D \Delta \zeta} \tag{8.7}$$

where D is the diameter of the cylinder.

The expressions (8.6) and (8.7) are the empirical equations. When $\alpha \to \pi/2$, then $\zeta \to z$ and the tangential force is small relative to the normal force. Thus, for a vertical cylinder we need to consider only the horizontal force, and therefore eqn (8.6) can be written as

$$\Delta F_n = \frac{1}{2} \rho U|U|DC_n \Delta z. \tag{8.8}$$

This is subsequently written as

$$\Delta F_D = \frac{1}{2} \rho C_D DU|U| \Delta z. \tag{8.9}$$

But we know from the linear theory, discussed in Chapter 4, that the expression for U is given by

$$U = \frac{Agk}{\sigma} \frac{\cosh k(z+h)}{\cosh(kh)} \cos(kx - \sigma t). \tag{8.10}$$

Therefore (8.9) can be written as

$$\Delta F_D = \left(\frac{1}{2} \rho D C_D \Delta z \right) \left(\frac{Agk \cosh k(z+h)}{\sigma \cosh(kh)} \right)^2$$

$$\times \cos(kx - \sigma t)|\cos(kx - \sigma t)|. \qquad (8.11)$$

Since the flow is time dependent, the inertial effects of the water must be included in the analysis. The mass of the fluid which is influenced by the cylinder is called the *added mass* and can be defined through the added-mass coefficient

$$C_M = \frac{\Delta F_I / \Delta z}{\frac{1}{4} \rho \pi D^2 \left(\dfrac{DU}{Dt} \right)}, \qquad (8.12)$$

where the force ΔF_I is the elemental inertia force of the fluid acting on the pile element. Thus the elemental force of inertia is given by

$$\Delta F_I = \frac{1}{4} \rho \pi D^2 C_M \left(Agk \frac{\cosh k(z+h)}{\cosh kh} \sin(kx - \sigma t) \right) \Delta z. \qquad (8.13)$$

The total normal force on the cylinder is obtained by adding the elemental drag force and elemental inertia force and integrating the sum over the wetted surface of the cylinder, from $z = -h$ to $z = \eta$.

Thus the total wave-induced force on the pile is

$$F = \int (\Delta F_D + \Delta F_I)$$

$$= \int_{z=-h}^{\eta} \frac{1}{2} \rho \left(\frac{Ag}{C} \frac{\cosh k(z+h)}{\cosh kh} \right)^2$$

$$\times \cos(kx - \sigma t)|\cos(kx - \sigma t)| D C_D \, dz$$

$$+ \int_{z=-h}^{\eta} \frac{1}{4} \rho \pi D^2 C_M \left(Agk \frac{\cosh k(z+h)}{\cosh kh} \sin(kx - \sigma t) \right) dz.$$

$$(8.14)$$

If it is assumed that the centre line of the pile is located at $x = 0$ and the diameter is much smaller than the wavelength, then after integration (8.14) can be written as

$$F = \frac{1}{4} \rho \left(\frac{Ag}{C} \right)^2 \frac{DC_D}{\cosh^2 kh}$$

$$\times \left[(h + \eta) + \left(\frac{1}{2k} \right) \sinh[2k(h + \eta)] \right] \cos \sigma t |\cos \sigma t|$$

$$- \frac{1}{4} \rho \pi D^2 A g C_M \frac{\sinh k(k + \eta)}{\cosh kh} \sin \sigma t. \tag{8.15}$$

It is noted that the coefficient of inertia C_M for a circular cylinder is unity. The moment about the base of the pile, resulting from the normal force F, is as follows:

$$M = - \int_{z=-h}^{\eta} (h + \eta)(\Delta F_D + \Delta F_I)$$

$$= - \frac{1}{4} \rho \left(\frac{Ag}{C} \right)^2 \frac{DC_D}{2\cosh^2(kh)} \left[\frac{(h + \eta)}{k} \sinh[2k(h + \eta)] + (h + \eta)^2 \right.$$

$$\left. + \left(\frac{1}{2k^2} \right) [1 - \cosh(2k(h + \eta))] \right] \cos(\sigma t)|\cos(\sigma t)|$$

$$+ \frac{1}{4} \rho \pi D^2 C_M \frac{Ag}{\cosh kh} \left[(h + \eta)\sinh[k(h + \eta)] \right.$$

$$\left. - \left(\frac{1}{k} \right) \{\cosh[k(h + \eta)] - 1\} \right] \sin \sigma t. \tag{8.16}$$

Thus the forces and the moments can be calculated from (8.15) and (8.16). The reader is referred to McCormick (1973) who has given a table of added-mass coefficients for certain structural components.

8.2.2 Drag and inertia

To see the relative importance of drag and inertia force components, we need to consider the value of the ratio $(\Delta F_I)_{max}/(\Delta F_D)_{max}$. Let us examine this ratio at $z = 0$:

$$\frac{(\Delta F_I)_{max}}{(\Delta F_D)_{max}} = \frac{\left(\rho C_M U \dfrac{\partial U}{\partial t}\right)_{max}}{\left(\frac{1}{2}\rho C_D A U |U|\right)_{max}}. \tag{8.17}$$

It is known from the small-amplitude wave theory presented in Chapter 4 that the maximum value of the inertia force occurs at the still water crossing where $\partial U/\partial t$ is a maximum, whereas the maximum drag force occurs at the wave crest.

Then we have

$$\left(\frac{\partial U}{\partial t}\right)_{max} = \frac{H}{2}\sigma^2 \coth kh \tag{8.18}$$

$$(U|U|)_{max} = (U^2)_{max} = \left(\frac{H}{2}\right)^2 \sigma^2 \coth^2 kh. \tag{8.19}$$

For the case of wave forces on a cylindrical pile, then we obtain from eqn (8.17) after using eqns (8.18) and (8.19)

$$\frac{(\Delta F_I)_{max}}{(\Delta F_D)_{max}} = \frac{C_M \pi D}{C_D H} \tanh kh$$

or

$$\left(\frac{H}{\pi D}\right) = \left(\frac{(\Delta F_D)_{max}}{(\Delta F_I)_{max}}\right)\left(\frac{C_M}{C_D}\right) \tanh kh. \tag{8.20}$$

To determine the relative importance of these two forces, we first plot the critical curve for which the two forces are equal.

We thus obtain that

$$\frac{H}{\pi D} = \frac{C_M}{C_D} \tanh kh. \tag{8.21}$$

The plot of this curve is shown in Fig. 8.2.

From Fig. 8.2, it is clear that the influence of the drag force is more than that of the inertia force above the critical curve and less below this critical

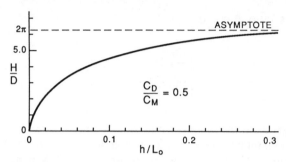

Fig. 8.2. $(H/\pi D)$ against kh for condition of equal maximum drag and inertia force components.

curve. It is to be noted that in shallow water ($kh \to 0$), the drag force tends to predominate the inertia force.

On the other hand, we can consider the following approach to see the relative importance of drag and inertia force components.

Consider that the total force is expressed in terms of a simple harmonic velocity given by

$$U = V_m \cos \sigma t. \tag{8.22}$$

Then the ratio of maximum inertia to drag force components is

$$\frac{(\Delta F_I)_{max}}{(\Delta F_D)_{max}} = \frac{C_M \pi D \sigma}{2 C_D U_M} = \pi^2 \left(\frac{C_M}{C_D}\right)\frac{1}{U_M T/D} = \pi^2 \frac{C_M}{C_D}\cdot\frac{1}{KC}. \tag{8.23}$$

Also this can be rewritten as

$$\frac{(\Delta F_I)_{max}}{(\Delta F_D)_{max}} = \frac{C_M \pi D}{2 C_D U_m/\sigma} = \frac{1}{2}\left(\frac{C_M}{C_D}\right)\cdot\left(\frac{\pi D}{L}\right). \tag{8.24}$$

Here KC is the Keulegan–Carpenter number and $\pi D/L$ is the diffraction parameter as defined previously.

Equation (8.23) suggests that for small and large values of Keulegan–Carpenter number (KC), the inertia and drag force components dominate, respectively. Also, eqn (8.24) confirms that for small and large values of diffraction parameter ($\pi D/L$), the drag and inertia force components dominate, respectively. Following Keulegan and Carpenter (1958), Figs 8.3 and 8.4 are redrawn for drag and inertia coefficients against the Keulegan–Carpenter number (KC) for forces measured at the node of a standing wave system.

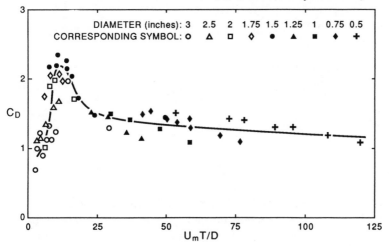

Fig. 8.3. Variation of C_D against KC (Keulegan and Carpenter (1958)).

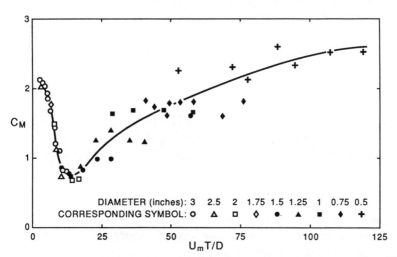

Fig. 8.4. Variation of inertia coefficients (C_M) with Keulegan–Carpenter number (KC) (Keulegan and Carpenter (1958)).

8.3 Froude–Krylov theory

The approximate wave forces on the structure can be calculated by a pressure–area method with the assumption that the structure is not there as far as the waves are concerned. This method is usually called the Froude-Krylov theory which has very limited application. The general formulae for the wave force calculations on a submerged object can be defined as follows:

$$F_x = C_H \iint_s pn_x \, ds \qquad (8.25)$$

and

$$F_z = C_V \iint_s pn_z \, ds \qquad (8.26)$$

in which C_H = the horizontal force coefficient, C_V = the vertical force coefficient, n_x, n_z = direction normals in the x- and y-directions respectively, ds = an elemental surface area of the submerged structure and F_x, F_z = the horizontal and vertical force components, respectively.

Here p represents the dynamic wave pressure based on the linear theory and is given by

$$p = \rho g \, \frac{H}{2} \, \frac{\cosh k(z + z_0)}{\cosh kh} \cos(kx - \sigma t). \qquad (8.27)$$

It is to be noted here that the coefficients C_H and C_V are the assigned values in accordance with the experimental data.

A brief discussion of the derivation of the Froude–Krylov forces on (a) a horizontal cylinder, (b) a vertical cylinder, and (c) a sphere is given below.

8.3.1 Horizontal cylinder

Consider a circular cylinder of length l and radius a being submerged horizontally such that a progressive wave is incident perpendicular to the axis of the cylinder a shown in Fig. 8.5.

A cylindrical polar coordinate system is used where the elevation of the axis of the cylinder is given by z_0 from the ocean floor. Thus the

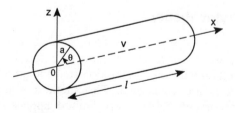

Fig. 8.5. A horizontal cylinder in waves.

coordinates (x, z) are given as

$$x = a \cos \theta \tag{8.28}$$

$$z = a \sin \theta \tag{8.29}$$

while the elemental area, ds, on the surface of the cylinder of length l is

$$ds = l(a\, d\theta) \tag{8.30}$$

where θ is the angle to the x-axis in the vertical plane.

The horizontal and vertical forces respectively can be expressed as

$$F_x = C_H \frac{\rho g H a l}{2 \cosh kh} \int_0^{2\pi} \cosh k(a \sin \theta + z_0)\cos(ka \cos \theta - \sigma t)\cos \theta\, d\theta \tag{8.31}$$

$$F_z = C_V \frac{\rho g H a l}{2 \cosh kh} \int_0^{2\pi} \cosh k(a \sin \theta + z_0)\cos(ka \cos \theta - \sigma t)\sin \theta\, d\theta. \tag{8.32}$$

After performing these integrals, we obtain

$$F_x = C_H \frac{\pi \rho g H k a^2 l}{2 \cosh kh} \cosh kz_0 \sin \sigma t \tag{8.33}$$

$$F_z = C_V \frac{\pi \rho g H k a^2 l}{2 \cosh kh} \sinh kz_0 \cos \sigma t. \tag{8.34}$$

8.3.2 Vertical cylinder

In this case of the vertical cylinder, only the horizontal component of the force on the curved surface will be present. Using the cylindrical polar coordinate (Fig. 8.6), the horizontal coordinate is

$$x = a \cos \theta \tag{8.35}$$

and

$$ds = (a\, d\theta)dz. \tag{8.36}$$

Assuming the length of the cylinder to be l and the vertical coordinate of the centre of the cylinder to be z_0 from the ocean floor, the horizontal

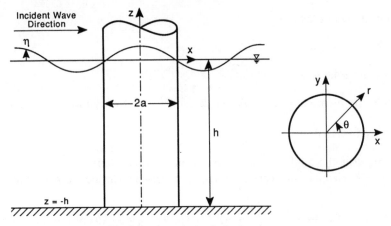

Fig. 8.6. A vertical cylinder in waves.

force is given by

$$F_x = C_H \frac{\rho g H a}{2\cosh kh} \int_{-l/2}^{l/2} \cosh k(z + z_0)\mathrm{d}z$$

$$\times \int_0^{2\pi} \cos(ka\cos\theta - \sigma t)\cos\theta\,\mathrm{d}\theta. \tag{8.37}$$

After performing the integration, we obtain

$$F_x = C_H \frac{\pi\rho g H a}{k\cosh kh}\left[\sinh k\left(z_0 + \frac{l}{2}\right) - \sinh k\left(z_0 - \frac{l}{2}\right)\right]J_1(ka)\sin\sigma t$$

$$\tag{8.38}$$

where $J_1(ka)$ is a Bessel function of the first kind and of order 1.

8.3.3 Sphere

Consider a fixed and fully submerged sphere which is in waves with its centre at an elevation of z_0 from the ocean floor as shown in Fig. 8.7.

With reference to Fig. 8.7, the relationship between the Cartesian and the spherical polar coordinates is

$$x = a\sin\theta\cos\alpha \tag{8.39}$$

$$z = a\cos\theta \tag{8.40}$$

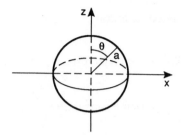

Fig. 8.7. A fully submerged sphere in waves.

in which a = radius of the sphere, θ = angle to z-axis, and α = angle on horizontal plane.

The horizontal and vertical forces in the x- and z-directions are respectively

$$F_x = C_H \iint_S (p \sin \theta \cos \alpha)(\mathrm{d}s) \tag{8.41}$$

$$F_z = C_V \iint_S (p \cos \theta)(\mathrm{d}s) \tag{8.42}$$

where $\mathrm{d}s$ is the elemental surface area and is given by

$$\mathrm{d}s = (a\,\mathrm{d}\theta)(a \sin \theta\,\mathrm{d}\alpha) = a^2 \sin \theta\,\mathrm{d}\theta\,\mathrm{d}\alpha. \tag{8.43}$$

Substituting the values of p and $\mathrm{d}s$ in terms of spherical polar coordinates from eqns (8.27) and (8.43) and using the limits of $\theta = 0$ to π and $\alpha = 0$ to 2π, the following expressions are obtained:

$$F_x = C_H \frac{\rho g H a^2}{2 \cosh kh} \int_0^\pi \cosh k(a \cos \theta + z_0)\sin^2 \theta$$

$$\times \left(\int_0^{2\pi} \cos(ka \sin \theta \cos \alpha - \sigma t)\cos \alpha\,\mathrm{d}\alpha \right)\mathrm{d}\theta \tag{8.44}$$

$$F_z = C_V \frac{\rho g H a^2}{2 \cosh kh} \int_0^\pi \cosh k(a \cos \theta + z_0)\sin \theta \cos \theta$$

$$\times \left(\int_0^{2\pi} \cos(ka \sin \theta \cos \alpha - \sigma t)\mathrm{d}\alpha \right)\mathrm{d}\theta. \tag{8.45}$$

250 Wave forces on offshore structures

The integrals are evaluated as follows:

$$\int_0^{2\pi} \cos(ka \sin\theta \cos\alpha - \sigma t)\cos\alpha \, d\alpha = 2\pi J_1(ka \sin\theta)\sin\sigma t \quad (8.46)$$

$$\int_0^{\pi} \cosh(ka \cos\theta)J_1(ka \sin\theta)\sin^2\theta \, d\theta = \frac{2}{3}ka \quad (8.47)$$

$$\int_0^{2\pi} \cos(ka \sin\theta \cos\alpha - \sigma t)d\alpha = 2\pi J_0(ka \sin\theta)\cos\sigma t \quad (8.48)$$

$$\int_0^{\pi} \sinh(ks \cos\theta)J_0(ka \sin\theta)\sin\theta \cos\theta \, d\theta = \frac{2}{3}ka \quad (8.49)$$

where the other integrals are zero. Here J_0 and J_1 are Bessel functions of zeroth and first order and of the first kind.

Using these values, we obtain from eqns (8.44) and (8.45),

$$F_x = C_H \frac{2\pi\rho gHka^3}{3\cosh kh}\cosh kz_0 \sin\sigma t \quad (8.50)$$

$$F_z = C_V \frac{2\pi\rho gHka^3}{3\cosh kh}\sinh kz_0 \cos\sigma t. \quad (8.51)$$

The force coefficients, C_H and C_V (see Chakrabarti (1987), which are used in the formula for the evaluation of wave forces on the structures, are in general functions of diffraction parameter ka. The Froude–Krylov theory fails to yield good results when ka values are large. In these cases, the diffraction theory must be used.

8.4 Diffraction theory

It has already been stated that if the structure is large relative to the wavelength, flow separation will not occur and consequently the drag force is negligible. In this case, the Morison equation is no longer applicable, and the flow field should be treated by the classical methods of potential flow. In particular, if the structure spans a significant portion of a wavelength, the incoming wave incident upon the surface of the structure undergoes significant scattering or diffraction. In this situation, the diffraction of waves must be taken into considertion in the wave load (force and moment) calculations.

We have considered two-dimensional flow in Chapter 4 and 5. In this section, the basic flow is considered to be three dimensional, oscillatory, irrotational and the fluid is incompressible. We have already seen in Chapter 2 that if the flow is irrotational, there exists a scalar potential, ϕ, the gradient of which may be defined as the fluid velocity. This potential function satisfies Laplace's equation which is in a rectangular Cartesian coordinate system $Oxyz$ as

$$\nabla^2\phi = \frac{\partial^2\phi}{\partial x^2} + \frac{\partial^2\phi}{\partial y^2} + \frac{\partial^2\phi}{\partial z^2} = 0 \tag{8.52}$$

where $\phi = (x, y, z, t)$ is the velocity potential and x, y, z are the coordinates and t is the time.

Figure 8.8 depicts the three-dimensional visualization of the flow situation with the boundaries of the problem.

The free surface of the wave is governed by two boundary conditions: the dynamic free-surface boundary condition and the kinematic free-surface boundary condition. The dynamic condition derives from the Bernoulli equation (eqn (2.36)) on the assumption that atmospheric pressure outside the fluid is constant:

$$-\frac{\partial\phi}{\partial t} + g\eta + \frac{1}{2}\left[\left(\frac{\partial\phi}{\partial x}\right)^2 + \left(\frac{\partial\phi}{\partial y}\right)^2 + \left(\frac{\partial\phi}{\partial z}\right)^2\right] = 0 \qquad \text{at} \qquad z = \eta. \tag{8.53}$$

The kinematic free-surface boundary condition states that a particle lying on the free surface at any instant of time will continue to remain on the free surface.

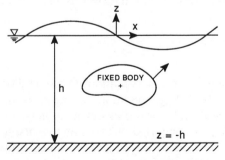

Fig. 8.8. Definition of the diffraction problem.

Mathematically,

$$w = \frac{d\eta}{dt} = \frac{\partial \eta}{\partial t} + \frac{\partial \eta}{\partial x} \frac{dx}{dt} + \frac{\partial \eta}{\partial y} \frac{dy}{dt}$$

or

$$-\frac{\partial \phi}{\partial z} = \frac{\partial \eta}{\partial t} + u \frac{\partial \eta}{\partial x} + v \frac{\partial \eta}{\partial y} = w.$$

This can be subsequently expressed as

$$-\frac{\partial \phi}{\partial z} = \frac{\partial \eta}{\partial t} - \frac{\partial \phi}{\partial x} \frac{\partial \eta}{\partial x} - \frac{\partial \phi}{\partial y} \frac{\partial \eta}{\partial y} \qquad \text{at} \qquad z = \eta \qquad (8.54)$$

where the velocity components are given by $\mathbf{V} = -\nabla \phi$, i.e.

$$u = -\frac{\partial \phi}{\partial x}, \qquad v = -\frac{\partial \phi}{\partial y} \qquad \text{and} \qquad w = -\frac{\partial \phi}{\partial z}. \qquad (8.55)$$

Assuming that the ocean floor is flat and horizontal, the boundary condition at the bottom states that the vertical velocity of the wave particle must be equal to zero:

$$\frac{\partial \phi}{\partial z} = 0 \qquad \text{at} \qquad z = -h. \qquad (8.56)$$

Next, the body-surface boundary condition states that the normal component of water particle velocity in contact with a boundary surface is equal to the velocity of the surface at that point.

If the surface is at rest the normal velocity must be zero:

$$\frac{\partial \phi}{\partial n} = 0 \qquad -h \le z \le \eta. \qquad (8.57)$$

In addition to the boundary conditions (8.53), (8.54), (8.56) and (8.57), we need another boundary condition far away from the structure to confirm that the scattered wave dies out at infinity. In essence the total velocity potential ϕ is composed of two components, namely ϕ_I which is the potential due to incident waves, and ϕ_s which is the potential due to waves scattered from the surface of the structure.

The surface boundary condition (8.57) can be simply written as

$$\frac{\partial \phi_I}{\partial n} = -\frac{\partial \phi_s}{\partial n}, \qquad -h \le z \le \eta. \tag{8.58}$$

Next, we define the far-field condition which is known as the Sommerfeld radiation condition.

We know the total potential is

$$\phi = \phi_I + \phi_s. \tag{8.59}$$

If we write

$$\phi_I = \text{Re}\{\Phi_I e^{-i\sigma t}\} \tag{8.60}$$

$$\phi_s = \text{Re}\{\Phi_s e^{-i\sigma t}\} \tag{8.61}$$

where Re stands for the real part of $\{\cdot\}$, and Φ_I and Φ_s are the complex incident and scattered potentials, respectively. Then from eqn (8.59) we have

$$\phi = \text{Re}\{(\Phi_I + \Phi_s)e^{-i\sigma t}\} = \text{Re}\{\Phi e^{-i\sigma t}\} \tag{8.62}$$

where

$$\Phi = \Phi_I + \Phi_s \tag{8.63}$$

is the total complex potential.

The Sommerfeld radiation condition can be stated as

$$\lim_{R \to \infty} \sqrt{R} \left(\frac{\partial}{\partial R} \pm i\lambda \right) \Phi_s = 0 \tag{8.64}$$

where λ = eigenvalue, $i = \sqrt{-1}$, and R = the radial distance.

In principle, once the total velocity potential is known and the pressures on the surface of the structure are known from Bernoulli's equation, the total force exerted on the body $\mathbf{F}$ and the moment of this force $\mathbf{M}$ in vector notation may be obtained from the following formulae:

$$\mathbf{F}(t) = -\iint_s p\mathbf{n}\, ds \tag{8.65}$$

$$\mathbf{M}(t) = -\iint_s p(\mathbf{r} \times \mathbf{n})\, ds \tag{8.66}$$

where $\mathbf{n}$ is the outward normal vector on s, p is the pressure and $\mathbf{r}$ is the vector from the point about which moments are taken.

The problem defined above is the complete boundary value problem which is highly nonlinear because of the nonlinear free-surface boundary conditions. In the following we shall discuss a linear diffraction theory mainly due to MacCamy and Fuchs (1954) applied to a large, vertical cylinder in waves.

8.4.1 Diffraction theory of MacCamy and Fuchs

Consider a fixed, vertical, circular surface-piercing cylinder which extends from the ocean floor as shown in Fig. 8.9.

This problem was initially solved by Havelock (1940) for the deep water case by using diffraction theory and then extended by MacCamy and Fuchs (1954) for the general intermediate depth case. The coordinates are chosen such that x is positive in the direction of wave propagation and z is positive in the upward direction. The origin is chosen at the centre of the cylinder at the still water level (SWL), corresponding to the water elevation, $\eta = A \sin(kx - \sigma t)$. The velocity potential for incident waves can be written as

$$\phi_I = \frac{gA}{\sigma} \frac{\cosh k(z+h)}{\cosh kh} \cos(kx - \sigma t). \tag{8.67}$$

Equation (8.67) can be expressed as

$$\phi_I = \mathrm{Re}\left\{ \frac{gA}{\sigma} \frac{\cosh k(z+h)}{\cosh kh} e^{i(kx - \sigma t)} \right\} = \mathrm{Re}\{\Phi_I e^{-i\sigma t}\}$$

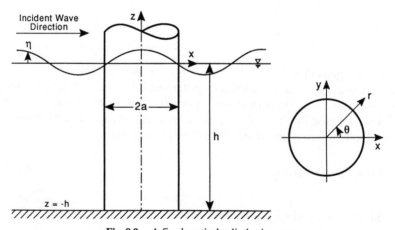

Fig. 8.9. A fixed vertical cylinder in waves.

where

$$\Phi_I = \frac{gA}{\sigma} \frac{\cosh k(z+h)}{\cosh kh} e^{ikx}. \tag{8.68}$$

If the cylindrical polar coordinates r and θ are used then

$$e^{ikx} = e^{ikr\cos\theta} = \sum_{m=0}^{\infty} \varepsilon_m i^m J_m(kr)\cos m\theta \tag{8.69}$$

where $J_m(kr)$ are Bessel functions of the first kind of orders $m = 0, 1, 2, \ldots, \varepsilon_0 = 1$ and $\varepsilon_m = 2$ for $m \geq 1$. For further reference see Wylie and Barrett (1982). Substituting eqn (8.69) into (8.68)

$$\Phi_I = \frac{gA}{\sigma} \frac{\cosh k(z+h)}{\cosh kh} \sum_{m=0}^{\infty} \varepsilon_m i^m J_m(kr)\cos m\theta. \tag{8.70}$$

We assume that the reflected wave which admits of a similar expression satisfying Laplace's equation and the boundary conditions and which radiates away may be written in the following symmetric form (about θ):

$$\Phi_S = \frac{gA}{\sigma} \frac{\cosh k(z+h)}{\cosh kh} \sum_{m=0}^{\infty} A_m H_m^{(1)}(kr)\cos m\theta \tag{8.71}$$

where the A_m are constants and $H_m^{(1)}(kr) = J_m(kr) + iY_m(kr)$ is a Hankel function of the first kind of order m, and Y_m is a Bessel function of the second kind.

The asymptotic form of a Hankel function of the first kind as r goes to infinity satisfies the radiation condition

$$\lim_{r \to \infty} \sqrt{r} \left(\frac{\partial}{\partial r} - ik \right) H_m^{(1)}(kr) = 0. \tag{8.72}$$

It is to be noted here that the asymptotic form of a Hankel function of the second kind, $H_m^{(2)}(kr) = J_m(kr) - iY_m(kr)$, satisfies the radiation condition

$$\lim_{r \to \infty} \sqrt{r} \left(\frac{\partial}{\partial r} + ik \right) H_m^{(2)}(kr) = 0. \tag{8.73}$$

256 Wave forces on offshore structures

The asymptotic form of $H_m^{(1)}(kr)$ as $r \to \infty$ is given by

$$H_m^{(1)}(kr) \simeq \left(\frac{2}{\pi kr}\right)^{1/2} \exp\left[i\left(kr - \frac{2m+1}{4}\pi\right)\right].$$
(8.74)

The total potential is the sum of the incident and scattered waves and is given by

$$\Phi = \frac{gA}{\sigma} \frac{\cosh k(z+h)}{\cosh kh} \sum_{m=0}^{\infty} (\varepsilon_m i^m J_m(kr) + A_m H_m^{(1)}(kr))\cos m\theta.$$
(8.75)

Using the body-surface boundary condition $\partial \Phi/\partial r = 0$ at $r = a$, we obtain

$$A_m = -\varepsilon_m i^m \frac{J_m'(ka)}{H_m^{(1)'}(ka)}$$
(8.76)

where the prime denotes differentiation with respect to the argument kr. The total potential is then given by

$$\Phi = \frac{gA}{\sigma} \frac{\cosh k(z+h)}{\cosh kh}$$

$$\times \sum_{m=0}^{\infty} \varepsilon_m i^m \left(J_m(kr) - \frac{J_m'(ka)}{H_m^{(1)'}(ka)} H_m^{(1)}(kr)\right)\cos m\theta.$$
(8.77)

Once the total velocity potential has been determined, all the quantities of physical interest may be obtained directly.

The water surface elevation η and dynamic pressure at the surface of the cylinder are calculated using the total velocity potential and the linearized Bernoulli equation

$$p = \rho \frac{\partial \phi}{\partial t} - \rho g z$$
(8.78)

where p is the fluid pressure and ρ is the density of the fluid. At the water surface $p = 0$ and $z = \eta$ and so from (8.78)

$$\eta = \frac{1}{g}\left(\frac{\partial \phi}{\partial t}\right)_{z=0} = \frac{1}{g}\frac{\partial}{\partial t} \text{Re}\{\Phi e^{-i\sigma t}\}|_{z=0}$$

or

$$\eta = -\frac{\sigma}{g} \operatorname{Re}\{i\Phi e^{-i\sigma t}\}|_{z=0}.$$ (8.79)

Now define

$$\eta = \operatorname{Re}\{\hat{\eta} e^{-i\sigma t}\}$$ (8.80)

where $\hat{\eta}$ is a complex water elevation.

Thus from eqns (8.79) and (8.80)

$$\hat{\eta} = -\frac{i\sigma}{g} \Phi|_{z=0}$$

$$= -A \sum_{m=0}^{\infty} \varepsilon_m i^{m+1} \left(J_m(kr) - \frac{J'_m(kr)}{H_m^{(1)'}(ka)} H_m^{(1)}(kr) \right) \cos m\theta.$$ (8.81)

Similarly, the dynamic pressure after neglecting the hydrostatic pressure term is

$$p = \rho \frac{\partial \phi}{\partial t} = \rho \frac{\partial}{\partial t} \operatorname{Re}\{\Phi e^{-i\sigma t}\}.$$

or

$$p = -\rho\sigma \operatorname{Re}\{i\Phi e^{-i\sigma t}\}.$$ (8.82)

Define now

$$p = \operatorname{Re}\{P e^{-i\sigma t}\}$$ (8.83)

where P is a complex dynamic pressure.

Then, evaluating P at the surface of the cylinder $r = a$,

$$P = -\rho\sigma i\Phi|_{r=a} = -\rho g A \frac{\cosh k(z+h)}{\cosh kh} \sum_{m=0}^{\infty} \frac{\varepsilon_m i^{m+1}}{H_m^{(1)'}(ka)}$$

$$\times \left[J_m(ka) H_m^{(1)'}(ka) - J'_m(ka) H_m^{(1)}(ka) \right] \cos m\theta.$$

(8.84)

Using Bessel's identity

$$J_m(ka)H_m^{(1)'}(ka) - J_m'(ka)H_m^{(1)}(ka) = \frac{2i}{\pi ka}, \tag{8.85}$$

eqn (8.84) may be written as

$$P = \frac{2\rho gA}{\pi ka} \frac{\cosh k(z+h)}{\cosh kh} \sum_{m=0}^{\infty} \frac{\varepsilon_m i^m}{H_m^{(1)'}(ka)} \cos m\theta. \tag{8.86}$$

The force in the horizontal x-direction per unit length of the cylinder is given by

$$f_x = \int pn_x \, ds = \int_{\theta=0}^{2\pi} \rho \frac{\partial \phi}{\partial t}(\cos \theta)(a \, d\theta)$$

$$= \frac{2\rho gA}{\pi k} \frac{\cosh k(z+h)}{\cosh kh}$$

$$\times \mathrm{Re}\left\{\int_0^{2\pi} \sum_{m=0}^{\infty} \frac{\varepsilon_m i^m}{H_m^{(1)'}(ka)} \cos m\theta \cos \theta \, e^{-i\sigma t} \, d\theta\right\}. \tag{8.87}$$

Owing to the orthogonality of $\{\cos m\theta\}$ in the range $0 \le \theta \le 2\pi$, i.e.

$$\int_0^{2\pi} \cos m\theta \cos n\theta \, d\theta = 0, \qquad m \neq n,$$

$$\int_0^{2\pi} \cos m\theta \cos n\theta \, d\theta \neq 0, \qquad m = n$$

we have only to consider one term in the series corresponding to $m = 1$ such that $\int_0^{2\pi} \cos^2\theta \, d\theta = \pi$. Therefore,

$$f_x = \frac{4\rho gA}{k}\left(\frac{\cosh k(z+h)}{\cosh kh}\right)\mathrm{Re}\left\{\frac{ie^{-i\sigma t}}{H_1^{(1)'}(ka)}\right\}$$

$$= \frac{4\rho gA}{k}\left(\frac{\cosh k(z+h)}{\cosh kh}\right)B(ka)\cos(\sigma t - \alpha) \tag{8.88}$$

where

$$B(ka) = \left[J_1'^2(ka) + Y_1'^2(ka)\right]^{-1/2} \tag{8.89}$$

and

$$\alpha = \tan^{-1}\left(\frac{J_1'(ka)}{Y_1'(ka)}\right). \tag{8.90}$$

The horizontal force per unit length may equivalently, be, written as the inertia part of the Morison equation

$$f_x = C_{M\rho}\pi a^2 \dot{U}_\alpha \tag{8.91}$$

in which

$$\dot{U}_\alpha = gkA \frac{\cosh k(z+h)}{\cosh kh} \cos(\sigma t - \alpha),$$

the water particle acceleration at an elevation $z + h$ from the bottom at a phase lag of α, and C_M is the effective inertia coefficient given by

$$C_M = \frac{4}{\pi(ka)^2\sqrt{J_1'^2(ka) + Y_1'^2(ka)}}. \tag{8.92}$$

8.5 Nonlinear diffraction theory: perturbation method

In the previous sections we discussed a linear diffraction theory, first formulated by MacCamy and Fuchs (1954), to calculate wave loads on a vertical circular cylinder. The cylinder extends from well above the water surface to the bottom of the sea. More recently a number of investigators including Mogridge and Jamieson (1976) have used this theory to obtain the wave loads on large submerged cylinders in the sea. In all these problems, the analyses were restricted to linear theory only, which has a very limited application in a practical field due to the character of the ocean waves often being nonlinear. To improve the correlation between experimental data and theory, it is necessary to include nonlinearity in the theory.

Taking into consideration this motivation, many investigators have directed their attention towards nonlinear theory in water waves. Chakrabarti (1975) obtained an expression for wave forces on a cylinder using Stokes' fifth-order theory, but the kinematic boundary condition was not satisfied in the vicinity of the cylinder. The results obtained were compared with the second-order theory of Yamaguchi and Tsuchiya (1974). Chakrabarti (1978) concluded that his theory and experimental data showed good agreement with their experimental data.

Raman and Venkatanarasaiah (1976) obtained a nonlinear solution of water waves using a perturbation technique. Unfortunately, their experimental results were obtained using a relatively small-diameter cylinder, for which the previous workers had assumed the viscous drag forces to be significant.

This section deals with the theory of second-order expressions to predict the wave forces on large offshore structures. The theory has been applied to surface-piercing circular caissons in waves. Following Stoker (1957), therefore, a mathematical model has been developed for the nonlinear wave structures which will correlate with nature. A perturbation technique has been used to solve the nonlinear diffraction problem. For the case of fixed offshore structures, closed-form analytical solutions have been obtained using second-order theory and the predictions compared with available experimental data.

8.5.1 Mathematical formulations

A rigid vertical cylinder of radius a is acted upon by a train of regular surface waves of height H, progressing in the positive x-direction, as shown in Fig. 8.10. It is assumed that the fluid is incompressible and the motion irrotational. When the amplitude is large, the small-amplitude theory does not hold good. In practice, the finite-amplitude wave theory, namely nonlinear wave theory, is of primary importance. In linear wave theory, the wave amplitudes to the second and higher orders are considered negligible, whereas in finite-amplitude wave theory these higher-order

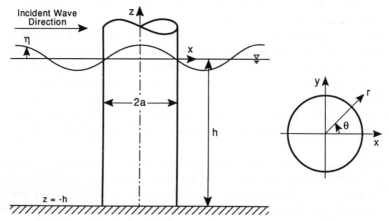

Fig. 8.10. Definition sketch of wave forces on a vertical circular cylinder.

terms are retained so as to give an accurate representation of the wave motion.

The motion is governed by the following equation:

$$\nabla^2 \phi = \frac{\partial^2 \phi}{\partial r^2} + \frac{1}{r}\frac{\partial \phi}{\partial r} + \frac{1}{r^2}\frac{\partial^2 \phi}{\partial \theta^2} + \frac{\partial^2 \phi}{\partial z^2} = 0 \tag{8.93}$$

in the region $a \leq r \leq \infty$; $-h \leq z \leq \eta$; $-\pi \leq \theta \leq \pi$. Equation (8.93) is Laplace's equation in cylindrical form and is derived from the fluid continuity equation. The boundary conditions associated with (8.93) are as follows:

Dynamic boundary condition:

$$\frac{\partial \phi}{\partial t} + g\eta + \frac{1}{2}\left[\left(\frac{\partial \phi}{\partial r}\right)^2 + \left(\frac{1}{r}\frac{\partial \phi}{\partial \theta}\right)^2 + \left(\frac{\partial \phi}{\partial z}\right)^2\right] = 0; \quad z = \eta, \quad r \geq a.$$
$$\tag{8.94}$$

Kinematic boundary condition:

$$\frac{\partial \eta}{\partial t} + \left(\frac{\partial \phi}{\partial r}\right)\left(\frac{\partial \eta}{\partial r}\right) + \frac{1}{r^2}\left(\frac{\partial \phi}{\partial \theta}\right)\left(\frac{\partial \eta}{\partial \theta}\right) = \frac{\partial \phi}{\partial z} \quad \text{on} \quad z = \eta; \quad r \geq a.$$
$$\tag{8.95}$$

Boundary condition on the body surface:

$$\frac{\partial \phi}{\partial r} = 0 \quad \text{on} \quad r = a; \quad -h \leq z \leq \eta. \tag{8.96}$$

Bottom boundary condition:

$$\frac{\partial \phi}{\partial z} = 0 \quad \text{on} \quad z = -h. \tag{8.97}$$

The additional boundary condition at infinity called the radiation boundary condition must be satisfied by the scattered potential Φ_s, were $\phi = \text{Re}\{(\Phi_I + \Phi_s)e^{-i\sigma t}\}$ in which Φ_I and Φ_s are defined to be the complex incident and complex scattered potentials respectively. This boundary condition states that the diffracted wave must vanish at infinity and mathematically can be defined by the Orr–Sommerfeld condition

$$\lim_{r \to \infty} \sqrt{r}\left(\frac{\partial}{\partial r} \pm ik\right)\Phi_s = 0 \tag{8.98}$$

where $\sqrt{-1} = i$.

The velocity components in cylindrical coordinates are $\mathbf{V} = \nabla\phi$, i.e.

$$v_r = \frac{\partial\phi}{\partial r}, \qquad v_\theta = \frac{1}{r}\frac{\partial\phi}{\partial\theta} \qquad \text{and} \qquad v_z = \frac{\partial\phi}{\partial z},$$

where ϕ is the total velocity potential, η is the height of the free surface, h is the depth of water below the still water level, v_r, v_θ and v_z are the velocity components, g is the acceleration due to gravity, k is the wave number $(= 2\pi/L)$, x, y, z are rectangular coordinates, r, θ, z are cylindrical coordinates, t is the time variable and L is the wavelength. The Cartesian form of (8.93) is

$$\nabla^2\phi = \frac{\partial^2\phi}{\partial x^2} + \frac{\partial^2\phi}{\partial y^2} + \frac{\partial^2\phi}{\partial z^2} = 0, \tag{8.99}$$

where $x = r\cos\theta$, $y = r\sin\theta$, $z = z$.

The procedure is to work with the complete form of the equations to assume that the solution can be represented in terms of a power series expansion of the parameter ε where $\varepsilon = kH/2$. Thus, expanding ϕ, η and p as a series in powers of ε yields

$$\phi = \sum_{n=1}^{\infty} \varepsilon^n \phi_n, \tag{8.100}$$

$$\eta = \sum_{n=1}^{\infty} \varepsilon^n \eta_n \tag{8.101}$$

$$p = \sum_{n=1}^{\infty} \varepsilon^n p_n. \tag{8.102}$$

The sum of the terms up to index n represents the nth-order theory for any particular quantity. p represents the pressure field, and ϕ, η and p are all functions of x, y, z and t. At the free surface, $z = \eta(x,y,t)$,

$$\phi(x,y,z,t) = \phi[x,y,\eta(x,y,t),t]. \tag{8.103}$$

Expanding (8.103) by Taylor's theorem about $z = 0$, we have

$$\phi(x,y,\eta,t) = \phi(x,y,0,t) + \eta\left(\frac{\partial \phi}{\partial z}\right)_{z=0} + \dots \quad (8.104)$$

Using (8.99) and (8.100), we have

$$\phi(x,y,\eta,t) = \varepsilon\phi_1(x,y,0,t) + \varepsilon^2\phi_2(x,y,0,t) + \dots$$

$$+ (\varepsilon\eta_1 + \varepsilon^2\eta_2)\left(\frac{\partial}{\partial z}(\varepsilon\phi_1 + \varepsilon^2\phi_2 + \dots)\right)_{z=0} + \dots$$

$$= \varepsilon\phi_1(x,y,0,t) + \varepsilon^2\left[\phi_2 + \eta_1\left(\frac{\partial \phi_1}{\partial z}\right)_{z=0}\right] + O(\varepsilon^3).$$

The modified velocity potential at the free surface is

$$\phi(x,y,\eta,t) = \varepsilon\phi_1(x,y,0,t) + \varepsilon^2\left[\phi_2(x,y,0,t)\right.$$

$$\left. + \eta_1\left(\frac{\partial \phi_1}{\partial z}\right)_{z=0}\right] + O(\varepsilon^3). \quad (8.105)$$

The differential equation which each of the velocity potential terms must satisfy is Laplace's equation, namely

$$\nabla^2\phi_n = \frac{\partial^2\phi_n}{\partial x^2} + \frac{\partial^2\phi_n}{\partial y^2} + \frac{\partial^2\phi_n}{\partial z^2} = 0. \quad (8.106)$$

This can be verified by substituting (8.100) into (8.99), collecting the coefficients of like powers of ε, and then equating to zero all coefficients of each power of ε.

The bottom boundary condition is

$$\frac{\partial \phi_n}{\partial z} = 0 \quad \text{on} \quad z = -h. \quad (8.107)$$

The boundary condition at the body surface is

$$\frac{\partial \phi_1}{\partial r} = 0, \quad (8.108)$$

$$\frac{\partial \phi_2}{\partial r} = 0. \quad (8.109)$$

Substituting the series into (8.94), the dynamic free-surface boundary condition is given by

$$\varepsilon\left(g\eta_1 + \frac{\partial\phi_1}{\partial t}\right) + \varepsilon^2\left\{g\eta_2 + \frac{\partial\phi_2}{\partial t} + \eta_1\frac{\partial^2\phi_1}{\partial t\partial z} + \frac{1}{2}\left[\left(\frac{\partial\phi_1}{\partial x}\right)^2 + \left(\frac{\partial\phi_1}{\partial y}\right)^2\right.\right.$$

$$\left.\left. + \left(\frac{\partial\phi_1}{\partial z}\right)^2\right]\right\} + O(\varepsilon^3) = 0 \qquad \text{on} \qquad z = 0. \tag{8.110}$$

The kinematic free-surface boundary condition is determined similarly from (8.95), and is

$$\varepsilon\left(\frac{\partial\eta_1}{\partial t} - \frac{\partial\phi_1}{\partial z}\right) + \varepsilon^2\left(\frac{\partial\eta_2}{\partial t} + \frac{\partial\phi_1}{\partial x}\frac{\partial\eta_1}{\partial x} + \frac{\partial\phi_1}{\partial y}\frac{\partial\phi_1}{\partial y}\right.$$

$$\left. - \frac{\partial\phi_2}{\partial z} + \eta_1\frac{\partial^2\phi_1}{\partial z^2}\right) + O(\varepsilon^3) = 0 \qquad \text{on} \qquad z = 0. \tag{8.111}$$

Thus, the first-order wave theory yields

$$\frac{\partial^2\phi_1}{\partial x^2} + \frac{\partial^2\phi_1}{\partial y^2} + \frac{\partial^2\phi_1}{\partial z^2} = 0, \tag{8.112}$$

together with the boundary conditions

$$g\eta_1 + \frac{\partial\phi_1}{\partial t} = 0 \qquad \text{on} \qquad z = 0, \tag{8.113}$$

$$\frac{\partial\eta_1}{\partial t} - \frac{\partial\phi_1}{\partial z} = 0 \qquad \text{on} \qquad z = 0, \tag{8.114}$$

$$\frac{\partial\phi_1}{\partial z} = 0 \qquad \text{on} \qquad z = -h, \tag{8.115}$$

$$\frac{\partial\phi_1}{\partial r} = 0 \qquad \text{on} \qquad r = a, \tag{8.116}$$

and

$$\lim_{r \to \infty} \sqrt{r} \left(\frac{\partial}{\partial r} \pm ik \right) \Phi_{s1} = 0 \tag{8.117}$$

where Φ_{s1} is the first-order complex scattered potential.

The second-order wave theory is

$$\frac{\partial^2 \phi_2}{\partial x^2} + \frac{\partial^2 \phi_2}{\partial y^2} + \frac{\partial^2 \phi_2}{\partial z^2} = 0, \tag{8.118}$$

with boundary conditions

$$g\eta_2 + \frac{\partial \phi_2}{\partial t} + \eta_1 \frac{\partial^2 \phi_1}{\partial t \partial z}$$
$$+ \frac{1}{2} \left[\left(\frac{\partial \phi_1}{\partial x} \right)^2 + \left(\frac{\partial \phi_1}{\partial y} \right)^2 + \left(\frac{\partial \phi_1}{\partial z} \right)^2 \right] = 0, \qquad z = 0, \tag{8.119}$$

$$\left(\frac{\partial \eta_2}{\partial t} + \frac{\partial \phi_1}{\partial x} \frac{\partial \eta_1}{\partial x} + \frac{\partial \phi_1}{\partial y} \frac{\partial \eta_1}{\partial y} - \frac{\partial \phi_2}{\partial z} + \eta_1 \frac{\partial^2 \phi_1}{\partial z^2} \right) = 0, \qquad z = 0 \tag{8.120}$$

$$\frac{\partial \phi_2}{\partial z} = 0 \qquad \text{on} \qquad z = -h \tag{8.121}$$

$$\frac{\partial \phi_2}{\partial r} = 0 \qquad \text{on} \qquad r = a, \tag{8.122}$$

and

$$\lim_{r \to \infty} \sqrt{r} \left(\frac{\partial}{\partial r} \pm i\lambda \right) \Phi_{s2} = 0 \tag{8.123}$$

where λ is an eigenvalue and Φ_{s2} is the second-order scattered complex potential.

Eliminating η_1 from (8.113) and (8.114), it is found that ϕ_1 satisfies the following equation:

$$\frac{\partial^2 \phi_1}{\partial t^2} + g \frac{\partial \phi_1}{\partial z} = 0 \qquad \text{on} \qquad z = 0. \tag{8.124}$$

Similarly, eliminating η_2 and η_1 from (8.119) and (6.120), and using the relations (6.113) and (6.114), we have

$$\frac{\partial^2 \phi_2}{\partial t^2} + g \frac{\partial \phi_2}{\partial z} = -\eta_1 \frac{\partial}{\partial z}\left(\frac{\partial^2 \phi_1}{\partial t^2} + g \frac{\partial \phi_1}{\partial z}\right) - \frac{\partial}{\partial t}\left[\left(\frac{\partial \phi_1}{\partial x}\right)^2\right.$$

$$\left. + \left(\frac{\partial \phi_1}{\partial y}\right)^2 + \left(\frac{\partial \phi_1}{\partial z}\right)^2\right], \qquad r \geq a. \tag{8.125}$$

The pressure $P(r, \theta, z, t)$ may be determined from Bernoulli's equation

$$\frac{P}{\rho} + gz + \frac{\partial \phi}{\partial t} + \frac{1}{2}\left[\left(\frac{\partial \phi}{\partial r}\right)^2 + \left(\frac{1}{r}\frac{\partial \phi}{\partial \theta}\right)^2 + \left(\frac{\partial \phi}{\partial z}\right)^2\right] = 0 \tag{8.126}$$

which, on substituting for ϕ as a series in powers of ε, leads to

$$P = -\rho gz - \varepsilon \rho \frac{\partial \phi_1}{\partial t} - \varepsilon^2 \rho \left[\frac{\partial \phi_2}{\partial t} + \frac{1}{2}\left(\left(\frac{\partial \phi_1}{\partial r}\right)^2\right.\right.$$

$$\left.\left. + \left(\frac{\partial \phi_1}{\partial z}\right)^2 + \frac{1}{r^2}\left(\frac{\partial \phi_1}{\partial \theta}\right)^2\right)\right] + O(\varepsilon^3). \tag{8.127}$$

The total horizontal force is

$$F_x = \int_0^{2\pi} \int_{-h}^{\eta} |P|_{r=a}(-a \cos \theta)\mathrm{d}z\, \mathrm{d}\theta, \tag{8.128}$$

where η is given by the perturbation expansion, (8.101).

Substituting (8.127) into (8.128) and writing the z-integral as the sum of $\int_{-h}^{0} + \int_{0}^{\eta}$, we obtain

$$F_x = a\rho \int_0^{2\pi} \left(\int_{-h}^0 \left(gz + \varepsilon \frac{\partial \phi_1}{\partial t} + \varepsilon^2 \left(\frac{\partial \phi_2}{\partial t} + \frac{1}{2} \left(\frac{\partial \phi_1}{\partial z} \right)^2 \right. \right. \right.$$
$$\left. \left. \left. + \frac{1}{2a^2} \left(\frac{\partial \phi_1}{\partial \theta} \right)^2 \right) \right) \right)_{r=a} dz + \int_0^{\varepsilon\eta_1 + \varepsilon^2 \eta_2} \left(gz + \varepsilon \frac{\partial \phi_1}{\partial t} \right.$$
$$\left. \left. + \varepsilon^2 \left(\frac{\partial \phi_2}{\partial t} + \frac{1}{2} \left(\frac{\partial \phi_1}{\partial z} \right)^2 + \frac{1}{2a^2} \left(\frac{\partial \phi_1}{\partial \theta} \right)^2 \right) \right) \right)_{r=a} dz \right) \cos \theta \, d\theta.$$

(8.129)

It is clear from (8.129) that the integral of gz, the hydrostatic term, up to $z = 0$ contains no $\cos \theta$ term and can be ignored. Also, the upper limit of the z-integral of the second-order terms may be taken at $z = 0$ instead of $z = \varepsilon\eta_1 + \varepsilon^2\eta_2$ since this would only introduce higher-order terms of ε^3 and higher. Thus, F_x may be written as

$$F_x = \varepsilon F_{x_1} + \varepsilon^2 F_{x_2},$$

(8.130)

where the first-order contribution is

$$\varepsilon F_{x_1} = a\rho \int_0^{2\pi} \left[\int_{-h}^0 \left(\varepsilon \frac{\partial \phi_1}{\partial t} \right)_{r=a} dz \right] \cos \theta \, d\theta$$

(8.131)

and the second-order contribution is

$$\varepsilon^2 F_{x_2} = a\rho \int_0^{2\pi} \left\{ \int_0^{\varepsilon\eta_1} \left(gz + \varepsilon \frac{\partial \phi_1}{\partial t} \right)_{r=a} dz \right.$$
$$\left. + \varepsilon^2 \int_{-h}^0 \left[\frac{\partial \phi_2}{\partial t} + \frac{1}{2} \left(\frac{\partial \phi_1}{\partial z} \right)^2 + \frac{1}{2a^2} \left(\frac{\partial \phi_1}{\partial \theta} \right)^2 \right]_{r=a} dz \right\} \cos \theta \, d\theta.$$

(8.132)

It is to be noted here that in the second-order force evaluation given in eqn (8.132), the first term on the right can be defined as the waterline force, the third term as the dynamic force and the second term as the quadratic force as stated by Lighthill (1979).

Using the results presented by Rahman and Heaps (1983), the non-dimensional forms of the first-order and second-order components of the total horizontal forces may now be written as

$$\frac{F_{x_1}}{\rho g D^3} = \frac{\tanh kh}{2(ka)^3} \cdot \frac{\cos(\sigma t - \alpha_1)}{|H_1^{(1)\prime}(ka)|} \tag{8.133}$$

and

$$\frac{F_{x_2}}{\rho g D^3} = \left(\frac{\tanh kh\, e^{-2i\sigma t}}{8(ka)} \int_0^\infty G(k_2)\,dk_2 + \text{complex conjugate} \right)$$

$$- \frac{1}{4\pi(ka)^4} \sum_{l=0}^{\infty} (-1)^l \left[\left(3 - \frac{2kh}{\sinh 2kh} \right) + \frac{l(l+1)}{a^2 k^2} \right.$$

$$\left. \times \left(1 + \frac{2kh}{\sinh 2kh} \right) \right] \times [C_l \cos 2\sigma t - S_l \sin 2\sigma t]$$

$$- \frac{1}{4\pi(ka)^4} \sum_{l=0}^{\infty} \left[\left(1 - \frac{l(l+1)}{a^2 k^2} \right) \left(1 + \frac{2kh}{\sinh 2kh} \right) E_l \right] \tag{8.134}$$

where

$$G(k_2 a) = \frac{\int_{ka}^\infty A_1(k_2 r) B_1(kr)(kr)\,d(kr)}{(k_2 a)\left[(k_2 a) - \dfrac{4ka \tanh kh}{\tanh k_2 h} \right] H_1^{(1)\prime}(k_2 a)}. \tag{8.135}$$

A second-order diffraction theory, developed by Rahman and Heaps (1983), has been presented and tested against the results of experimental data. Closed-form integral solutions satisfying all necessary hydrodynamic conditions were obtained by using the perturbation method.

In Fig. 8.11, both the first-order and second-order solutions (see Rahman (1988)) are compared with the force measurements of Chakrabarti (1975), which are generally seen to be closer to the second-order predictions. It is worth mentioning here that a complete second-order diffraction solution for an axisymmetric body using the Green function method was recently obtained by Kim and Yue (1989, 1990). The interested reader is referred to their work. In the following we briefly discuss the development of the Green function method from a mathematical viewpoint.

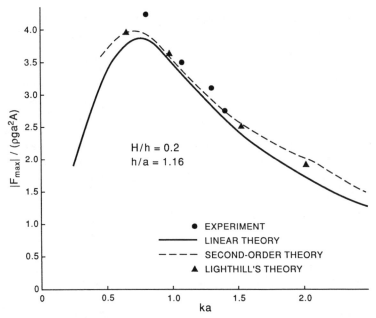

Fig. 8.11. Comparison of linear and second-order wave forces with experimental data of Chakrabarti (1975).

8.6 Green function method

In the preceding section an analytical treatment of the boundary value problems arising in physical situations has been presented. Many analytical techniques to find the exact closed-form solution of partial differential equations within simple geometrical configurations have been discussed. However, in many practical situations, a closed-form analytical solution is not always available when the geometrical configuration is not simple but complex. It is then necessary to resort to numerical methods in order to deal with the problems that cannot be solved analytically. With the advent of today's high-speed electronic computer, numerical methods are becoming increasingly popular and useful in practical applications.

This section is, therefore, devoted to the computer-implemented simulation model which includes the Green function method (boundary integral method). The Green function method is used to find the wave loads on arbitrary-shaped large off-shore structures.

To predict the wave loading on large offshore structures of arbitrary geometry, it is almost a formidable task to determine the exact closed-form solution. For such problems, numerical treatment is absolutely necessary. Garrison and his co-workers (1971, and 1972, 1977) Hogben and Standing

(1975) have utilized the boundary integral method known as the Green function method to evaluate the wave loading on arbitrary structures. In describing this approach the following fundamental result of potential theory is needed. The scattered velocity potential $\Phi_s(\phi_s = \mathrm{Re}\{\Phi_s e^{-i\sigma t}\})$ may be represented as due to a continuous distribution of point sources over the submerged body surface. If the potential of the fluid due to a point source of unit strength located at the point (ξ, η, ζ) is known, then owing to the linearity of the problem this may be applied to any required strength and then superimposed with any number of other wave sources. The velocity potential due to the entire continuous distribution of sources over the body surface is then given as (see Rahman (1988))

$$\Phi_s(x, y, z) = \frac{1}{4\pi} \iint\limits_s f(\xi, \eta, \zeta) G(x, y, z; \xi, \eta, \zeta)\,ds \qquad (8.136)$$

where $f(\xi, \eta, \zeta)$ represents a source strength located at the point (ξ, η, ζ) on the body surface, and ds is an elementary area on the submerged surface. The function G represents a Green-function for the general field point (x, y, z) within the fluid due to a source of unit strength at (ξ, η, ζ).

An expression of the Green function which satisfies Laplace's equation and the appropriate boundary conditions such as the free-surface, bottom, and radiation boundary conditions (except the body-surface boundary condition) is given by Wehausen and Laitone (1960):

$$G(x, y, z; \xi, \eta, \zeta) = \frac{1}{R} + \frac{1}{R_1}$$

$$+ 2PV \int_0^\infty \frac{(\mu + \nu)e^{-\mu h}\cosh \mu(z + h)\cosh \mu(\zeta + h)}{\sinh \mu h - \nu \cosh \mu h} J_0(\mu r)\,d\mu$$

$$+ i\frac{2\pi(k^2 - \nu^2)\cosh k(z + h)\cosh k(\zeta + h)}{(k^2 - \nu^2)h + \nu} J_0(kr) \qquad (8.137)$$

where PV represents principal value and

$$R = \sqrt{(x - \xi)^2 + (y - \eta)^2 + (z - \xi)^2} \qquad (8.138)$$

$$R_1 = \sqrt{(x - \xi)^2 + (y - \eta)^2 + (z + 2h + \zeta)^2} \qquad (8.139)$$

$$r = \sqrt{(x - \xi)^2 + (y - \eta)^2} \qquad (8.140)$$

$$\nu = \frac{\sigma^2}{g} = k \tanh kh. \qquad (8.141)$$

The improper integral in the Green function may be replaced by an infinite series

$$G(x, y, z; \xi, \eta, \zeta)$$

$$= \frac{2\pi(\nu^2 - k^2)}{(k^2 - \nu^2)h + \nu} \cosh k(z + h)\cosh k(\zeta + h)$$

$$\times [Y_0(kr) - iJ_0(kr)]$$

$$+ 4 \sum_{m=1}^{\infty} \frac{\mu_m^2 + \nu^2}{(\mu_m^2 + \nu^2)h - \nu} \cos[\mu_m(z + h)]$$

$$\times \cos[\mu_m(\xi + h)]K_0(\mu_m r) \qquad (8.142)$$

where Y_0 is a Bessel function of the second kind of order 0; K_0 is a modified Bessel function of the second kind of order 0; μ_m are the roots of the equation (see Abramowitz and Stegun (1965))

$$\mu_m \tan \mu_m h + \nu = 0 \qquad (8.143)$$

with μ_0 the imaginary root ($\mu_0 = -ik$), and, for $m \geq 1$, μ_m the positive real roots in ascending order.

The source strength distribution function $f(\xi, \eta, \zeta)$ may be determined by applying the body-surface boundary condition which yields

$$-f(x, y, z) + \frac{1}{2\pi} \iint_s f(\xi, \eta, \zeta) \frac{\partial G}{\partial n}(x, y, z; \xi, \eta, \zeta)ds$$

$$= 2U_n(x, y, z) \quad (8.144)$$

in which $-U_n(x, y, z)$ is the normal fluid velocity due to the incident wave. Equation (8.144) may be rewritten as

$$-f_i + \alpha_{ij}f_j = 2U_{ni} \qquad (8.145)$$

where

$$\alpha_{ij} = \frac{1}{2\pi} \iint_{\Delta s_j} \frac{\partial G}{\partial n}(x_i, y_i, z_i; \xi_j, \eta_j, \zeta_j)ds. \qquad (8.146)$$

In arriving at eqn (8.145), the surface of the structure has been discretized into a finite number of facets; the point (x_i, y_i, z_i) refers to the centroid of the ith facet of the surface area Δs_i. The quantity $\partial G/\partial n$

represents the derivative of the Green function in the outward normal direction:

$$\frac{\partial G}{\partial n} = \frac{\partial G}{\partial x} n_x + \frac{\partial G}{\partial y} n_y + \frac{\partial G}{\partial z} n_z. \tag{8.147}$$

Once the source distribution f is known, the scattered potential Φ_s is obtained from

$$\Phi_{si} = \beta_{ij} f_j \tag{8.148}$$

where

$$\beta_{ij} = \frac{1}{4\pi} \iint_{\Delta s_j} G(x_i, y_i, z_i; \xi_j, \eta_j, \zeta_j) \mathrm{d}s. \tag{8.149}$$

The hydrodynamic pressure on the submerged surface of the structure is calculated from the linearized Bernoulli equation

$$p(x, y, z, t) = -\rho \frac{\partial \phi}{\partial t}$$

$$= \mathrm{Re}\{-\mathrm{i}\rho\sigma[\Phi_I(x, y, z) + \Phi_s(x, y, z)]\mathrm{e}^{-\mathrm{i}\sigma t}\}. \tag{8.150}$$

Then the hydrodynamic forces acting on the submerged structure may be determined by integrating the pressure on the immerged surface. The horizontal components are given by

$$F_x(t) = \iint_s p(x, y, z, t) n_x \, \mathrm{d}s \tag{8.151}$$

and

$$F_y(t) = \iint_s p(x, y, z, t) n_y \, \mathrm{d}s. \tag{8.152}$$

Calculations of wave forces on a number of large, fixed structures by the Green function method have been presented by Garrison and Stacey (1977). Figure 8.12 shows a vertical cylindrical caisson sitting at the bottom of the sea. The numerical calculations were carried out by the Green function method. Figure 8.13 shows a comparison between the exact and the numerical results for a caisson with a height to radius ratio of unity (Garrison and Stacey (1977)).

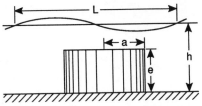

Fig. 8.12. Fixed, vertical cylindrical caisson in waves (from Garrison and Stacey (1977).

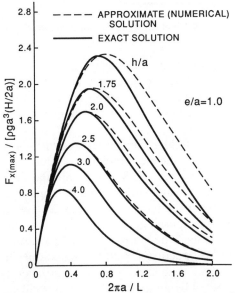

Fig. 8.13. Horizontal inertia coefficients for vertical caisson for $e/a = 1$.

8.7 Dynamics of floating structures

An offshore structure which is free to move in waves as shown in Fig. 8.14 undergoes six independent degrees of motion–three translational and three rotational.

Assuming the origin of the set of axes $OXYZ$ situated at the centre of gravity of the structure, the translational motions are described along these axes. The longitudinal motion along X is called surge, the transverse motion along Y is sway, and the vertical motion along Z is defined as heave. The rotational motion (moment) about X is called roll, about Y is called pitch and about Z is called yaw. Floating structures, such as articulated towers, moored tankers, tension leg platforms, etc., are very much in use in practice. Engineers want to know the stress distributions on such structures; and to determine these stresses, the motions of the

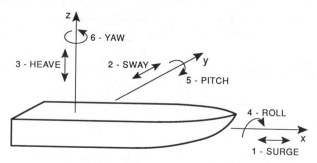

Fig. 8.14. Sketch of a floating ship with six degrees of motion.

structures must be known in addition to the wave forces on them. The equation of motion becomes nonlinear because of the presence of nonlinear damping, restoring and exciting forces. However, through more reasonable assumptions, these nonlinearities are removed to yield a linearized equation.

We assume that the fluid is incompressible and the flow is rotational. The problem reduces to the determination of the velocity potential, ϕ, which satisfies Laplace's equation, $\nabla^2\phi = 0$, with the fluid velocity vector $\mathbf{V}$ defined by $\mathbf{V} = \text{grad } \phi$. Assuming that the amplitude of the waves is small, the scattering process of the incident waves in the pressure of the floating structure can be decomposed and treated as the sum of two distinct mechanisms as described by Sarpkaya and Issacson (1981): the fluid motion produced by a structure forced to oscillate in otherwise still water and the interaction of a regular wave with a restrained body. The floating structure will be forced by an incident plane to oscillate in six degrees of freedom as explained earlier. For each mode of oscillations, the resulting small periodic motion, of angular frequency σ, may be expressed in the form $X_j e^{-i\sigma t}$ where X_j is the complex amplitude for the mode of motion j, with $j = 1, 2, 3, \ldots, 6$ corresponding to surge, sway, heave, roll, pitch and yaw respectively (see Fig. 8.14). The velocity potential at a point in the fluid in the presence of an oscillating structure in waves may be written as

$$\phi = \text{Re}\left\{ \left(\Phi_I + \Phi_s + \sum_{j=1}^{6} X_j \Phi_j \right) e^{-i\sigma t} \right\} \tag{8.153}$$

where Φ_I is the incident wave potential, Φ_s is the scattered potential, Φ_j the radiated or forced potentials due to the oscillation of the structure and $\text{Re}\{\cdot\}$ stands for the real part. Here Φ_I, and Φ_s and Φ_j are all functions of x, y, z and X_j is the independent parameter.

The complex potentials of eqn (8.153) must satisfy Laplace's equation and the following boundary conditions:

- The linearized kinematic and dynamic boundary conditions, at the free surface, combined to yield

$$\frac{\partial \Phi_j}{\partial z} - \frac{\sigma^2}{g} \Phi_j = 0 \quad \text{at} \quad z = 0, \quad j = 1, 2, \ldots, 6. \tag{8.154}$$

Equation (8.154) must also be satisfied by the Φ_I and Φ_s potentials.
- The vanishing of the vertical velocity component on the ocean floor:

$$\frac{\partial \Phi_j}{\partial z} = 0 \quad \text{at} \quad z = -h, \quad j = 1, 2, \ldots, 6, I, s. \tag{8.155}$$

- The radiation condition to ensure that the scattered and radiated potentials correspond to outgoing waves:

$$\lim_{r \to \infty} \sqrt{r} \left(\frac{\partial \Phi_j}{\partial r} - ik\Phi_j \right) = 0, \quad j = 1, 2, 3, \ldots, 6, s. \tag{8.156}$$

There is one more boundary condition which has to be satisfied by Φ_j at the surface of the structure. This boundary condition to be imposed on the floating body, s_0, is obtained by equating the normal derivative of (8.153) to the normal component of the body velocity (V_{nj}) at the point on the structure. Since the amplitudes X_j are independent parameters, it follows that

$$\frac{\partial}{\partial n} \left(X_j \Phi_j e^{-i\sigma t} \right) = V_{nj} = \frac{\partial}{\partial t} \left(X_j n_j e^{-i\sigma t} \right)$$

or

$$\frac{\partial \Phi_j}{\partial n} = (-i\sigma) n_j, \quad j = 1, 2, \ldots, 6. \tag{8.157}$$

in which

$$n_1 = n_x, \quad n_2 = n_y, \quad n_3 = n_z$$

$$n_4 = (yn_z - zn_y), \quad n_5 = (zn_x - xn_z), \quad n_6 = (xn_y - yn_x)$$

$$\tag{8.158}$$

where n_x, n_y and n_z are the direction cosines with respect to the x, y and z directions respectively.

The forced motion potentials Φ_j, $j = 1, 2, \ldots, 6$, are usually known as solutions of the radiation problem.

The remaining potentials, namely the incident and scattered potentials, are independent of the body motion and may be defined with the body fixed in position. The appropriate boundary condition on the body surface is given by

$$\frac{\partial \Phi_s}{\partial n} = -\frac{\partial \Phi_I}{\partial n} \quad \text{on} \quad S_0. \tag{8.159}$$

This problem is usually known as a wave diffraction problem.

This boundary value problem can be solved with the Green function method. Referring to Wehausen and Laitone (1960), the unknown potentials $\Phi_j(x, y, z)$ are expressed in terms of a surface distribution of sources

$$\Phi_j(x, y, z) = \frac{1}{4\pi} \iint_{S_0} f_j(\xi, \eta, \zeta) G(x, y, z; \xi, \eta, \zeta) \, ds$$

$$\text{for} \quad j = 1, 2, \ldots, 6 \tag{8.160}$$

where $f_j(\xi, \eta, \zeta)$ is a source distribution function at (ξ, η, ζ), a point on S_0, and $G(x, y, z; \xi, \eta, \zeta)$ is a Green function for the field point (x, y, z) due to a source of a unit strength at (ξ, η, ζ).

The Green function satisfies the boundary conditions (8.154), (8.155) and (8.156). It remains for $f_j(\xi, \eta, \zeta)$ to be chosen so that the body-surface boundary condition (8.157) is satisfied. This source strength distribution function, for each mode of oscillations j, may be obtained using a method described by Isaacson (1982). Then the unknown potentials $\Phi_j(x, y, z)$ ($j = 1, 2, \ldots, 6$) can be obtained from eqn (8.160).

8.7.1 Damping and added mass

The body motion amplitudes, X_j, are obtained from the equation of motion for a freely floating offshore structure:

$$\sum_{j=1}^{6} \left[-\sigma^2 (M_{ij} + a_{ij}) - i\sigma b_{ij} + c_{ij} \right] X_j = F_i^e \quad \text{for} \quad i = 1, 2, \ldots, 6$$

$$\tag{8.161}$$

where M_{ij} is the mass matrix, c_{ij} the hydrodynamic stiffness matrix and F_i^e the exciting forces associated with the potentials $(\Phi_I + \Phi_s)$ and identical to the wave force components for a fixed structure. Referring to Newman (1977), the added-mass coefficient, a_{ij}, and the damping coefficient, b_{ij}, are defined in terms of the force components associated with the radiated potentials, F_{ij}^R.

$$a_{ij} = \frac{1}{\sigma^2} \operatorname{Re}\{F_{ij}^{(R)}\} \tag{8.162}$$

$$b_{ij} = \frac{1}{\sigma} I_m\{F_{ij}^{(R)}\} \tag{8.163}$$

where $I_m\{\cdot\}$ indicates the imaginary part.

Once the source strength distribution functions f_j and the amplitudes of motion have been determined, an expression for the potential ϕ at any point in the fluid can be derived. Thus, consequently, the linear dynamic pressure due to ϕ can be computed from the linearized Bernoulli equation given by

$$p = -\rho\left(\frac{\partial\phi}{\partial t} + gz\right)$$

$$= -\rho \operatorname{Re}\left\{\left(\Phi_I + \Phi_s + \sum_{j=1}^{6} X_j\Phi_j\right)(-i\sigma)e^{-i\sigma t}\right\} - \rho gz, \tag{8.164}$$

Once the pressure distribution on the surface of the structure is known, the forces may be determined by integrating the fluid pressure over the wetted surface S_0 as follows:

$$\mathbf{F} = -\rho g \iint_{S_0} z\mathbf{n}\,ds - \rho \operatorname{Re}\left\{(-i\sigma)e^{-i\sigma t} \iint_{S_0} (\Phi_I + \Phi_s)\mathbf{n}\,ds\right\}$$

$$- \rho \operatorname{Re}\left\{\sum_{j=1}^{6} (-i\sigma)X_j e^{-i\sigma t} \iint_{S_0} \Phi_j\mathbf{n}\,ds\right\}. \tag{8.165}$$

The three integrals in eqn (8.165) distinctly represent three different contributions to the total force. The first integral is due to the hydrostatic force, the second integral yields the exciting force and the third integral is known as the radiated force in terms of the *damping* and *added-mass* coefficients.

Define the force due to the radiated potential as (the third integral coefficient of $X_j e^{-i\sigma t}$):

$$F_{ij}^{(R)} = -\rho(-i\sigma) \iint_{S_0} \Phi_j n_i \, ds = -\rho \iint_{S_0} \Phi_j \frac{\partial \Phi_i}{\partial n} \, ds. \tag{8.166}$$

The last result is due to boundary condition (8.157). This force can be determined provided Φ_j is known.

From eqn (8.161), the force coefficient due to the motion of the structure is given by

$$F_{ij}^{(R)} = -\sigma^2 a_{ij} - i\sigma b_{ij}. \tag{8.167}$$

Once $F_{ij}^{(R)}$ is determined from eqn (8.166), the added-mass and damping coefficients are known from eqn (8.167) as the real and imaginary parts of $F_{ij}^{(R)}$ respectively as given in eqns (8.162) and (8.163). The following section presents the results of a floating ice floe in cylindrical form. These results were mainly due to Masson and LeBlond (1989).

8.7.2 Response amplitude operators (RAOs) for a cylindrical ice floe

Consider the cylindrical ice flow as shown in Fig. 8.15. Assume the floe to be symmetric about the z-axis which is vertically upwards. Because of its symmetry, the floating cylinder will be forced by an incident plane wave to oscillate in three degrees of freedom: surge, back and forth in the direction of incident wave number $\mathbf{k}$; heave, up and down; and pitch, about an axis parallel to the wave crest. Owing to this oscillation, each mode may be expressed in the form $X_j e^{-i\sigma t}$ with angular frequency σ and where X_j is the complex amplitude for the mode of motion j, with $j = 1$, 2 and 3 corresponding to surge, heave and pitch respectively as shown in Fig. 8.15.

The velocity component which is also harmonic can be obtained by superposition of five components (see eqn (8.153)):

$$\phi = \text{Re}\left\{ \left(\Phi_I + \Phi_s + \sum_{j=1}^{3} X_j \Phi_j \right) e^{-i\sigma t} \right\}. \tag{8.168}$$

The symbols have their usual meaning.

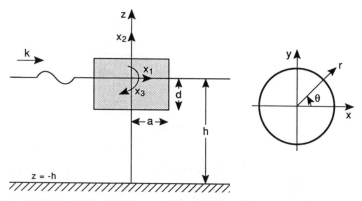

Fig. 8.15. Cylindrical flow of radius $a = D/2$ and draft d in water of depth h. The three modes of motion have amplitudes of X_j with subscript 1 for surge, 2 for heave and 3 for pitch. Here, (x, y, z) form a Cartesian coordinate system and (r, θ, z) the corresponding cylindrical coordinate system.

The potential associated with an incident plane wave of amplitude $A \, (= H/2)$ in water of finite depth is usually given as

$$\Phi_I = \frac{gA}{\sigma} \frac{\cosh k(z+h)}{\cosh kh} \exp(ikx) \tag{8.169}$$

where x is measured in the direction of wave propagation and z vertically upwards from the still water level. Using the analysis presented by Masson and LeBlond (1989) and following the numerical development of the preceding section, the equation of motion for a freely floating cylindrical floe is given by (see eqn (8.161))

$$\sum_{j=1}^{3} \left[-\sigma^2 (M_{ij} + a_{ij}) - i\sigma b_{ij} + c_{ij} \right] X_j = F_i^{(e)} \qquad \text{for} \qquad i = 1, 2, 3. \tag{8.170}$$

The floe response amplitudes X_j, $j = 1, 2, 3$, are presented in the form of non-dimensional response-amplitude operators (RAOs) defined here as X_j/A for surge and heave and $LX_3/(360A)$ for pitch with X_3 in degrees. The response varies with incident wave-length, the geometry and dimensions of the floe, as well as with the water depth. Figure 8.16 shows the response of a cylindrical floe, to which a triangular keel of draft 3 m has

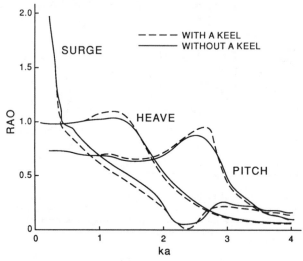

Fig. 8.16. Response-amplitude operators of a cylindrical floe with (dashed lines) and without (solid lines) a keel. The floe has a radius $a = 10$ m and draft $d = 2$ m and a keel of draft 3 m (water depth of 30 m).

been appended. It appears from this figure that the presence of the keel has little effect on the overall characteristics of the response curves.

8.8 Exercises

1. Verify the horizontal force and overturning moment given in eqs (8.15) and (8.16), respectively. Plot these two profiles by developing a computer program.

2. Develop a FORTRAN program and plot the Morison coefficient, C_M, and the phase angle, α, given in eqns (8.90) and (8.92), respectively.

3. Two vertical cylinders separated by 100 m horizontally and having diameters of 2 m and 10 m, respectively, are placed in ocean waves. The wave characteristics are given by $H = 20$ m, $T = 5$ sec, $h = 30$ m. (a)Calculate the maximum total wave force on the two cylinders when the angle of the wave incident on the cylinders is 0°, 30°, 45°, 60° and 90°.
(b) What would be the overturning moment for the direction α_{max} of maximum force?

4. Using results of the MacCamy–Fuchs theory, discuss the reasons for the decrease in C_M with increasing diffraction coefficient ka.

5. A circular cylinder of radius a and length l is held fixed in a horizontal plane at an elevation h_0 above the bottom in water of depth h. A linear wave of height H and period T is propagating in the x-direction. Considering only the inertia force, develop expressions for the time-varying components of forces in the x- and y-directions, $F_x(t)$ and $F_y(t)$, and the moment $M_\alpha(t)$ about the z-axis.

6. Simplify eqns (8.15) and (8.16) for the cases of shallow and deep water. Discuss the variations of drag and inertia force and moment components with wave period for these cases.

7. Consider a fixed vertical cylinder of length l and radius a in a wave of height H and period T. The water depth is h. The horizontal forces acting on this cylinder are given by eqn (8.38) as the Froude–Krylov force. Compare this result with that obtained by diffraction theory (eqn (8.88)).

References

Abramowitz, M. and Stegun, I. A. (Editors) (1965). *Handbook of Mathematical Functions*, Dover, New York.

Chakrabarti, S. K. (1975). Second order wave forces on large vertical cylinders. *Proc. ASCE*, 101, WW3, 311–417.

Chakrabarti, S. K. (1978). Comments on 'Second order wave effects on large diameter vertical cylinders' *J. Ship Res.*, 22, 226–268.

Chakrabarti, S. K. (1987). *Hydrodynamics of Offshore Structures*, Computational Mechanics Publications, Southampton.

Garrison, C. J. and Chow, P. Y. (1972). Wave forces on submerged bodies. *J. Waterw., Harbors Coastal Eng. Div., ASCE*, 98, WW3, 375–392.

Garrison, C. J. and Rao, V. S. (1971). Interaction of waves with submerged objects. *J. Waterw., Harbors Coastal Eng. Div., ASCE*, 97, WW2, 259–277.

Garrison, C. J. and Stacey, R. (1977). Wave loads on North sea gravity platforms: A comparison of theory and experiment. *Proc. Offshore Technology Conf., Houston*, Paper No. OTC 2794, Vol I, 513–524.

Havelock, T. H. (1940). The pressure of water waves upon a fixed obstacle. *Proc. R. Soc.* A963, 175–190.

Hogben, N. and Standing, R. G. (1975). Experience in computing wave loads on large bodies. *Proc. Offshore Technology Conf., Houston*, Paper No. OCT 2189, 11, 413–431.

Isaacson, M. (1982). Fixed and floating axisymmetric structures in waves. *J. Waterw., Harbors Coastal Ocean Div. ASCE* 108, WW2, 180–1149.

Keulegan, G. H. and Carpenter, L. H. (1958). Forces on cylinders and plates in an oscillating fluid. *J. Res. NBS*, 60(5), 423–440.

Kim, M. H. and Yue, D. K. P. (1989). The complete second-order diffraction solution for an axisymmetric body. Part 1. Monochromatic incident waves. *J. Fluid Mech.*, 200, 235–264.

Kim, M. H. and Yue, D. K. P. (1990). The complete second-order diffraction solution for an axisymmetric body. Part 2. Biochromatic incident waves and body motions. *J. Fluid Mech.*, 211, 557–593.

Lighthill, M. J. (1979). Waves and hydrodynamic loading. *Behavior of Offshore Structures Conference, Imperial College, London*, pp. 1–40.

MacCamy, R. C. and Fuchs, R. A. (1954). Wave forces on piles: A diffraction theory. *US Army Beach Erosion Board, Tech. Memo*, No. 69, 17 pp.

McCormick, M. E. (1973). *Ocean Engineering Wave Mechanics*, John Wiley and Sons, New York.

Masson, D. and LeBlond, P. H. (1989). Spectral evolution of wind generated surface gravity waves in a dispersed ice-field. *J. Fluid Mech.*, 202, 43–81.

Mogrdige, G. R. and Jamieson, W. W. (1976). Wave loads on large circular cylinders: A design method. *National Research Council, Ottawa, Memo*, MH-111.

Morison, J. R., O'Brien, M. P., Johnson, J. W. and Schaaf, S. A. (1950). The forces exerted by surface waves on piles. *J. Petrol Technol.*, 189, 149–154.

Newman, J. N. (1977). *Marine Hydrodynamics*, MIT Press, Cambridge, Massachusetts.

Rahman, M. (1988). *The Hydrodynamics of Waves and Tides, with Applications*, Computational Mechanics Publications, Southampton.

Rahman, M. and Heaps, H. S. (1983). Wave forces on offshore structures: Nonlinear wave diffraction by large cylinders. *J. Oceanogr.*, 13, 2225–2235.

Raman, H. and Venkatanarasaiah, P. (1976). Forces due to nonlinear water waves on vertical cylinders. *Proc. ASCE*, 102, WW3, 301–316.

Sarpkaya, T. and Isaacson, M. (1981). *Mechanics of Wave Forces on Offshore Structures*, Van Nostrand Reinhold, New York.

Stoker, J. J. (1957). *Water Waves*, Interscience, New York.

Wehausen, J. V. and Laitone, E. V. (1960). Surface waves. *Handb. Phys.*, 9, 446–778.

Wylie, C. R. and Barrett, L. C. (1982). *Advanced Engineering Mathematics*, McGraw-Hill, New York.

Yamaguchi, M. and Tsuchiya, Y. (1974). Nonlinear effect of waves on wave pressure and wave force on a large cylindrical pile. (in Japanese). *Proc. Civ. Eng. Soc. in Japan*, No. 229, September.

9
Nonlinear long waves in shallow water

9.1 Introduction

In Chapter 5 the Stokes perturbation procedure for finite-amplitude waves was developed. In that method the perturbation parameter was used as $kH/2$, the relative wave steepness. However, in very shallow water (see Chapter 5) the Stokes perturbation techniques fails to hold good which will be clear from the following analysis.

let us consider, for the sake of simplicity, the following momentum equation:

$$\frac{\partial u}{\partial t} + u\frac{\partial u}{\partial x} + v\frac{\partial u}{\partial y} + w\frac{\partial u}{\partial z} = -\frac{1}{\rho}\frac{\partial p}{\partial x} + \nu \nabla^2 u. \tag{9.1}$$

We know from the Stokes expansion that the condition for linearity is $k(H/2) \ll 1$. Let us check this linearity assumption by comparing a nonlinear term, say $u(\partial u/\partial x)$, with a linear term, say $\partial u/\partial t$, both evaluated at the free surface $z = 0$. Then the ratio

$$\left(\frac{u\dfrac{\partial u}{\partial x}}{\dfrac{\partial u}{\partial t}} \right) = O\left(\frac{U^2/L}{U/T} \right) = 0\left(\frac{U}{L/T} \right) = \left(\frac{|U|}{C} \right)$$

where L is the characteristic wavelength, T is the characteristic wave period, U is the characteristic fluid velocity, and C is the characteristic phase velocity.

We know from Chapter 4 that the expressions for U and C are

$$|U| = \frac{gkH}{2\sigma}\frac{\cosh k(z+h)}{\cosh kh}|e^{ikx}| \tag{9.2}$$

and

$$C = \left(\frac{g}{k} \tanh kh\right)^{1/2}.$$ (9.3)

Thus

$$\left(\frac{|U|}{C}\right)_{z=0} = \left(\frac{\dfrac{gkH}{2\sigma} \dfrac{\cosh k(z+h)}{\cosh kh}|e^{ikx}|}{\left(\dfrac{g}{k} \tanh kh\right)^{1/2}}\right)_{z=0} = \frac{(kH/2)}{\tanh kh}.$$

For the deep water wave, $kh \to \infty$, so $(|U|/C)_{z=0} = (kH/2) = kA$; and for the shallow water wave, $kh \to 0$, so $(|U|/C)_{z=0} = (H/2)/h = A/h$.

From this analysis it is clear that the nonlinear theory of the Stokes expansion for $kA < 1$ is valid for the deep water waves only. However, for the nonlinear theory for the shallow water waves, both $kh \ll 1$ and $A/h \ll 1$ must hold.

Let us define these two parameters as

$$\gamma = kh \ll 1 \quad \text{and} \quad \delta = \frac{A}{h} \ll 1.$$ (9.4)

Here the second restriction is a much severe one for many practical problems. Therefore, it is essential to investigate a nonlinear theory of shallow water waves. Historically, two different theories were developed, one by Airy and the second by Boussinesq (1877), which led to opposite conclusions. However, these differences were resolved by a fundamental paper by Ursell (1953). In particular, Ursell (1953) showed that the ratio

$$U_r = \left(\frac{\delta}{\gamma^2}\right) = \frac{A/h}{(kh)^2} = \frac{kA}{(kh)^3},$$ (9.5)

which is known as the Ursell number, plays an important role in deciding the choice of approximation.

9.2 Mathematical developments

The following mathematical equations for the constant depth case are due to Benney (1966) and Peregrine (1967). This analysis is made for the three-dimensional case so that the limiting cases for one and two dimensions can be retrieved easily.

Because of the presence of two parameters, it will be convenient for the sake of clarity to use dimensionless variables. The mathematical equations in three dimensions can be written as (Benney (1962) and Mei (1983)).

Laplace's equation:

$$\frac{\partial^2 \Phi}{\partial X^2} + \frac{\partial^2 \Phi}{\partial Y^2} + \frac{\partial^2 \Phi}{\partial Z^2} = 0 \qquad -h < Z < \tilde{\eta}. \tag{9.6}$$

Dynamic free-surface boundary condition:

$$\frac{\partial \Phi}{\partial T} + g\tilde{\eta} + \frac{1}{2}\left[\left(\frac{\partial \Phi}{\partial X}\right)^2 + \left(\frac{\partial \Phi}{\partial Y}\right)^2 + \left(\frac{\partial \Phi}{\partial Z}\right)^2\right] \quad \text{at} \quad Z = \tilde{\eta}. \tag{9.7}$$

Kinematic free-surface boundary condition:

$$\frac{\partial \tilde{\eta}}{\partial T} + \frac{\partial \Phi}{\partial X}\frac{\partial \tilde{\eta}}{\partial X} + \frac{\partial \Phi}{\partial Y}\frac{\partial \tilde{\eta}}{\partial Y} = \frac{\partial \Phi}{\partial Z} \quad \text{at} \quad Z = \tilde{\eta}. \tag{9.8}$$

Bottom boundary condition:

$$\frac{\partial \Phi}{\partial Z} = 0 \quad \text{at} \quad Z = -h. \tag{9.9}$$

Here Φ = velocity potential, $\tilde{\eta}$ = wave elevation, X, Y, Z = Cartesian coordinates, g = acceleration due to gravity, and T = time.

The linear theory suggests that the following dimensionless variables are appropriate to make the above equations dimensionless:

$$x = kX, \quad y = kY, \quad z = Z/h, \quad t = k(gh)^{1/2}T,$$

$$\eta = \tilde{\eta}/A, \qquad \phi = \frac{\Phi}{\left(\dfrac{A}{kh}\right)(gh)^{1/2}}. \tag{9.10}$$

The velocity components are normalized as

$$U = \frac{\partial \Phi}{\partial X} = \frac{A}{h}(gh)^{1/2}\frac{\partial \phi}{\partial x} = \frac{A}{h}\sqrt{gh}\,u$$

$$V = \frac{\partial \Phi}{\partial Y} = \frac{A}{h}\sqrt{gh}\,\frac{\partial \Phi}{\partial y} = \frac{A}{h}\sqrt{gh}\,v$$

$$W = \frac{\partial \Phi}{\partial Z} = \left(\frac{1}{kh}\right)\left(\frac{A}{h}\right)\sqrt{gh}\,\frac{\partial \phi}{\partial z} = \left(\frac{1}{kh}\right)\left(\frac{A}{h}\right)\sqrt{gh}\,w. \tag{9.11}$$

It is to be noted here that the difference in scaling for horizontal and vertical components are required by continuity.

The normalized equations and boundary conditions are:

Laplace's equation:

$$\gamma^2 \left(\frac{\partial^2 \phi}{\partial x^2} + \frac{\partial^2 \phi}{\partial y^2} \right) + \frac{\partial^2 \phi}{\partial z^2} = 0, \qquad -1 < z < \delta\eta. \tag{9.12}$$

Dynamic boundary condition:

$$\gamma^2 \left(\frac{\partial \phi}{\partial t} + \eta \right) + \frac{\delta}{2} \left\{ \gamma^2 \left[\left(\frac{\partial \phi}{\partial x} \right)^2 + \left(\frac{\partial \phi}{\partial y} \right)^2 \right] + \left(\frac{\partial \phi}{\partial z} \right)^2 \right\} = 0, \qquad z = \delta\eta. \tag{9.13}$$

Kinematic boundary condition:

$$\gamma^2 \left[\frac{\partial \eta}{\partial t} + \delta \left(\frac{\partial \phi}{\partial x} \frac{\partial \eta}{\partial x} + \frac{\partial \phi}{\partial y} \frac{\partial \eta}{\partial y} \right) \right] = \frac{\partial \phi}{\partial z}, \qquad z = \delta\eta. \tag{9.14}$$

Bottom boundary condition:

$$\frac{\partial \phi}{\partial z} = 0 \quad \text{at} \quad z = -1. \tag{9.15}$$

Keeping δ arbitrary, first assume that $\gamma = kh \ll 1$.

Since ϕ is analytic, we can expand it as a power series in the vertical coordinate so that

$$\phi(x, y, z, t) = \sum_{n=0}^{\infty} (z + 1)^n \phi_n(x, y, t). \tag{9.16}$$

Define

$$\nabla_h \equiv \left(\frac{\partial}{\partial x}, \frac{\partial}{\partial y} \right)$$

as the horizontal gradient and physically,

$$k \nabla_h \equiv k \left(\frac{\partial}{\partial x}, \frac{\partial}{\partial y} \right) \equiv \nabla_H.$$

Then we have

$$\nabla_h \phi = \sum_{n=0}^{\infty} (z+1)^n \nabla_h \phi_n$$

$$\nabla_h^2 \phi = \sum_{n=0}^{\infty} (z+1)^n \nabla_h^2 \phi_n$$

$$\frac{\partial \phi}{\partial z} = \sum_{n=0}^{\infty} n(z+1)^{n-1} \phi_n = \sum_{n=1}^{\infty} n(z+1)^{n-1} \phi_n$$

$$= \sum_{n=0}^{\infty} (n+1)(z+1)^n \phi_{n+1}$$

$$\frac{\partial^2 \phi}{\partial z^2} = \sum_{n=0}^{\infty} n(n+1)(z+1)^{n-1} \phi_{n+1}$$

$$= \sum_{n=1}^{\infty} n(n+1)(z+1)^{n-1} \phi_{n+1}$$

$$= \sum_{n=0}^{\infty} (n+1)(n+2)(z+1)^n \phi_{n+2}.$$

Using these definitions in Laplace's equation (9.12), we obtain

$$\gamma^2 \nabla_h^2 \phi + \frac{\partial^2 \phi}{\partial z^2} = \gamma^2 \sum_{n=0}^{\infty} (z+1)^n \nabla_h^2 \phi_n$$

$$+ \sum_{n=0}^{\infty} (n+1)(n+2)(z+1)^n \phi_{n+1} = 0.$$

Since z is arbitrary $-1 \le z < \delta\eta$, the coefficient of each power of $(z+1)$ must vanish yielding a recurrence relation

$$\phi_{n+2} = \frac{-\gamma^2 \nabla_h^2 \phi_n}{(n+1)(n+2)}, \qquad n = 0, 1, 2, \dots . \tag{9.17}$$

But the bottom boundary condition implies that $\partial\phi/\partial z = 0$ at $z = -1$. That is

$$\sum_{n=0}^{\infty} n(z+1)^{n-1} \phi_n(x, y, t) = 0 \quad \text{at} \quad z = -1$$

or

$$\sum_{n=1}^{\infty} n(z+1)^{n-1}\phi_n = 0 \quad \text{at} \quad z = -1$$

or

$$\sum_{n=0}^{\infty} (n+1)(z+1)^n \phi_{n+1} = 0 \quad \text{at} \quad z = -1$$

corresponding to $n = 0$, $\phi_1 = 0$.

It is clear from eqn (9.17) that

$$\phi_1 = \phi_3 = \phi_5 = \ldots = \phi_{2n+1} = 0. \tag{9.18}$$

Also from eqn (9.17), we have

$$n = 0: \quad \phi_2 = \frac{-\gamma^2 \nabla_h^2 \phi_0}{1 \cdot 2}$$

$$n = 2: \quad \phi_4 = \frac{-\gamma^2 \nabla_h^2 \phi_2}{3 \cdot 4} = \frac{(-\gamma^2)(-\gamma^2)\nabla_h^2 \nabla_h^2 \phi_0}{1 \cdot 2 \cdot 3 \cdot 4} = \frac{\gamma^4 \nabla_h^2 \nabla_h^2 \phi_0}{4!}$$

$$n = 4: \quad \phi_6 = \frac{-\gamma^2}{5 \cdot 6} \nabla_h^2 \phi_4 = \frac{-\gamma^6 \nabla_h^2 \nabla_h^2 \nabla_h^2 \phi_0}{6!}.$$

If we assume that $\phi_0 = O(\phi) = O(1)$ then it may be concluded that $\phi_2 = O(\gamma^2)$, $\phi_4 = O(\gamma^4)$, $\phi_6 = O(\gamma^6)$ and so on. Thus with an error of $O(\gamma^6)$, the velocity potential is

$$\phi = \phi_0 - \frac{\gamma^2}{2}(z+1)^2 \nabla_h^2 \phi_0 + \frac{\gamma^4}{24}(z+1)^4 \nabla_h^4 \phi_0 + O(\gamma^6). \tag{9.19}$$

Now we will investigate the dynamic and kinematic free-surface boundary conditions. Put

$$\zeta = 1 + \delta\eta. \tag{9.20}$$

With this substitution, eqn (9.13) reduces to

$$\gamma^2 \left(\frac{\partial \phi_0}{\partial t} - \frac{\gamma^2}{2} \zeta^2 \nabla_h^2 \frac{\partial \phi_0}{\partial t} + \eta \right)$$

$$+ \frac{1}{2} \delta\gamma^2 \left[(\nabla_h \phi_0)^2 - \gamma^2 \zeta^2 \nabla_h \phi_0 \cdot \nabla_h^2 (\nabla_h \phi_0) \right]$$

$$+ \frac{1}{2} \delta\gamma^4 \zeta^2 (\nabla_h^2 \phi_0)^2 = o(\gamma^6). \tag{9.21}$$

Equation (9.14) can be written as

$$\gamma^2 \left[\frac{1}{\delta} \frac{\partial \zeta}{\partial t} + \nabla_h \zeta \cdot \left(\nabla_h \phi_0 - \frac{\gamma^2}{2} \zeta^2 \, \nabla_h^2 (\nabla_h \phi_0) \right) \right]$$

$$= -\gamma^2 \zeta \, \nabla_h^2 \phi_0 + \frac{\gamma^4}{6} \zeta^3 \, \nabla_h^4 \phi_0 + o(\gamma^6). \tag{9.22}$$

Now define the horizontal velocity as

$$\mathbf{u}_0 = \nabla_h \phi_0. \tag{9.23}$$

Then taking the gradient of eqn (9.21) we obtain (dynamic BC)

$$\frac{\partial \mathbf{u}_0}{\partial t} + \delta \mathbf{u}_0 (\nabla_h \cdot \mathbf{u}_0) + \frac{1}{\delta} \nabla_h \zeta$$

$$+ \gamma^2 \nabla_h \left(-\frac{\delta}{2} \zeta^2 \mathbf{u}_0 \cdot \nabla_h^2 \mathbf{u}_0 + \frac{\delta}{2} \zeta^2 (\nabla_h \cdot \mathbf{u}_0)^2 \right.$$

$$\left. -\frac{1}{2} \zeta^2 \frac{\partial}{\partial t} (\nabla_h \cdot \mathbf{u}_0) \right) = O(\gamma^4). \tag{9.24}$$

Also eqn (9.22) can be simplified as (kinematic BC)

$$\frac{1}{\delta} \frac{\partial \zeta}{\partial t} + \nabla_h \zeta \cdot \left(\mathbf{u}_0 - \frac{\gamma^2}{2} \zeta^2 \, \nabla_h^2 \mathbf{u}_0 \right)$$

$$+ \zeta \, \nabla_h \cdot \mathbf{u}_0 - \frac{\gamma^2}{6} \zeta^3 \, \nabla_h^2 (\nabla_h \cdot \mathbf{u}_0) = O(\gamma^4). \tag{9.25}$$

Once η and $\mathbf{u}_0 = (u, v)$ are solved, the actual velocity components are given by

$$(u, v) = \nabla_h \phi = \mathbf{u}_0 - \frac{\gamma^2}{2} (z + 1)^2 \, \nabla_h \nabla_h \cdot \mathbf{u}_0 + O(\gamma^4) \tag{9.26}$$

$$w = \frac{\partial \phi}{\partial z} = -\gamma^2 (z + 1) \nabla_h^2 \phi_0 = -\gamma^2 (z + 1) \nabla_h \cdot \mathbf{u}_0 + O(\gamma^4). \tag{9.27}$$

290 Nonlinear long waves in shallow water

The pressure field is obtained from Bernoulli's equation whose dimensionless form is exactly

$$-p = z + \delta\left\{ \frac{\partial \phi}{\partial t} + \frac{\delta}{2}\left[(\nabla_h \phi)^2 + \frac{1}{2}\left(\frac{\partial \phi}{\partial z} \right)^2 \right] \right\} \tag{9.28}$$

where p has been normalized by $\rho g h$ $\left(p = \dfrac{P}{\rho g h} \right)$.

The approximate pressure field may be obtained by substituting eqns (9.24) and (9.26) into eqn (9.28).

$$-p = z + \delta\left[\left(\frac{\partial \phi_0}{\partial t} - \frac{\gamma^2}{2}(z+1)^2 \nabla_h \cdot \mathbf{u_0} \right) \right.$$

$$\left. + \frac{\delta}{2}\left[\mathbf{u}_0^2 - \gamma^2(z+1)^2 \mathbf{u_0} \cdot \nabla_h^2 \mathbf{u_0} + \gamma^2(z+1)^2(\nabla_h \cdot \mathbf{u_0})^2 \right] \right] + O(\gamma^4).$$

Equation (9.21) can be used to eliminate $\partial \phi_0 / \partial t$ so that

$$p = (\delta\eta - z) - \frac{\gamma^2}{2}\left[\zeta^2 - (z+1)^2 \right]$$

$$\times \left\{ \nabla_h \cdot \mathbf{u_0} + \delta\left[\mathbf{u_0} \cdot \nabla_h^2 \mathbf{u_0} - (\nabla_h \cdot \mathbf{u_0})^2 \right] \right\} + O(\gamma^4). \tag{9.29}$$

Next we can introduce the depth-averaged horizontal velocity defined by

$$\langle \mathbf{u} \rangle = \frac{1}{\zeta} \int_{-1}^{\delta\eta} (\nabla_h \phi) \mathrm{d}z = \frac{1}{\zeta} \int_{-1}^{\delta\eta} \left(\mathbf{u_0} - \frac{\gamma^2}{2}(z+1)^2 \nabla_h^2 \mathbf{u_0} + \dots \right) \mathrm{d}z$$

$$= \mathbf{u_0} - \frac{\gamma^2}{6} \zeta^2 \nabla_h^2 \mathbf{u_0} + O(\gamma^4) \tag{9.30}$$

which can be easily inverted to yield

$$\mathbf{u_0} = \langle \mathbf{u} \rangle + \frac{\gamma^2}{6} \zeta^2 \nabla_h^2 \langle \mathbf{u} \rangle + O(\gamma^4). \tag{9.31}$$

Substituting eqn (9.31) into eqn (9.25) it follows that

$$\frac{\partial \zeta}{\partial t} + \delta \nabla_h \cdot (\zeta \langle \mathbf{u} \rangle) = O(\gamma^4) \tag{9.32}$$

which is just the depth-averaged continuity equation.

Now expressing $\mathbf{u}_0$ in terms of $\langle\mathbf{u}\rangle$, eqn (9.24) can be rewritten as

$$\frac{\partial\langle\mathbf{u}\rangle}{\partial t} + \delta\langle\mathbf{u}\rangle\cdot(\nabla_h\cdot\langle\mathbf{u}\rangle) + \frac{1}{\delta}\,\nabla_h\zeta + \frac{\gamma^2}{6}\,\frac{\partial}{\partial t}(\zeta^2\,\nabla_h^2\langle\mathbf{u}\rangle)$$

$$+ \gamma^2\,\nabla_h\left(-\frac{\delta}{3}\,\zeta^2\langle\mathbf{u}\rangle\cdot\nabla_h^2\langle\mathbf{u}\rangle + \frac{\delta}{2}\,\zeta^2(\nabla_h\cdot\langle\mathbf{u}\rangle)^2\right.$$

$$\left. -\frac{\zeta^2}{2}\,\nabla_h\cdot\frac{\partial}{\partial t}\langle\mathbf{u}\rangle\right) = O(\gamma^4). \tag{9.33}$$

These equations are valid for δ only. Extension to δ^2 is straightforward but tedious. Expressing eqns (9.32), (9.33) and (9.29) in physical variables, we obtain

$$\frac{\partial}{\partial T}(h + \tilde{\eta}) + \nabla_H\cdot((h + \tilde{\eta})\langle\mathbf{U}\rangle) = 0, \tag{9.34}$$

$$\frac{\partial}{\partial T}\langle\mathbf{U}\rangle + \langle\mathbf{U}\rangle\cdot\nabla_H\langle\mathbf{U}\rangle + g\,\nabla_H(h + \tilde{\eta})$$

$$+ \frac{1}{6}\,\frac{\partial}{\partial T}((h + \tilde{\eta})\nabla_H^2\langle\mathbf{U}\rangle)$$

$$+ \nabla_H\left(-\frac{2}{3}(h + \tilde{\eta})^2\langle\mathbf{U}\rangle\cdot\nabla_H^2\langle\mathbf{U}\rangle\right.$$

$$+ \frac{1}{2}(h + \tilde{\eta})^2(\nabla_H\cdot\langle\mathbf{U}\rangle)^2$$

$$\left. -\frac{1}{2}(h + \tilde{\eta})^2\,\nabla_H\cdot\frac{\partial}{\partial T}\langle\mathbf{U}\rangle\right) = 0 \tag{9.35}$$

and

$$P = \rho g(\tilde{\eta} - Z) - \frac{\rho}{2}\left[(h + \tilde{\eta})^2 - (Z + h)^2\right]\left(\nabla_H\cdot\frac{\partial}{\partial T}\langle\mathbf{U}\rangle\right.$$

$$\left. + \left[\langle\mathbf{U}\rangle\cdot\nabla_H^2\langle\mathbf{U}\rangle - (\nabla_H\cdot\langle\mathbf{U}\rangle)^2\right]\right) \tag{9.36}$$

where

$$\langle\mathbf{U}\rangle = \frac{A}{h}\,\sqrt{gh}\,\langle\mathbf{u}\rangle, \quad \nabla_H = \left(\frac{\partial}{\partial X}, \frac{\partial}{\partial Y}\right) = k\left(\frac{\partial}{\partial x}, \frac{\partial}{\partial y}\right) = k\,\nabla_h.$$

Limiting cases

(1) Airy's theory for very long waves: $\gamma \to 0$ *and* $\delta = O(1)$

Airy's theory gives the leading-order approximation for very long waves of finite amplitude, because for very long waves $\gamma = kh \to 0$ and $\delta = A/h = O(1)$. Thus omitting terms proportional to γ^2 for eqns (9.32), (9.33) and (9.29), we obtain in physical variables

$$\frac{\partial}{\partial T}(h + \tilde{\eta}) + \nabla_H \cdot ((h + \tilde{\eta})\langle \mathbf{U} \rangle) = 0 \tag{9.37}$$

$$\frac{\partial}{\partial T}\langle \mathbf{U} \rangle + \langle \mathbf{U} \rangle \cdot \nabla_H \langle \mathbf{U} \rangle + g \nabla_H (h + \tilde{\eta}) = 0 \tag{9.38}$$

$$P = \rho g(\tilde{\eta} - Z). \tag{9.39}$$

These equations are valid for variable depth $h(x, y)$ also. A distinguishing feature of Airy's approximation is that the pressure is hydrostatic.

(2) Boussinesq theory: $O(\delta) = O(\gamma^2) < 1$

For weakly nonlinear and moderately long waves in shallow water $\delta = A/h < 1$ and $\gamma^2 = (kh)^2 < 1$. And in that situation eqns (9.32), (9.33) and (9.29) reduce to

$$\frac{\partial \eta}{\partial t} + \nabla_h \cdot ((1 + \delta\eta)\langle \mathbf{u} \rangle) = 0 \tag{9.40}$$

$$\frac{\partial}{\partial t}\langle \mathbf{u} \rangle + \delta\langle \mathbf{u} \rangle \cdot (\nabla_h \cdot \langle \mathbf{u} \rangle) + \nabla_h \eta - \frac{\gamma^2}{3}\nabla_h^2 \cdot \frac{\partial}{\partial t}\langle \mathbf{u} \rangle = 0 \tag{9.41}$$

$$p = \delta\eta - z - \frac{\gamma^2}{2}(-z^2 - 2z)\nabla_h \cdot \frac{\partial}{\partial t}\langle \mathbf{u} \rangle. \tag{9.42}$$

In physical variables these can be written as

$$\frac{\partial \tilde{\eta}}{\partial T} + \nabla_H \cdot ((\tilde{\eta} + h)\langle \mathbf{U} \rangle) = 0 \tag{9.43}$$

$$\frac{\partial}{\partial T}\langle \mathbf{U} \rangle + \langle \mathbf{U} \rangle \cdot \nabla_H \langle \mathbf{U} \rangle + g \nabla_H \tilde{\eta} - \frac{h^2}{3}\nabla_H^2 \frac{\partial}{\partial T}\langle \mathbf{U} \rangle = 0 \tag{9.44}$$

$$P = \rho g(\tilde{\eta} - Z) + \frac{\rho}{2}(2Zh + Z^2)\nabla_H \cdot \frac{\partial}{\partial T}\langle \mathbf{U} \rangle. \tag{9.45}$$

Equations (9.40) and (9.41) or equivalently (9.43) and (9.44) are called the Boussinesq equations. Here the pressure term is no longer hydrostatic. Moreover, the Airy and Boussinesq theories differ by the linear term multiplied by γ^2 in eqn (9.42).

Remark

The two parameters $\delta = (A/h)$ and $\gamma = kh$ can be treated as nonlinearity and dispersive parameters. The Boussinesq equation accounts for the effects of nonlinearity δ and dispersion γ^2 to the leading order. However, when $\delta \gg \gamma^2$ they reduce to Airy's equations which are valid for all δ; when $\delta \ll \gamma^2$ they reduce to the linearized approximation with weak dispersion. When $\delta \to 0$ and $\gamma^2 \to 0$ the classical linearized theory can be recovered. The foregoing analysis is valid for the constant depth case. For the variable depth case, the reader is referred to the standard textbooks including Mei (1983).

9.3 Nonlinear dispersive long waves of permanent nature

Mathematical solutions of the form $\exp[ik(x - ct)]$ always represent wave propagating at a constant speed c called the phase speed. These waves propagate without change of form and are called progressive waves. We know that the role of nonlinearity is to steepen a crest while dispersion tends to disperse into waves of different wavelengths. Thus progressive waves have the role of dynamical equilibrium to balance these two effects.

Normalized Boussinesq equations (9.40) and (9.41) can be written for one-dimensional waves:

$$\frac{\partial \eta}{\partial t} + \delta \frac{\partial}{\partial x}\left(\eta \frac{\partial \phi}{\partial x}\right) + \frac{\partial^2 \phi}{\partial x^2} = 0 \tag{9.46}$$

$$\frac{\partial \phi}{\partial t} + \eta + \frac{\delta}{2}\left(\frac{\partial \phi}{\partial x}\right)^2 = \frac{\gamma^2}{3}\frac{\partial^2 \phi}{\partial x \partial t}. \tag{9.47}$$

Eliminating η from (9.46) and (9.47) we get a single equation in ϕ:

$$\frac{\partial^2 \phi}{\partial t^2} - \frac{\partial^2 \phi}{\partial x^2} = \frac{\gamma^2}{3}\frac{\partial^4 \phi}{\partial x^2 \partial t^2} - \delta \frac{\partial}{\partial t}\left[\left(\frac{\partial \phi}{\partial x}\right)^2 + \frac{1}{2}\left(\frac{\partial \phi}{\partial t}\right)^2\right]. \tag{9.48}$$

In evaluating this we ignored terms smaller than $O(\delta, \gamma^2)$.

Rewriting eqn (9.48) in subscript notation for convenience, we obtain

$$\phi_{tt} - \phi_{xx} = \frac{\gamma^2}{3} \phi_{xxtt} - \delta \left(\delta_x^2 + \tfrac{1}{2} \phi_t^2 \right)_t.$$ (9.49)

This is the differential equation which we want to examine for possible progressive wave solution. We now seek a solution of the following form

$$\phi = f(\xi) \quad \text{with} \quad \xi = x - ct.$$ (9.50)

Substituting this form of solution into eqn (9.49), we obtain the following ordinary differential equation in f:

$$(c^2 - 1)f'' = \frac{\gamma^2}{3} c^2 f^{IV} + \delta c \left(1 + \frac{c^2}{2} \right)(f'^2)'.$$ (9.51)

This equation implies that $c = 1 + O(\delta, \gamma^2)$. Integrating once with respect to ξ yields

$$(c^2 - 1)f' + B_1 = \frac{\gamma^2}{3} c^2 f''' + \delta c \left(1 + \frac{c^2}{2} \right) f'^2.$$ (9.52)

But we know $\eta = -\phi_t = cf'$, and $c = O(1)$ where $c = C/\sqrt{gh}$. Hence without loss of generality eqn (9.52) can be rewritten as

$$(c^2 - 1)\eta + B_1 = \frac{\gamma^2}{3} \eta'' + \frac{3}{2} \delta \eta^2.$$ (9.53)

Multiplying this equation by η' and integrating with respect to ξ again yields

$$(c^2 - 1)\frac{\eta^2}{2} + B_1 \eta + B_2 = \frac{\gamma^2}{6} \eta'^2 + \frac{\delta}{2} \eta^3$$

which can be rewritten as

$$-\frac{\delta}{2} \eta^3 + (c^2 - 1)\frac{\eta^2}{2} + B_1 \eta + B_2 = \frac{\gamma^2}{6} \eta'^2$$ (9.54)

where the integration constants B_1 and B_2 are both of $O(\delta, \gamma^2)$.

In the following we shall investigate two important analytical solutions of physical interest: (A) solitary waves, (b) cnoidal waves.

9.3.1 Solitary waves

A solitary wave has a single crest whose amplitude diminishes to zero as $|\xi| \to \infty$. Therefore, as $|\xi| \to \infty$, $\eta = 0$, $\eta' = 0$ and $\eta'' = 0$ so that $B_1 = 0$, $B_2 = 0$. Equation (9.54) simply becomes

$$(\eta')^2 = 3\frac{\delta}{\gamma^2}\left(\frac{c^2-1}{\delta} - \eta\right)\eta^2. \tag{9.55}$$

Since the left-hand side of (9.55) is positive, the right-hand side must be positive also for a physical solution.

Therefore, we must have $c > 1$ or in a physical variable, $C > \sqrt{gh}$, which is known as the supercritical wave speed. Moreover, $\eta \le \left(\dfrac{c^2-1}{\delta}\right)$ must also hold good. Hence $(c^2 - 1)/\delta$ is just the maximum amplitude of the crest which is unity because of normalization. Therefore,

$$\frac{c^2-1}{\delta} = 1 \quad \text{or} \quad c^2 = 1 + \delta. \tag{9.56}$$

In dimensional form (9.56) becomes

$$C = \sqrt{gh}\,(1+A/h)^{1/2} \quad \text{or} \quad C = [g(h+A)]^{1/2}. \tag{9.57}$$

This relation was first found by Rayleigh.

Putting eqn (9.56) into eqn (9.55) yields

$$\left(\frac{d\eta}{d\xi}\right) = \frac{(3\delta)^{1/2}}{\gamma}\,\eta(1-\eta)^{1/2}$$

which can be integrated to give

$$\frac{(3\delta)^{1/2}}{\gamma}(\xi-\xi_0) = -2\tanh^{-1}(1-\eta)^{1/2}$$

which simplifies to

$$\eta = \mathrm{sech}^2\left[\frac{(3\delta)^{1/2}}{2\gamma}(\xi-\xi_0)\right] \tag{9.58}$$

where ξ_0 is an integration constant and can be chosen to be zero. In dimensional form the surface profile can be written as

$$\tilde{\eta} = A\,\mathrm{sech}^2\left[\frac{\sqrt{3}}{2}\left(\frac{A}{h^3}\right)^{1/2}(X-CT)\right]. \tag{9.59}$$

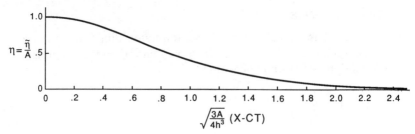

Fig. 9.1. Dimensionless free-surface profile of a solitary wave.

This is called the solitary wave of Boussinesq (1871, 1872). The solitary wave form is shown in Fig. 9.1. The A therefore represents the height of the wave and h the depth at infinity.

Solitary waves can be easily generated in a long tank by giving any kind of impulse. For applications of the solitary wave theory, the reader is referred to the extensive work of Munk (1949). Numerical solutions of solitary waves can be found in the work of Byatt-Smith (1970). Fenton (1972) extended the expansion procedure of Benjamin and Lighthill (1954) to the ninth order. In a series of remarkable papers, Longuet-Higgins (1973, 1974) and his associates made a very thorough investigation of solitary waves (see Byatt-Smith and Longuet-Higgins (1976)).

9.3.2 Cnoidal waves

Besides the solitary waves discussed in the previous section, the solution of eqn (9.54) may admit periodic progressive waves also. Korteweg and de Vries (1895) noted this periodic wave in the study of shallow water wave theory. These waves reduce to the solitary waves presented above under certain simplifying boundary conditions. We shall present here the analytical solutions of such periodic progressive waves.

Rewrite eqn (9.54) as follows:

$$\left(\frac{\gamma^2}{3\delta}\right)\eta'^2 = -\eta^3 + \frac{c^2-1}{\delta}\eta^2 + A_1\eta + A_2 \tag{9.60}$$

where A_1 and A_2 are the two redefined integration constants. Because the right-hand side of eqn (9.60) is a cubic polynomial, it can be factorized in the following manner:

$$\left(\frac{\gamma^2}{3\delta}\right)\eta'^2 = (\eta - \eta_1)(\eta - \eta_2)(\eta_3 - \eta) = P_3(\eta) \tag{9.61}$$

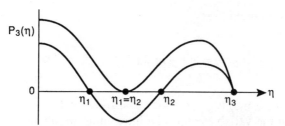

Fig. 9.2. Sketch of $P_3(\eta)$ to show three zeros.

in which $\eta_1 < \eta_2 < \eta_3$ are the three real zeros of the third-order polynomial P_3, as shown in Fig. 9.2.

It can be easily verified that η must lie between the two zeros η_3 and η_2 which correspond to the heights of the crest and the trough, respectively.

Let

$$\eta_3 - \eta_2 = H. \tag{9.62}$$

Equation (9.61) can be integrated in terms of an elliptic integral by introducing $\eta = \eta_3 \cos^2 \theta + \eta_2 \sin^2 \theta$ with

$$\theta = \theta(\eta). \tag{9.63}$$

Then

$$\eta' = (\eta_3 - \eta_2)(-2\sin\theta\cos\theta)\theta'. \tag{9.64}$$

Then inserting eqns (9.63) and (9.64) into eqn (9.61) yields

$$4\theta'^2(\eta_3 - \eta_2)^2 \sin^2\theta \cos^2\theta$$

$$= \left(\frac{3\delta}{\gamma^2}\right)\left[\eta_3 \cos^2\theta + \eta_2 \sin^2\theta - \eta_1\right]$$

$$\times \left[\eta_3 \cos^2\theta - \eta_2(1 - \sin^2\theta)\right] \times \left[\eta_3(1 - \cos^2\theta) - \eta_2 \sin^2\theta\right]$$

$$= \left(\frac{3\delta}{\gamma^2}\right)\left[(\eta_3 - \eta_1) - (\eta_3 - \eta_2)\sin^2\theta\right]$$

$$\times \cos^2\theta(\eta_3 - \eta_2) \times \sin^2\theta(\eta_3 - \eta_2)$$

$$= \left(\frac{3\delta}{\gamma^2}\right)(\eta_3 - \eta_2)^2 \cos^2\theta \sin^2\theta\left[(\eta_3 - \eta_1) - (\eta_3 - \eta_2)\sin^2\theta\right].$$

Simplifying we obtain

$$\theta'^2 = \left(\frac{3\delta}{4\gamma^2}\right)(\eta_3 - \eta_1)(1 - m\sin^2\theta) \tag{9.65}$$

where $m = (\eta_3 - \eta_2)/(\eta_3 - \eta_1)$ such that $0 < m < 1$.

Separating the variables and then integrating we obtain

$$\int_0^\theta \frac{d\theta}{(1 - m\sin^2\theta)^{1/2}} = \pm\frac{(3\delta)^{1/2}}{2\gamma}(\eta_3 - \eta_1)^{1/2}(\eta - \eta_0) = f(\theta, m). \tag{9.66}$$

This integral is known as the incomplete Jacobian elliptic integral with modulus m and amplitude θ. More explicitly, we denote

$$\cos\theta = C_n\left(\frac{(3\delta)^{1/2}}{2\gamma}(\eta_3 - \eta_1)^{1/2}(\eta - \eta_0)\right),$$

$$\sin\theta = S_n\left(\frac{(3\delta)^{1/2}}{2\gamma}(\eta_3 - \eta_1)^{1/2}(\eta - \eta_0)\right)$$

where C_n and S_n are called the cosine-elliptic and the sine-elliptic functions respectively.

This surface height is then given by

$$\eta = \eta_3\cos^2\theta + \eta_2\sin^2\theta = \eta_2 + (\eta_3 - \eta_2)\cos^2\theta$$

$$= \eta_2 + (\eta_3 - \eta_2)C_n^2\left(\frac{(3\delta)^{1/2}}{2\gamma}(\eta_3 - \eta_1)^{1/2}(\eta - \eta_0)\right). \tag{9.67}$$

This can be written in physical terms as

$$\tilde{\eta} = \tilde{\eta}_2 + (\tilde{\eta}_3 - \tilde{\eta}_2)C_n^2\left(\frac{3^{1/2}}{2}\frac{(\tilde{\eta}_3 - \tilde{\eta}_1)^{1/2}}{h^{3/2}}(X - CT - X_0)\right). \tag{9.68}$$

Korteweg and de Vries (1895) coined the word cnoidal for the function C_n. Thus eqn (9.67) or (9.68) is now called the *CNOIDAL WAVE*.

Since $\cos\theta$ is periodic with period 2π, the Jacobian elliptic function C_n is a periodic function with a period of $4K$, where K is a complete elliptic integral of the first kind defined by

$$K = f(\pi/2, m) = \int_0^{\pi/2} \frac{d\theta}{(1 - m\sin^2\theta)^{1/2}}. \tag{9.69}$$

Since the function $\cos^2\theta$ has the period π, so C_n^2 must have the period $2K$ such that the dimensionless wavelength λ of a cnoidal wave is given by

$$\frac{(3\delta)^{1/2}}{2\gamma}(\eta_3 - \eta_1)^{1/2}\lambda = 2K$$

or

$$\lambda = \frac{4K\gamma}{[3\delta(\eta_3 - \eta_1)]^{1/2}}. \tag{9.70}$$

The wavelength depends on the amplitude m and therefore eqn (9.67) may be written as

$$\eta = \eta_2 + (\eta_3 - \eta_2)C_n^2\left(\frac{2K}{\lambda}(\eta - \eta_0)\right). \tag{9.71}$$

Also the wave speed is found from eqn (9.60) in terms of η_1, η_2 and η_3 as $(c^2 - 1)/\delta = \eta_1 + \eta_2 + \eta_3$ or

$$c^2 = 1 + \delta(\eta_1 + \eta_2 + \eta_3). \tag{9.72}$$

The cnoidal wave is specified by three parameters η_3, η_2 and η_1. However, for engineering applications, it is more convenient if these parameters are replaced by the wavelength λ, the wave height H and the mean depth. In the following we present this analysis.

The mean depth is defined in such a way that the net area occupied by fluid within a wavelength is zero. Therefore, $\int_0^\lambda \eta \, d\eta = 0$ which simply implies that

$$\int_0^\pi (\eta_3 \cos^2\theta + \eta_2 \sin^2\theta)\frac{d\eta}{d\theta} d\theta = 0.$$

300 Nonlinear long waves in shallow water

However, from eqn (9.65), after dropping the constant multiplier, we have

$$\frac{d\eta}{d\theta} = \frac{1}{(1 - m \sin^2 \theta)^{1/2}}.$$

Also

$$\eta_3 \cos^2 \theta + \eta_2 \sin^2 \theta = \eta_3 - (\eta_3 - \eta_2)\sin^2 \theta$$
$$= \eta_1 + (\eta_3 - \eta_1) - (\eta_3 - \eta_2)\sin^2 \theta$$
$$= \eta_1 + (\eta_3 - \eta_1)(1 - m \sin^2 \theta).$$

Therefore the above integral is written as

$$\int_0^{\pi/2} \frac{\eta_1 + (\eta_3 - \eta_1)(1 - m \sin^2 \theta)}{(1 - m \sin^2 \theta)^{1/2}} \, d\theta = 0 \tag{9.73}$$

or

$$\eta_1 K(m) + (\eta_3 - \eta_1)E(m) = 0 \tag{9.74}$$

where $E(m) = \int_0^{\pi/2}(1 - m \sin^2 \theta)^{1/2} \, d\theta$ is an elliptic integral of the second kind.

Equation (9.74) yields

$$\eta_1 K(m) + \frac{\eta_3 - \eta_2}{m} E(m) = 0,$$

or

$$\eta_1 K(m) + \frac{H}{m} E(m) = 0.$$

Hence

$$\eta_1 = -\frac{HE}{mK} \tag{9.75}$$

because $H = \eta_3 - \eta_2$. Then from eqn (9.74)

$$\eta_3 = \eta_1\left(1 - \frac{K}{E}\right) = -\frac{HE}{mK}\left(1 - \frac{K}{E}\right) \quad \text{or} \quad \eta_3 = \frac{H}{m}\left(1 - \frac{E}{K}\right). \tag{9.76}$$

Also we know

$$\eta_3 - \eta_2 = H,$$

$$\eta_2 = \eta_3 - H = \frac{H}{m}\left(1 - \frac{E}{K}\right) - H = \frac{H}{m}\left(1 - m - \frac{E}{K}\right). \tag{9.77}$$

These expressions may then be inserted to obtain the dimensionless wave speed

$$c^2 = 1 + \delta(\eta_1 + \eta_2 + \eta_3) = 1 + \delta\frac{H}{m}\left(2 - m - \frac{3E}{K}\right). \tag{9.78}$$

From eqn (9.70) the dimensionless wavelength is then

$$\lambda = \frac{4K\gamma}{(3\delta)^{1/2}}\left(\frac{m}{H}\right)^{1/2}. \tag{9.79}$$

The dimensionless wave period is then

$$p = \frac{\lambda}{c} = \frac{\left(4\gamma/(3\delta)^{1/2}\right)\left(\dfrac{m}{H}\right)^{1/2}K}{\left[1 + \delta\left(\dfrac{H}{m}\right)\left(2 - m - \dfrac{3E}{K}\right)\right]^{1/2}}. \tag{9.80}$$

To obtain the physical variables the following transformation is necessary:

$$x = kX, \quad t = k(gh)^{1/2}T, \quad c = (gh)^{1/2}C, \quad \lambda = k\Lambda$$

$$H = \frac{\hat{H}}{A}, \quad \gamma = kh, \quad \delta = \frac{A}{h}$$

where X, T, C, Λ and $\hat{H}$ are the physical variables.

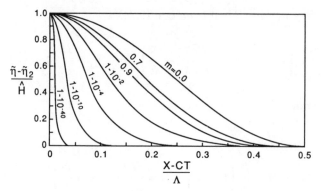

Fig. 9.3. Surface profile of cnoidal waves.

In physical variables, then, we have

$$C^2 = gh\left[1 + \frac{\hat{H}}{h}\frac{1}{m}\left(2 - m - \frac{3E}{K}\right)\right] \tag{9.81}$$

$$\Lambda = 4Kh\left(\frac{mh}{3\hat{H}}\right)^{1/2} \tag{9.82}$$

$$P = \left(\frac{h}{g}\right)^{1/2}\frac{4K\left(mh/3\hat{H}\right)^{1/2}}{\left[1 + \frac{\hat{H}}{h}\frac{1}{m}\left(2 - m - \frac{3E}{K}\right)\right]^{1/2}} \tag{9.83}$$

$$\tilde{\eta} = \tilde{\eta}_2 + \hat{H}C_n^2\left(\frac{2K}{\Lambda}(X - CT)\right). \tag{9.84}$$

Wiegel (1960) has plotted the wave profile $\tilde{\eta}$ for various values of m ranging from $m = 0$ to 1 and these results are reproduced in Figs 9.3 and 9.4. Limiting cases of cnoidal waves reduce to some interesting results as follows:

Case (i) $m \to 1$

In this situation

$$m = \frac{\eta_3 - \eta_2}{\eta_3 - \eta_1} \to 1$$

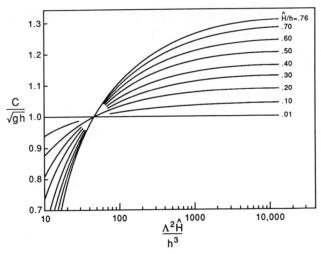

Fig. 9.4. Dispersion relation of cnoidal waves.

provided $\eta_2 \to \eta_1$. Consequently $E(1) = 1$ and $K(1) \to \infty$. Then $\Lambda \to \infty$ and $C^2 v = \mathrm{sech}^2 v$. Also the ratio K/λ in eqn (9.79) approaches a finite limit so that the wave elevation becomes

$$\tilde{\eta} = \hat{H}\, \mathrm{sech}^2 \left[\frac{3^{1/2}}{2} \left(\frac{\hat{H}}{h^3} \right)^{1/2} (X - CT) \right]$$

which is the form of the solitary wave given in eqn (9.59). Thus it is evident that the solitary wave is the limit of the cnoidal wave with infinite wavelength.

The wave speed then becomes $C^2 = gh(1 + \hat{H}/h)$.

Case (ii) $m \to 0$

In this situation $\eta_3 - \eta_2 = H \to 0$. This means the given wave is infinitesimal. It can be easily verified that in this case $C^2 \to gh$, $C_n(v) \to \cos v$, $K \to \pi/2$, and the wave elevation is given by

$$\tilde{\eta} = \tilde{\eta}_2 + \hat{H} \cos^2 \left[\frac{\pi}{\Lambda} (X - CT) \right].$$

Now $\tilde{\eta}_2 = -A = -\tfrac{1}{2}\hat{H}$ so that

$$\tilde{\eta} = \frac{\hat{H}}{2} \cos \left[\frac{2\pi}{\Lambda} (X - CT) \right]$$

which is the linearized sinusoidal wave and the wave speed $C = \sqrt{gh}$. Thus cnoidal wave theory spans the range from sinusoidal or Airy theory in deep water to solitary wave theory in shallow water.

9.4 The Korteweg–de Vries (KdV) equation

The nonlinear permanent wave given by eqn (9.48) may be rewritten in subscript notation as

$$\phi_{tt} - \phi_{xx} = \frac{\gamma^2}{3} \phi_{xxtt} - \delta \left(\phi_x^2 + \frac{1}{2} \phi_t^2 \right)_t. \tag{9.85}$$

Use the following transformations:

$$\mu = x - t \quad \text{and} \quad \tau = \delta t. \tag{9.86}$$

In terms of these transformations, the derivatives become

$$\frac{\partial}{\partial x} = \frac{\partial}{\partial \mu}$$

$$\frac{\partial}{\partial t} = \frac{\partial \mu}{\partial t} \frac{\partial}{\partial \mu} + \frac{\partial \tau}{\partial t} \frac{\partial}{\partial \tau} = -\frac{\partial}{\partial \mu} + \delta \frac{\partial}{\partial \tau}$$

$$\frac{\partial^2}{\partial x^2} = \frac{\partial^2}{\partial \mu^2}$$

$$\frac{\partial^2}{\partial t^2} = \frac{\partial^2}{\partial \mu^2} - 2\delta \frac{\partial^2}{\partial \mu \partial \tau} + \delta^2 \frac{\partial^2}{\partial \tau^2}.$$

By substituting these transformations into eqn (9.85), we obtain

$$\left(\phi_{\mu\mu} - 2\delta \phi_{\mu\tau} + \delta^2 \phi_{\tau\tau} \right) - \left(\phi_{\mu\mu} \right)$$

$$= \frac{\delta^2}{3} \left(\phi_{\mu\mu\mu\mu} - 2\delta \phi_{\mu\mu\mu\tau} + \delta^2 \phi_{\mu\mu\tau\tau} \right) - \delta \left(-\frac{\partial}{\partial \mu} + \delta \frac{\partial}{\partial \tau} \right)$$

$$\times \left(\phi_\mu^2 + \frac{1}{2} (-\phi_\mu + \delta\phi_\tau)^2 \right)$$

which can be immediately written as

$$-\delta(2\phi_{\mu\tau}) = \frac{\gamma^2}{3}\phi_{\mu\mu\mu\mu} + \frac{3\delta}{2}(\phi_\mu^2)_\mu + O(\delta,\gamma^2)$$

or

$$(\phi_\mu)_\tau + \frac{3}{4}(\phi_\mu^2)_\mu + \frac{\gamma^2}{6\delta}\phi_{\mu\mu\mu\mu} = O(\delta,\gamma^2).$$

If we substitute $\phi_\mu = u$, then the above equation reduces to

$$u_\tau + \frac{3}{2}uu_\mu + \frac{\gamma^2}{6\delta}u_{\mu\mu\mu} = 0. \tag{9.87}$$

This equation is commonly called the Korteweg–de Vries (KdV) equation. In physical variables, it can be written as

$$\frac{\partial\tilde\eta}{\partial T} + \sqrt{gh}\left(1 + \frac{3}{2}\frac{\tilde\eta}{h}\right)\frac{\partial\tilde\eta}{\partial X} + \frac{h^2}{6}\sqrt{gh}\frac{\partial^3\tilde\eta}{\partial X^3} = 0. \tag{9.88}$$

In 1985, Korteweg and de Vries derived the nonlinear equation (9.88) for long water waves of phase velocity $C_0 = \sqrt{gh}$ in a channel of constant depth h which has the remarkable alternative form

$$\tilde\eta_T + C_0\left(1 + \frac{3}{2}\frac{\tilde\eta}{h}\right)\tilde\eta_X + \omega\tilde\eta_{XXX} = 0$$

where $\omega = \frac{1}{6}C_0h^2$ is a constant for fairly long waves. The dispersion relation (9.3) for a dispersive surface wave on water of constant depth h is

$$\sigma = \sqrt{gh\tanh kh} = C_0k\sqrt{1 - \frac{1}{3}k^2h^2} \simeq C_0k\left(1 - \frac{1}{6}k^2h^2\right).$$

The phase and group velocities of the waves can be obtained, respectively, as $C_p = \sigma/k = C_0 - \omega k^2$ and $C_g = d\sigma/dk = C_0 - 3\omega k^2$. This is the simplest nonlinear equation for dispersive waves, and combines the nonlinearity and dispersion. The most remarkable feature is that the dispersive term in the KdV equation thus allows the solitary and periodic waves which are not found in any shallow water wave theory. We shall discuss the general solution of the KdV equation (9.87) in Chapter 10.

9.5 The validity of different wave theories

So far we have learned about a number of wave theories. They are (a) linear theory (Airy), (b) nonlinear theory (Stokes), (c) solitary wave theory (Boussinesq) and (d) cnoidal wave theory (Boussinesq). It is thus important to determine the validity of the various water wave theories which may be applicable for certain practical problems. This validity is usually determined by two aspects, namely the physical and mathematical validity. By the physical validity we mean how well the mathematical prediction of the various theories agrees with the actual measurements. Therefore, the mathematical theory which governs the physics of the problem must be compatible. The mathematical theory is the boundary value problem which must satisfy the field equation along with all the relevant boundary conditions. It is found that all the theories mentioned above satisfy the bottom boundary condition exactly. However, the solitary and cnoidal wave theories only approximately satisfy Laplace's equation within the fluid. The analytical validity of these theories was examined by Dean (1970, 1974). Figure 9.5 shows the results of the comparison. This figure provides the regions of the best fit of these theories. It has been observed that the cnoidal wave theory does well in shallow water while the Stokes fifth-order theory proved to be more applicable in deep water. The linear wave theory seems to be well fitted for the intermediate water depths.

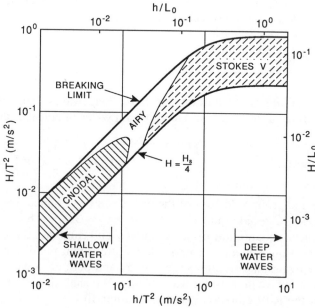

Fig. 9.5.   Comparison of different wave theories providing best fit (from Dean (1970, 1974)).

9.6 Exercises

1. Determine the region of validity for the second-order Stokes theory. Find the value of the Ursell parameter which is more restrictive.

2. What is the difference between Airy's theory and Boussinesq's theory in shallow water waves? It is to be noted that Airy's equations are obtained by letting $\gamma = kh \to 0$ and $\delta = A/h = O(1)$; and Boussinesq's equations are obtained by letting $O(\delta) = O(\gamma^2) < 1$. Explain the role of these two parameters.

3. In the perturbation expansion of the shallow water wave, two small parameters, namely $\delta = A/h$ and $\gamma = kh$, play a very important role in deciding the Airy and Boussinesq theory. The effect of nonlinearity is accounted by δ whereas the dispersion effect is accounted by γ^2. What justification can be made for this choice? Discuss.

4. What is the velocity of the solitary wave given in eqn (9.59)? Plot the wave profile for different wave celerity.

References

Airy, G. B. (1845). Tides and waves. *Encycl. Metrop.*, Art. 192, 241–396.

Benjamin, T. B. and Lighthill, M. J. (1954). On cnoidal waves and bores. *Proc. R. Soc.*, A224, 448–460.

Benney, D. J. (1962). Nonlinear gravity wave interactions. *J. Fluid Mech.*, 14, 557–584.

Benney, D. J. (1966). Long nonlinear waves in fluid flows. *J. Math. Phys.*, 45, 52–63.

Boussinesq, J. (1871). Theorie de l'intumescence liquide, appelee onde solitaire ou de translation se prapageant dans un canal rectangulaire. *CR Acad. Sci., Paris,* 72, 755–759.

Boussinesq, J. (1872). Theorie des ondes et des remous qui se propagent le long d'un canal rectangulaire horizontal, en communiquant au liquide contenu dans ce canal des vitesses sensiblement pareilles de la surface au fond *J. Math. Pures Appl.*, 17, 55–108.

Boussinesq, J. (1877). *Mem. Pres. Acad. Sci.*, 3rd edition, Paris, XXIII, p. 46.

Byatt-Smith, J. G. B. (1970). An exact integral equation for steady surface waves. *Proc. R. Soc.*, A315, 405–418.

Byatt-Smith, J. G. B. and Longuet-Higgins, M. S. (1976). On the speed and profile of steep solitary waves. *Proc. R. Soc.*, A350, 175–189.

Dean, R. G. (1970). Relative validity of water wave theories. *J. Waterw., Harbors Div., ASCE*, 96(WW1), 105–109.

Dean, R. G. (1974). Evaluation and development of water wave theories for engineering application, Vols 1 and 2, *Spec. Rep.* 1, US Army, *Coastal Engineering Research Center*, Fort Belvoir, Virginia.

Fenton, J. D. (1972). A ninth order solution for the solitary wave. *J. Fluid Mech.*, 53, 257–271.

Korteweg, D. J. and de Vries, G. (1895). On the change of form of long waves advancing in a rectangular canal, and on a new type of long stationary waves. *Philos. Mag.*, Ser 5, 39, 422–443.

Longuet-Higgins, M. S. (1973). On the form of highest progressive and standing waves in deep water. *Proc. R. Soc.*, A331, 445–456.

Longuet-Higgins, M. S. (1974). On the mass, momentum, energy and circulation of a solitary wave. *Proc. R. Soc.*, A337, 1–13.

Mei, C. C. (1983). *The Applied Dynamics of Ocean Surface Waves*, Wiley Interscience, New York.

Munk, W. H. (1949). The solitary wave theory and its applications to surf problems. *Ann. New York Acad. Sci.*, 51, 376–423.

Peregrine, D. H. (1967). Long waves on a beach. *J. Fluid. Mech.*, 27, 815–827.

Ursell, F. (1953). The long-wave paradox in the theory of gravity waves. *Proc. Cambridge Philos. Soc.*, 49, 685–694.

Wiegel, R. L. (1960). A presentation of Cnoidal wave theory for practical application. *J. Fluid Mech.*, 7, 273–286.

10
Inverse scattering transforms and the theory of solitons

10.1 Introduction

The theory of solitons and the method of inverse scattering transforms have received extreme interest from applied mathematicians and physicists in recent years, due to the most recent and fascinating achievements in the domain of applied mathematics. The KdV equation which leads to the theory of solitons has been the object of intensive scientific research in different branches of physics. We have already seen in Chapter 9, that under certain periodic boundary conditions, the KdV equation admits solitary wave solutions. These solitary waves which travel at different speeds may pass through one another without any change in the original structures; the only effect of the interaction is a change of phase. These phenomena are obvious in particle physics, such as electrons, protons, etc. In 1965, Zabusky and Kruskal coined the word soliton for solitary waves.

In 1967, Gardner et al. published a remarkable paper on the analytical solution to the initial value problem due to a disturbance of finite length in an infinite domain $|X| < \infty$. Many survey papers on the study of nonlinear waves may be found in Ablowitz and Segur (1981), Karpman (1975), Gardner et al. (1974), Lamb (1945), Scott et al. (1973), Miura (1976), Whitham (1974) and Bullough and Caudrey (1980).

The solitons can be of great use in practice. If we consider a pulse which carries a bit of information with it such that it suffers a significant dispersion, then on reaching the destination, the pulse may be so spread out and blurred that the information may be totally unintelligible. However, if the pulse travels as a soliton, then it can carry the information over a large distance without being distorted and without suffering any significant loss in its intensity.

Most survey papers on the Kortweg–de Vries equation start with a quotation from J. Scott-Russell's 'Report on Waves' in 1844 describing his famous chase on horseback behind a wave in a channel. We reproduce here his interesting and exciting description of the solitary wave which he observed accidentally.

'I was observing the motion of a boat which was rapidily drawn along a narrow channel by a pair of horses, when the boat suddenly stopped—not so the mass of water in the channel which it had put in the motion; it accumulated round the prow of the vessel in a state of violent agitation, then suddenly leaving it behind, rolled forward with great velocity assuming the form of a large solitary elevation, a rounded, smooth and well defined heap of water, which continued its course along the channel apparently without change of form or diminution of speed. I followed it on horseback, and overtook it still rolling at a rate of some eight to nine miles an hour, preserving its original figure some thirty feet long and a foot to a foot and half in height. Its height gradually diminished and after a chase of one or two miles I lost it in the windings of the channel. Such, in the month of August, 1834, was my first chance interview with that singular and beautiful phenomenon.'

10.2 Mathematical development

In the following discussion, we obtain the standard form of the Korteweg–de Vries (KdV) equation by making the transformation

$$u = -4v$$

$$\mu = \frac{\gamma}{(6\varepsilon)^{1/2}} x \tag{10.1}$$

$$\tau = \frac{\gamma}{(6\varepsilon)^{1/2}} t$$

to eqn (8.87) presented in Chapter 9.

The KdV equation then can be obtained as

$$v_t - 6vv_x + v_{xxx} = 0. \tag{10.2}$$

The numerical factor in front of the second term does not have any particular significance.

Assume the initial condition $v(x,0) = v_0(x)$ to be bounded and three times continuously differentiable and as in Chapter 9 consider the following boundary conditions.

$v(x,t)$, along with its derivatives, tends to zero as

$$|x| \to \infty. \tag{10.3}$$

Under these boundary conditions, the solitary wave solution of eqn (10.2) is

$$v(x,t) = -\frac{b^2}{2}\left(\text{sech}^2\left[\frac{b}{2}(x - b^2 t)\right]\right) \tag{10.4}$$

where $b^2/2$ is the amplitude of the solitary wave.

10.3 The Schrödinger equation and its properties

The transformed problem (10.2) turns out to centre around the steady one-dimensional Schrödinger equation

$$\Psi_{xx} + (\lambda - v)\Psi = 0 \tag{10.5}$$

where $v = v(x, t)$ with t being a parameter.

In quantum mechanics, the function Ψ in eqn (10.5) is the wave function of a moving particle under an external field whose potential energy $v(x, t)$ is prescribed. Here the problem is to find the eigenvalues λ and the corresponding eigenfunctions.

Landau and Lifschitz (1958) have discussed the general properties of the Schrödinger equation (10.5). We shall, however, record below only those properties which will be needed in this investigation.

(i) The eigenvalues of λ may be discrete or discontinuous or both in a given problem.

(ii) The discrete eigenvalues are negative and correspond to the stable states of finite motion of the particle. We denote the discrete eigenvalues by

$$\lambda = \lambda_n = -k_n^2, \qquad n = 1, 2, 3, \ldots \tag{10.6}$$

where $k_i > 0$ such that the corresponding eigenfunctions Ψ_n vanish at infinity. Their asymptotic behavior at infinity is

$$\Psi_n \sim b_n e^{-k_n x} \qquad x \to \infty, \qquad \Psi_n \sim d_n e^{k_n x} \qquad x \to -\infty \tag{10.7}$$

where b_n and d_n are the normalization constraints chosen so that

$$\int_{-\infty}^{\infty} \Psi_n^2 \, dx = 1 \qquad \text{for all} \qquad n. \tag{10.8}$$

In general, k_n, b_n and d_n depend on the parameter t.

(iii) The continuous eigenvalues correspond to infinite motion in which the particle reaches infinity. The potential field $v(x)$ may be neglected at sufficiently large distances and the particle may be regarded as free. In this situation, the energy of a free particle is positive, which suggests that the continuous eigenvalues are positive,

$$\lambda = k^2, \quad k > 0. \tag{10.9}$$

(iv) None of the eigenvalues is degenerate. That means one and only one eigenfunction corresponds to each discrete eigenvalue. This can be easily shown as follows.

If possible, let there be two eigenfunctions Ψ_1 and Ψ_2 corresponding to only a discrete eigenvalue λ. Then from eqn (10.5) we have

$$\frac{\Psi_{1xx}}{\Psi_1} = v - \lambda = \frac{\Psi_{2xx}}{\Psi_2} \quad \text{or} \quad \Psi_{1xx}\Psi_2 - \Psi_{2xx}\Psi_1 = 0.$$

Integrating once, we obtain $\Psi_{1x}\Psi_2 - \Psi_{2x}\Psi_1 = \text{constant} = 0$ in view of the boundary condition $\Psi_{1,2} \to 0$ as $|x| \to \infty$. Therefore, we have $(\Psi_{1x}/\Psi_1) = (\Psi_{2x}/\Psi_2)$ which yields the following after integration: $\Psi_1 = A\Psi_2$, where A is a constant. This means that the two eigenfunctions are the same except for a constant factor. Thus, the discrete eigenvalues are not degenerate. However, the continuous eigenvalues are degenerate.

(v) Let Ψ_n correspond to λ_n when we arrange the discrete eigenvalues in order of magnitude: $\lambda_1 < \lambda_2 < \lambda_3 < \ldots < \lambda_m$. Then the $(n + 1)$th eigenfunction vanishes n times in the finite domain of the x-axis.

(vi) Consider a potential such that $v(x) \to 0$ as $|x| \to \infty$. Then, when $|x| \to \infty$, the Schrödinger equation assumes the following asymptotic form: $\Psi_{xx} + \lambda\Psi = 0$. For discrete eigenvalues, it takes the form $\Psi_{xx} - k^2\Psi = 0$ and its two independent solutions are

$$\Psi = c_{\pm} \exp(\pm kx). \tag{10.10}$$

Clearly, when $x \to \infty$, the physical solution is $\exp(-kx)$, and when $x \to -\infty$, the physical solution is $\exp(kx)$. For the continuous eigenvalues, the Schrödinger equation takes the form $\Psi_{xx} + k^2\Psi = 0$ and two independent solutions are

$$\Psi = b_{\pm} \exp(\pm ikx). \tag{10.11}$$

Here the solution $\exp(ikx)$ corresponds to the particle moving in the positive x-direction, where the solution $\exp(-ikx)$ relates to the particle moving in the negative x-direction.

(vii) When $x \to \infty$ and $v(x) \to 0$, we can take the wave function of a continuous eigenvalue asymptotically as a linear combination of two plane waves $\exp(\pm ikx)$ and similarly when $x \to -\infty$. Consider a plane wave coming from $x = \infty$. Then the asymptotic behavior of the eigenfunction is

$$\Psi \sim \exp(-ikx) + b(k)\exp(ikx) \qquad \text{as} \qquad x \to \infty,$$

and

$$\Psi \sim a(k)\exp(-ikx) \qquad \text{as} \qquad x \to -\infty. \tag{10.12}$$

Clearly, for each of these k, Ψ describes a direct scattering problem where the complex number $b(k)$ and $a(k)$ depending on the wave number are, respectively, the reflection and the transmission coefficients. Both the coefficients depend on the parameter t. Here we have taken the incident wave with amplitude unity.

In the case of discrete eigenvalues k_m the eigenfunction $\Psi_m \to 0$ as $|x| \to \infty$. The eigenfunction is square integrable between $-\infty$ and ∞. We effect the normalization according to the following rule:

$$\int_{-\infty}^{\infty} \Psi_m^2 \, dx = 1. \tag{10.13}$$

The law of conservation of energy states that the energy of the incident wave = the energy of the reflected wave + the energy of the transmitted wave. That means

$$1 = |b|^2 + |a|^2 \tag{10.14}$$

where k_m, $b(k)$ and $a(k)$ constitute the scattering parameters of the wave.

(viii) As is known the Schrödinger equation is of order 2 and as such there exist two independent solutions to this equation. Let us assume that Ψ and ϕ are the two independent solutions of the equation. Hence $\Psi_{xx} + (\lambda - v)\Psi = 0$, $\phi_{xx} + (\lambda - v)\phi = 0$. On substitution of $\phi = \Psi Y$ into the ϕ equation we obtain $(Y_{xx}/Y_x) + (2\Psi_x/\Psi) = 0$ which on integration yields $Y_x = A/\Psi^2$ and integrating again $Y = A\int (dx/\Psi^2) + B$. Therefore the desired second independent solution is $\phi = A\Psi\int (dx/\Psi^2) + B\Psi$. Neglect the second term on the right as it will be included in the Ψ term. Therefore, the second solution is $\phi = \Psi \int_0^x (dx/\Psi^2)$.

(ix) Let the potential function $v(x)$ be even. Then the Schrödinger equation is invariant under the transformation $x \to -x$. Thus $\Psi(x)$ and $\Psi(-x)$ are both eigenfunctions but they differ by a constant multiplier such that $\Psi(-x) = c\Psi(x)$. If we effect the transformation once more, then $\Psi(x) = c\Psi(-x) = c^2\Psi(x)$ so that $c = \pm 1$ and $\Psi(-x) = \pm \Psi(x)$.

Therefore, if the potential is symmetrical about $x = 0$, the steady-state wave functions are either even or odd. This is the outcome of our assumption that the eigenvalues of the Schrödinger operator corresponding to the discrete eigenvalues are nondegenerate.

10.4 The KdV equation and the Schrödinger equation

The KdV equation is given by

$$v_t - 6vv_x + v_{xxx} = 0 \tag{10.15}$$

which can be written in conservation form as

$$T_t + X_x = 0 \tag{10.16}$$

where $T = v$ and $X = -3v^2 + v_{xx}$.

Assume that v is periodic in x, or that v and its derivatives vanish sufficiently rapidly at $x = \pm\infty$; then integrating the conservation law (10.16) we get

$$\int_{-\infty}^{\infty} \left(\frac{\partial T}{\partial t} + \frac{\partial X}{\partial x} \right) dx = 0,$$

$$\frac{\partial}{\partial t} \int_{-\infty}^{\infty} T \, dx + [X]_{-\infty}^{\infty} = 0,$$

$$\frac{\partial}{\partial t} \int_{-\infty}^{\infty} T \, dx = 0$$

which implies that

$$I = \int T \, dx = \text{independent of time.} \tag{10.17}$$

Thus, $I = \int v \, dx$, where the limits of integration are $\pm\infty$ or two ends of a period in x, is a time-invariant functional of the solution of the KdV equation. A time-invariant functional is usually called an integral equation. Miura, Gardner and Kruskal (1968) proved in a remarkable paper that the KdV equation has an infinity of polynomial conservation laws. We have already seen the first one. The second one is derived by multiplying eqn (10.15) by v so that in this case

$$T = \tfrac{1}{2}v^2, \qquad X = \tfrac{1}{3}v^3 + vv_{xx} - \tfrac{1}{2}v_x^2. \tag{10.18}$$

The third was derived by Whitham (1967a, b)

$$T = \tfrac{1}{3} v^3 - v_x^2, \qquad X = \tfrac{1}{4} v^4 + v^2 v_{xx} - 2vv_x^2 - 2v_x u_{xxx} + v_{xx}^2. \quad (10.19)$$

Since each conservation law gives an integral of the equation, it follows that the KdV equation has infinitely many integrals of the form (10.17). Next, we show the relationship between the Schrödinger equation and the KdV equation.

Let us substitute

$$v = w^2 + w_x \qquad\qquad (10.20)$$

into the KdV equation (10.15). Then we have the following:

$$2w(w_t - 6w^2 w_x + w_{xxx}) + (w_t - 6w^2 w_x + w_{xxx})_x = 0. \qquad (10.21)$$

Equation (10.21) will be true provided w satisfies the associated KdV equation

$$w_t - 6w^2 w_x + w_{xxx} = 0. \qquad\qquad (10.22)$$

Thus it is clear that if w evolves according to (10.22), v defined by (10.20) evolves according to the KdV equation. Hence the associated KdV equation can also give infinitely many conservation laws of the form (10.17).

The first three of these conservation laws are the following:

$$T = w, \qquad X = \tfrac{1}{3} w^3 + w_{xx} \qquad\qquad (10.23)$$

$$T = \tfrac{1}{2} w^2, \qquad X = \tfrac{1}{4} w^4 + ww_{xx} - \tfrac{1}{2} w_x^2 \qquad\qquad (10.24)$$

and

$$T = \tfrac{1}{4} w^4 - \tfrac{3}{2} w_x^2, \qquad X = \tfrac{1}{6} w^6 + w^3 w_{xx} - 3w^2 w_x^2 - 3w_x w_{xxx} + \tfrac{3}{2} w_{xx}^2. \qquad (10.25)$$

Thus the integrals of the associated KdV equation are obtained from these conservation laws. It is then clear that eqn (10.20) provides a relation between the above-mentioned integrals of the KdV equation (10.15) and those of the associated KdV equation (10.22). The relationship between the KdV equation and the associated KdV equation is one to one. However, the integral $\int wdx$ of the associated KdV equation cannot be obtained from an integral of the KdV equation by the transformation (10.20).

It is known from the theory of elementary differential equations that eqn (10.20) is a Riccati equation provided v is a prescribed function. This is a nonlinear first-order differential equation which can be linearized by the well-known transformation

$$w = \frac{\Psi_x}{\Psi}. \tag{10.26}$$

The equation for Ψ is the one-dimensional Schrödinger equation

$$\Psi_{xx} - v\Psi = 0 \tag{10.27}$$

with the energy level terms missing. Note that the KdV equation is invariant under the transformation

$$t \to t', \qquad x \to x' - 6ct', \qquad v \to v' + c. \tag{10.28}$$

The energy levels in eqn (10.27) can be introduced. Thus this leads to a proof that the eigenvalues of the Schrödinger equation

$$\Psi_{xx} + (\lambda - v)\Psi = 0, \tag{10.29}$$

where v evolves according to the KdV equation, are time-independent functionals of v. We give the proof below.

10.5 Time independence of the eigenvalues of the Schrödinger equation: determination of scattering parameters

A remarkable and surprising discovery of Gardner *et al.* (1967) is that the eigenvalues of the Schrödinger equation (10.29) do not vary with time if $v(x, t)$ in eqn (10.29) satisfies the KdV equation (10.15) and vanishes sufficiently fast as $|x| \to \infty$. This result can be proved as follows.

Solving v from eqn (10.29), we have

$$v = \frac{\Psi_{xx}}{\psi} + \lambda. \tag{10.30}$$

Substituting eqn (10.30) into eqn (10.15), we get

$$\lambda_t \Psi^2 + (\Psi R_x - \Psi_x R)_x = 0 \tag{10.31}$$

where

$$R = \Psi_t + \Psi_{xxx} - 6\lambda\Psi_x - 3\frac{\Psi_x\Psi_{xx}}{\Psi}$$

$$= \Psi_t + \Psi_{xxx} - 3(v + \lambda)\Psi_x. \tag{10.32}$$

Integrating eqn (10.31) with respect to x from $x = -\infty$ to ∞ and considering the normalized wave function, we have

$$\lambda_t = 0 \tag{10.33}$$

in view of the boundary conditions, i.e.

$$\Psi = 0, |x| \to \infty, \frac{\partial\Psi}{\partial x} = 0, |x| \to \infty.$$

Therefore $\lambda = \lambda_n$ where λ_n is a constant.

Consequently, the eigenvalue $\lambda = \lambda_n$ may be determined from the initial data of $v(x,0) = v_0(x)$ prescribed a priori for solving the KdV equation. For the continuous eigenvalues, λ can be assumed constant in t, and hence eqn (10.33) is valid.

Substituting eqn (10.33) into eqn (10.31), we obtain for discrete as well as continuous eigenvalues:

$$\Psi R_{xx} - \Psi_{xx}R = 0. \tag{10.34}$$

Putting eqn (10.30) into eqn (10.34), we have

$$R_{xx} + (\lambda - v)R = 0. \tag{10.35}$$

Therefore, R also satisfies the Schrödinger equation and the general solution of (10.35) can be obtained as

$$R = c(t)\Psi + D(t)\Psi \int_0^x \frac{dx}{\Psi^2}. \tag{10.36}$$

If Ψ is the eigenfunction of the discrete eigenvalue λ, $1/\Psi^2$ is exponentially large as $|x| \to \infty$. Therefore for the boundedness of R, we must take

$$D = 0 \tag{10.37}$$

and then

$$R = c(t)\Psi. \tag{10.38}$$

Putting eqn (10.32) into eqn (10.38), we finally obtain

$$c(t)\Psi^2 = \Psi\Psi_t + \Psi\Psi_{xxx} - 6\lambda\Psi\Psi_x - 3\Psi_x\Psi_{xx}. \tag{10.39}$$

Now, integrating eqn (10.39) with respect to x from $x = -\infty$ to ∞ and using the boundary conditions on Ψ and its derivatives and the normality condition

$$\frac{\partial}{\partial t} \int_{-\infty}^{\infty} \Psi^2 \, dx = 0,$$

we obtain

$$c(t) = 0. \tag{10.40}$$

So eqn (10.39) reduces to

$$\Psi_t + \Psi_{xxx} - 6\lambda\Psi_x - 3\frac{\Psi_x\Psi_{xx}}{\Psi} = 0. \tag{10.41}$$

It is known from eqn (10.41) that when $x \to \infty$, it can be written for the discrete eigenvalue case as,

$$\Psi \sim c_m(t)\exp(-k_m x), \qquad \lambda_m = -k_m^2, \qquad k_m > 0. \tag{10.42}$$

Substituting eqn (10.42) into eqn (10.41), we obtain

$$\frac{d}{dt} c_m = 4k_m^3 c_m$$

which on integration yields

$$c_m(k_m, t) = c_m(k_m, 0)\exp(4k_m^3 t). \tag{10.43}$$

Let us consider the continuous eigenvalue k. For a steady plane wave coming from $x = \infty$, it can be written from eqn (10.11) as

$$\Psi \sim a(k,t)\exp(-ikx), \qquad x \to -\infty \tag{10.44}$$

where a is the transmission coefficient. Substituting eqn (10.44) into eqn (10.36), where R is given by eqn (10.32), gives us

$$a_t + 4ik^3 a - ca = \frac{D}{a} \int_0^x \exp(2ikx) \, dx \tag{10.45}$$

where the left-hand side is entirely a function of t and the right-hand side contains a function of x as a factor. Equation (10.45) will only be true if

$$D = 0 \tag{10.46}$$

and, consequently, we have

$$a_t - (c - 4ik^3)a = 0 \tag{10.47}$$

where c is a function of t alone.

Let us consider the case when $x \to \infty$. The asymptotic behaviour of the plane wave is given by

$$\Psi \sim \exp(-ikx) + b(k,t)\exp(ikx) \tag{10.48}$$

where the amplitude of the incident wave is taken to be unity, and $b(k,t)$ is the reflection coefficient.

Substitution of eqn (10.48) into eqn (10.36) yields

$$\exp(ikx)(b_t - 4ik^3b - cb) + \exp(-ikx)(4ik^3 - c) = 0. \tag{10.49}$$

This equation will be true provided the coefficients of $\exp(ikx)$ and $\exp(-ikx)$ are zero separately. That means

$$c = 4ik^3 \tag{10.50}$$

$$b_t - 4ik^3b - cb = 0. \tag{10.51}$$

Substituting the value of c from eqn (10.50) in eqns (10.47) and (10.51), after integration

$$a(k,t) = a(k,0) \tag{10.52}$$

$$b(k,t) = b(k,0)\exp(8ik^3t). \tag{10.53}$$

Thus eqns (10.43), (10.52) and (10.53) determine the evolution of scattering parameters c_m, a and b with respect to t in terms of their values at $t = 0$. The constants $c_m(k_m, 0)$, $a(k,0)$ and $b(k,0)$ can be determined by solving the Schrödinger equation with potential $v_0(x)$, the prescribed initial value for the solution of the KdV equation. In this way, the solution of the direct scattering problem is determined.

10.6 Inverse scattering problem

In a series of surprising discoveries, Gardner *et al.* (1967, 1974) developed a method of solution for the KdV equation known as the method of the inverse scattering transform. Following their method, we shall solve the inverse scattering problem to determine the potential $v(x,t)$ from knowledge of the scattering data. This procedure was initiated by Gel'fand and Levitan (1955) and Kay and Moses (1956) and then further developed by Gardner *et al.* (1974).

According to these authors, the desired solution of the KdV equation can be obtained from the equation

$$v(x,t) = -2\frac{\mathrm{d}}{\mathrm{d}x}K(x,x) \tag{10.54}$$

where $K(x,y)$ satisfies the Gel'fand–Levitan equation

$$K(x,y) + B(x,y) + \int_x^\infty B(y+z)K(x,z)\mathrm{d}z = 0 \tag{10.55}$$

and the kernel B is given by

$$B(\zeta) = \sum_{m=1}^N c_m^2(k_m,t)\exp(-k_m\zeta) + \frac{1}{2\pi}\int_{-\infty}^\infty b(k,t)\exp(ik\zeta)\mathrm{d}k \tag{10.56}$$

where N represents the N discrete eigenvalues of the Schrödinger equation. Equation (10.55) is sometimes called the Marchenko equation (see Lamb (1945)). In eqn (10.56) the first term on the right-hand side represents the contribution of the discrete part of the spectrum, whereas the second term represents the contribution of the continuous part of the spectrum. Here the spectrum is defined as the collection of all eigenvalues. The basis of eqns (10.54) to (10.56) is a well-developed subject in mathematical physics. We shall not discuss their evolution here primarily because they are too lengthy. However, interested readers may consult Lamb (1945), or Ablowitz and Segur (1981) for a thorough discussion.

Using eqns (10.43), (10.52) and (10.53), we can show t dependence explicitly in eqn (10.56):

$$B(\zeta) = \sum_{m=1}^N c_m^2(k_m,0)\exp(8k_m^3 t - k_m\zeta)$$

$$+ \frac{1}{2\pi}\int_{-\infty}^\infty b(k,0)\exp[i(8k^3 t + k\zeta)]\mathrm{d}k. \tag{10.57}$$

Before we proceed to solve the Gel'fand–Levitan equation (10.55) to determine the kernel $K(x,y)$, we must determine first the scattering parameters $c_m(k_m,0)$, $a(k,0)$ and $b(k,0)$ by solving the direct scattering problem of the Schrödinger equation with the initial data $v_0(x)$ as the potential.

10.7 Summary of solution by the inverse scattering transform

Collecting the main results of the preceding sections, we arrive at the following method of solving the KdV equation. Our problem is to determine the solution $v(x,t)$ of the KdV equation

$$v_t - 6vv_t + v_{xxx} = 0, \qquad x \in (-\infty, \infty), \qquad t \geq 0 \tag{10.58}$$

with the given initial data

$$v(x,0) = v_0(x). \tag{10.59}$$

Associated with that problem, we consider the Schrödinger equation

$$\Psi_{xx} + (\lambda - v)\Psi = 0, \qquad x \in (-\infty, \infty). \tag{10.60}$$

For $t = 0$ we can compute the spectrum, which consists of a finite number of discrete eigenvalues $\lambda = -k_n^2$ and a continuous part $\lambda = k^2$. We can further compute the scattering parameters $c_m(k_m,0)$, $a(k,0)$ and $b(k,0)$. Then the evolution of the scattering parameters can be computed from

$$c_m(k_m,t) = c_m(k_m,0)\exp(4k_m^3 t)$$

$$a(k,t) = a(k,0)$$

$$b(k,t) = b(k,0)\exp(8ik^3 t). \tag{10.61}$$

The potential of the Schrödinger equation can be recovered from the scattering data at any time $t > 0$ by solving the inverse scattering problem. To do this, we need the following function

$$B(\zeta) = \sum_{m=1}^{N} c_m^2(k_m,0)\exp(8k_m^3 t - k_m \zeta)$$

$$+ \frac{1}{2\pi} \int_{-\infty}^{\infty} b(k,0)\exp[i(8k^3 t + k\zeta)]dk \tag{10.62}$$

and the Gel'fand–Levitan integral equation

$$K(x,y) + B(x,y) + \int_x^\infty B(y+z)K(x,z)\mathrm{d}z = 0. \tag{10.63}$$

the solution of the initial value problem for the KdV equation is obtained from the formula

$$v(x,t) = -2\,\frac{\mathrm{d}}{\mathrm{d}x}\,K(x,x). \tag{10.64}$$

It is noted here that the original problem for the nonlinear partial differential equation (10.58) is transformed and reduced in this way to the problem of solving a one-dimensional linear integral equation.

10.8 Soliton solution of the KdV equation

To understand the inverse scattering method fully, we shall consider first in this section simple cases of the one-soliton solution of the KdV equation and then extend the investigation to the two-soliton solution. It can be easily seen from this discussion that corresponding to one discrete eigenvalue of the Schrödinger equation, there exists only one soliton solution of the KdV equation, and similarly corresponding to two discrete eigenvalues two-soliton solutions exist. Because of the lengthy involvement of the solution process we have omitted the N-soliton solutions of the KdV equation. Interested readers are referred to Bhatnagar (1979) for a fuller account of these solutions.

10.8.1 One-soliton solution

We select this case simply to verify the inverse scattering method just presented above. We have already seen that the solution (10.4) of the KdV equation (10.2) is given by

$$v(x,t) = -\frac{b^2}{2}\,\mathrm{sech}^2\!\left(\frac{b}{2}(x-b^2t)\right) \tag{10.65}$$

and this solution corresponds to the initial condition

$$v(x,0) = v_0(x) = -\frac{b^2}{2}\,\mathrm{sech}^2\!\left(\frac{b}{2}x\right). \tag{10.66}$$

Our aim is to obtain the solution (10.65) by the inverse scattering method starting with the initial condition (10.66).

We know that in the direct scattering problem, we have to solve the eigenvalue problem associated with the Schrödinger equation:

$$\Psi_{xx} + \left(\lambda + \frac{b^2}{2}\,\text{sech}^2\,\frac{bx}{2}\right)\Psi = 0. \tag{10.67}$$

Following the work of Kay and Moses (1956), it can be shown that the potential (10.66) which appears in eqn (10.67) is reflectionless, which simply implies that the reflection coefficient $b(k,0) = 0$ for the continuous eigenvalues. Assuming this to be true, we proceed to determine the discrete eigenvalues. Landau and Lifschitz (1958) described the method of solution of eqn (10.67) in terms of hypergeometric functions. It can be easily seen by the following transformations

$$\frac{b}{2}x = X, \qquad \frac{b^3}{8}t = T, \frac{4}{b^2}v = V(X,T), \qquad \frac{4\lambda}{b^2} = \Lambda \tag{10.68}$$

that eqn (10.67) can be transformed to

$$\Psi_{XX} + (\Lambda + 2\,\text{sech}^2\,X)\Psi = 0 \tag{10.69}$$

where

$$V_0 = -2\,\text{sech}^2\,X \tag{10.70}$$

is the initial potential. Let us substitute

$$\Psi = w\,\text{sech}^p\,X \tag{10.71}$$

in eqn (10.69) to obtain

$$w_{XX} - 2p(\tanh X)w_X + w[(2-p-p^2)\text{sech}^2\,X + \Lambda + p^2] = 0. \tag{10.72}$$

If we want the coefficient of w to be independent of X, $2-p-p^2 = 0$ must be chosen, so that $p = 1,\ -2$ are the two roots of the above equation. If $p = 1$ is chosen, with this choice of p, eqn (10.72) reduces to

$$w_{XX} - 2(\tanh X)w_X + (\Lambda + 1)w = 0 \tag{10.73}$$

where

$$\Psi = w \operatorname{sech} X. \tag{10.74}$$

Equation (10.73) can be reduced to the standard form of the hypergeometric equation

$$X(1-X)y_{XX} + [\gamma - (\alpha + \beta + 1)X]y_X - \alpha\beta y = 0. \tag{10.75}$$

Put

$$\xi = \sinh^2 X. \tag{10.76}$$

Thus we have the following differential equation in ξ:

$$\xi(1+\xi)w_{\xi\xi} + \tfrac{1}{2}w_\xi + \tfrac{1}{4}(1-k^2)w = 0 \tag{10.77}$$

$$w = (1+\xi)^{1/2}\Psi \tag{10.78}$$

$$\Lambda = -k^2. \tag{10.79}$$

Changing $\xi \to -\xi$ in eqn (10.77), it can be easily verified that the two independent solutions of (10.77) (see Abramowitz and Stegum (1965, p. 563) may be written as

$$w_1 = F\left(-\frac{1}{2} + \frac{k}{2}, -\frac{1}{2} - \frac{k}{2}; \frac{1}{2}; -\xi\right) \tag{10.80}$$

and

$$w_2 = \sqrt{\xi}\, F\left(\frac{k}{2}, -\frac{k}{2}; \frac{3}{2}; -\xi\right). \tag{10.81}$$

We know that $V_0(x) = -2\operatorname{sech}^2 X$ is an even function of X; therefore the solution Ψ of the Schrödinger equation (10.69) can be even or odd in X. Moreover, it is noted that ξ defined by eqn (10.76) is an even function of X whereas $\sqrt{\xi}$ is an odd function of X. We also note from eqn (10.76) that $\xi \to \infty$ as $X \to \pm\infty$. Since $\Psi \to 0$ as $|X| \to \infty$ (i.e. as $\xi \to \infty$), hence

$$\frac{w}{\sqrt{1+\xi}} \to 0 \quad \text{as} \quad \xi \to \infty.$$

From this analysis, it is clear that the w_1 integral must be even and the w_2 integral must be odd. At the same time, for w to be bounded at the singular point $\xi = -1$ of eqn (10.77), eqns (10.80) and (10.81) should reduce to polynomials. The two independent solutions in terms of Ψ then can be written as

$$\Psi_1 = (1+\xi)^{-1/2} F\left(-\frac{1}{2}+\frac{k}{2}, -\frac{1}{2}-\frac{k}{2}; \frac{1}{2}; -\xi\right) \tag{10.82}$$

$$\Psi_2 = \left(\frac{\xi}{1+\xi}\right)^{1/2} F\left(\frac{k}{2}, -\frac{k}{2}; \frac{3}{2}, -\xi\right). \tag{10.83}$$

The values of k for which w_1 and w_2 become polynomials, or in other words, Ψ_1 and Ψ_2 tend to zero as $\xi \to \infty$, correspond to discrete eigenvalues.

It is clear from eqn (10.83) that

$$\lim_{\xi \to \infty} \left(\frac{\xi}{1+\xi}\right)^{1/2} = 1$$

which implies that eqn (10.83) is not admissible even if we choose $k = 0$. However, eqn (10.82) will give us an admissible solution only when

$$-\frac{1}{2}+\frac{k}{2} = 0 \quad \text{or} \quad k = 1 \quad \text{or} \quad \Lambda = -1. \tag{10.84}$$

Thus the Schrödinger equation has only one eigenvalue $k = 1$ which is discrete.

The eigenfunction corresponding to the reflectionless potential (10.70) can be written as

$$\Psi = \frac{A}{(1+\xi)^{1/2}} = \frac{A}{\cosh X} \tag{10.85}$$

where A is a constant and is determined from the normality condition of Ψ as follows:

$$\int_{-\infty}^{\infty} \Psi^2 \, dx = 1 \quad \text{or} \quad A^2 \int_{-\infty}^{\infty} \operatorname{sech}^2 x \, dx = 1, \quad 2A^2 = 1, \quad A = \frac{1}{\sqrt{2}}$$

and the normalized eigenfunction is then given by

$$\Psi = \frac{1}{\sqrt{2}} \operatorname{sech} X. \tag{10.86}$$

Using eqn (10.10), when $k = 1$, it is seen that the constant c can be extracted:

$$c(0) = \lim_{X \to \infty} (\Psi(X)\exp(X)) = \lim_{X \to \infty} \left(\frac{\sqrt{2}\,e^X}{e^X + e^{-X}} \right) = \sqrt{2}. \qquad (10.87)$$

Since the potential equation (10.70) is reflectionless

$$b(k,0) = 0, \qquad a(k,0) = 1. \qquad (10.88)$$

Hence from eqns (10.43) and (10.56),

$$c(T) = c(0)\exp(4T) = \sqrt{2}\,\exp(4T) \qquad (10.89)$$

$$B(\xi,T) = 2\exp(8T - \xi). \qquad (10.90)$$

The Gel'fand–Levitan integral equation reduces to

$$K(X,Y) + 2\exp(8T - X - Y)$$

$$+ 2\exp(8T - Y) \int_X^\infty \exp(-Z)K(X,Z)\mathrm{d}Z = 0. \qquad (10.91)$$

Consider the dependence of eqn (10.91) on Y; then without loss of generality we can assume

$$K(X,Y) = L(X)\exp(-Y). \qquad (10.92)$$

This assumption removes the Y dependence completely from eqn (10.91) and this equation reduces to

$$L(X) + 2\exp(8T - X) + 2\exp(8T)L(X)\int_X^\infty \exp(-2Z)\mathrm{d}Z = 0.$$

Therefore

$$L(X) = \frac{-2\exp(8T - X)}{1 + \exp(-2X + 8T)}$$

which can be written as

$$L(X) = \frac{-2\exp(X)}{1 + \exp(2X - 8T)} \qquad (10.93)$$

and consequently eqn (10.92) becomes

$$K(X,Y) = \frac{-2\exp(X-Y)}{1+\exp(2X-8T)}.$$

(10.94)

Therefore the solutions of the KdV equation satisfying the initial condition (10.70) become

$$V(X,T) = -2\frac{\mathrm{d}}{\mathrm{d}X}K(X,X)$$

$$= -2\frac{\mathrm{d}}{\mathrm{d}X}\left(\frac{-2\exp(X-Y)}{1+\exp(2X-8T)}\right) = -2\operatorname{sech}^2(X-4T)$$

or in our original variables

$$v(x,t) = -\frac{b^2}{2}\operatorname{sech}^2\left[\frac{b}{2}(x-b^2t)\right]$$

(10.95)

which is the same as the solitary wave solution (10.65).

10.8.2 Two-soliton solution

We now extend the investigtion to the two-soliton solution. Consider, as an initial condition for the KdV equation,

$$v_0(x) = -6\operatorname{sech}^2 x$$

(10.96)

in which the amplitude and width of the wave do not match according to the solitary wave solution (10.65) for $t = 0$. In this case, we need to solve the following eigenvalue problem for the Schrödinger equation:

$$\Psi_{xx} + (\lambda + 6\operatorname{sech}^2 x)\Psi = 0.$$

(10.97)

The potential $v_0(x)$ is reflectionless and therefore will only give discrete eigenvalues. Let

$$\Psi = w\operatorname{sech}^p x.$$

(10.98)

On substitution of eqn (10.98) in eqn (10.97) and choosing p such that the coefficient of w is independent of x, we have $p = 2, 3$. Choose $p = 2$ and consequently we have

$$\Psi = w\operatorname{sech}^2 x$$

(10.99)

$$w_{xx} - 4(\tanh x)w_x + (4+\lambda)w = 0.$$

(10.100)

Now, changing the independent variable x to ξ by the transformation

$$\xi = \sinh^2 x, \tag{10.101}$$

eqn (10.100) may be written as

$$\xi(1+\xi)w_{\xi\xi} + (\tfrac{1}{2} - \xi)w_\xi + \tfrac{1}{4}(4 - k^2)w = 0 \tag{10.102}$$

where

$$\lambda = -k^2.$$

The initial potential (10.96) is an even function of x, so the two independent solutions of (10.102) in even and odd functions may be written as

$$w_1 = F\left(-1 + \frac{k}{2}, -1 - \frac{k}{2}; \frac{1}{2}; -\xi\right) \tag{10.103}$$

and

$$w_2 = \sqrt{\xi}\, F\left(-\frac{1}{2} + \frac{k}{2}, -\frac{1}{2} - \frac{k}{2}; \frac{3}{2}; -\xi\right) \tag{10.104}$$

where F represents the hypergeometric function with the appropriate arguments.

Therefore the even and odd solutions of the Schrödinger equation (10.97) are

$$\Psi_1 = \frac{1}{\cosh^2 x}\, F\left(-1 + \frac{k}{2}, -1 - \frac{k}{2}; \frac{1}{2}; -\sinh^2 x\right) \tag{10.105}$$

$$\Psi_2 = \frac{\sinh x}{\cosh^2 x}\, F\left(-\frac{1}{2} + \frac{k}{2}, -\frac{1}{2} - \frac{k}{2}; \frac{3}{2}; -\sinh^2 x\right). \tag{10.106}$$

The boundary conditions Ψ_1 and $\Psi_2 \to 0$ as $|x| \to \infty$ can only be satisfied when the hypergeometric functions in (10.105) and (10.106) reduce to polynomials. Equation (10.105) yields the admissible solution provided

$$k = 2(1 - n) > 0$$

where

$$n = 1, 2, 3, \ldots .$$

This gives only one value $k = 2$ corresponding to $n = 0$ for which

$$\Psi_1 = \frac{1}{\cosh^2 x}, \qquad k = 2 \tag{10.107}$$

Equation (10.106) reduces to the polynomial, when

$$k = 1 - 2n > 0, n = 0, 1, 2, \ldots .$$

Hence again only one value $k = 1$ is obtained corresponding to $n = 0$. The admissible solution is then

$$\Psi_2 = \frac{\sinh x}{\cosh^2 x}, \qquad k = 1.$$

The normalization functions corresponding to $k = 2$ and $k = 1$ respectively are

$$\Psi_1 = \frac{\sqrt{3}}{2} \operatorname{sech}^2 x, \qquad k_1 = 2 \tag{10.108}$$

$$\Psi_2 = \sqrt{3/2} \operatorname{sech}^2 x \sinh x, \qquad k_2 = 1. \tag{10.109}$$

With knowledge of Ψ_1 and Ψ_2, we can determine the scattering parameters $c_1(0)$ and $c_2(0)$ as follows:

$$c_1(0) = \lim_{x \to \infty} \{\Psi \exp(2x)\} = \lim_{x \to \infty} \left\{ \left(\frac{\sqrt{3}}{2} \operatorname{sech}^2 x \right)(\exp(2x)) \right\} = 2\sqrt{3} \tag{10.110}$$

and

$$c_2(0) = \lim_{x \to \infty} \{\Psi_2 \exp(x)\} = \lim_{x \to \infty} \{(\sqrt{3/2} \operatorname{sech}^2 x \sinh x)\exp(x)\} = \sqrt{6}. \tag{10.111}$$

Hence from eqn (10.43),

$$c_1(t) = c_1(0)\exp(4k_1^3 t) = 2\sqrt{3} \exp(32t) \tag{10.112}$$

$$c_2(t) = c_2(0)\exp(4k_2^3 t) = \sqrt{6} \exp(4t). \tag{10.113}$$

Next, the kernel $B(\xi)$ in eqn (10.56) is given by

$$B(\xi) = 12\exp(64t - 2\xi) + 6\exp(8t - \xi). \tag{10.114}$$

Then the Gel'fand–Levitan integral equation (10.55) can be written as

$$K(x,y) + [12\exp(64t - 2x - 2y) + 6\exp(8t - x - y)]$$

$$+ \int_x^\infty \{[12\exp(64t - 2y - 2z)$$

$$+ 6\exp(8t - y - z)]K(x,z)\}\mathrm{d}z = 0. \tag{10.115}$$

To remove the y dependence in the above equation, put

$$K(x,y) = A_1 L_1(x)\exp(-2y) + A_2 L_2(x)\exp(-y) \tag{10.116}$$

where

$$A_1 = 12\exp(64t)$$

$$A_2 = 6\exp(8t). \tag{10.117}$$

Using this information in eqn (10.115), two equations are obtained determining $L_1(x)$ and $L_2(x)$. These are derived by equating the coefficients of $\exp(-2y)$ and $\exp(-y)$ separately to zero and performing the necessary integration

$$\left(1 + \frac{A_1}{4}\exp(-4x)\right)L_1 + \left(\frac{A_2}{3}\exp(-3x)\right)L_2 = -\exp(-2x)$$

$$\left(\frac{A_1}{3}\exp(-3x)\right)L_1 + \left(1 + \frac{A_2}{2}\exp(-2x)\right)L_2 = -\exp(-x).$$

Solving these equations for L_1 and L_2, we obtain

$$L_1 = -\frac{1}{B}\left(\exp(-2x) + \frac{A_2}{6}\exp(-4x)\right)$$

$$L_2 = \frac{1}{B}\left(-\exp(-x) + \frac{A_1}{12}\exp(-5x)\right) \tag{10.118}$$

where

$$B = 1 + \frac{A_1}{4}\exp(-4x) + \frac{A_2}{2}\exp(-2x) + \frac{A_1 A_2}{72}\exp(-6x).$$

Substituting these values of L_1 and L_2 in eqn (10.116), we obtain

$$K(x,x) = -6\frac{\exp(8t-2x) + 2\exp(64t-4x) + \exp(72t-6x)}{1 + 3\exp(8t-2x) + 3\exp(64t-4x) + \exp(72t-6x)}.$$

$$(10.119)$$

Therefore, the solution of the KdV equation satisfying the initial condition (10.96) is

$$v(x,t) = -2\frac{\mathrm{d}}{\mathrm{d}x}K(x,x)$$

$$= -12\frac{3 + 4\cosh(8t-2x) + \cosh(64t-4x)}{[\cosh(36t-3x) + 3\cosh(28t-x)]^2}. \qquad (10.120)$$

Consider now the asymptotic behaviour of eqn (10.120) as $x \to \pm\infty$ to investigate the existence of soliton solutions corresponding to eigenvalues $k_1 = 2$ and $k_2 = 1$. Corresponding to the eigenvalue $k_1 = 2$, let us see

$$\xi = x - 4k_1^2 t = x - 16t \qquad (10.121)$$

in eqn (10.120) and take the limit as $t \to \pm\infty$ keeping ξ fixed. From eqn (10.120) after introducing (10.121),

$$v(x,t) = -12\left(\frac{N}{D}\right) \qquad (10.122)$$

is obtained, where

$$N = 3 + 4\cosh(-24t - 2\xi) + \cosh(-4\xi)$$

$$D = [\cosh(-12t - 3\xi) + 3\cosh(12t - \xi)]^2.$$

When $t \to -\infty$, N and D behave as

$$N = 3 + \cosh(-4\xi) + 2\exp(-24t)\exp(-2\xi)$$

$$D = \exp(-24t)[\tfrac{1}{2}\exp(-3\xi) + \tfrac{3}{2}\exp(\xi)]^2.$$

Therefore,

$$\lim_{t \to -\infty, \, \xi \text{ fixed}} v(x,t) = -96 \, \frac{\exp(-2\xi)}{[\exp(-3\xi) + 3\exp(\xi)]^2}$$

$$= -96 \, \frac{1}{[\exp(-2\xi) + 3\exp(2\xi)]^2}$$

$$= -96 \, \frac{\exp(4\xi)}{[1 + 3\exp(4\xi)]^2}.$$

Now put

$$3 = \exp(-4\xi_1) \tag{10.123}$$

which gives

$$\lim_{t \to -\infty, \, \xi \text{ fixed}} v(x,t) = -8\,\mathrm{sech}^2 \, 2(\xi - \xi_1)$$

$$= -8\,\mathrm{sech}^2[2(x - 16t - \xi_1)] \tag{10.124}$$

plus

$$\lim_{t \to \infty, \, \xi \text{ fixed}} v(x,t) = -96 \, \frac{\exp(-4\zeta)}{[1 + 3\exp(-4\xi)]^2}$$

$$= -8\,\mathrm{sech}^2 \, 2(\xi - \xi_1')$$

$$= -8\,\mathrm{sech}^2[2(x - 16t - \xi_1')] \tag{10.125}$$

where $\exp(4\xi_1') = 3$.

In a similar way, let us set corresponding to $k_2 = 1$

$$\xi = x - 4k_2^2 t = x - 4t \tag{10.126}$$

and take the limit as $t \to \pm\infty$ keeping ξ fixed. In this case,

$$v(x,t) = -12\left(\frac{N}{D}\right)$$

where

$$N = 3 + 4\cosh(-2\xi) + \cosh(48t - 4\xi)$$

$$D = [\cosh(24t - 3\xi) + 3\cosh(24t - \xi)]^2.$$

Taking the limit, we have

$$\lim_{t \to -\infty, \xi \text{ fixed}} v(x,t) = -24 \frac{\exp(-2\xi)}{[1 + 3\exp(-2\xi)]^2}$$

$$= -2\operatorname{sech}^2(\xi - \xi_2)$$

$$= -2\operatorname{sech}^2(x - 4t - \xi_2) \tag{10.127}$$

where $\exp(2\xi_2) = 3$ and

$$\lim_{t \to \infty, \xi \text{ fixed}} v(x,t) = -24 \frac{\exp(2\xi)}{[1 + 3\exp(-2\xi)]^2}$$

$$= -2\operatorname{sech}^2(\xi - \xi_2')$$

$$= -2\operatorname{sech}^2(x - 4t - \xi_2') \tag{10.128}$$

where $\exp(-2\xi_2') = 3$.

It is to be noted here that eqns (10.124) and (10.125) represent the same soliton travelling from $x = -\infty$ except that they differ in phase. Similarly, it can be noticed that eqns (10.127) and (10.128) represent the same soliton travelling from $x = -\infty$ to $x = \infty$ with the exception that they differ in phase. (Compare with the standard form of soliton solution

$$v(x,t) = -\frac{b^2}{2} \operatorname{sech}^2\left(\frac{b}{2}(x - b^2 t)\right);$$

for the solution (100.124), $b = 4$, whereas for the solution (10.127), $b = 2$). Therefore it can be concluded that to each eigenvalue of the Schrödinger equation (10.97), there exists a soliton solution. Thus the solution (10.120) represents a double wave solution which breaks into two solitons both as $t \to -\infty$ and $t \to +\infty$. The effect of nonlinear interaction between these two solitons according to the KdV equation simply displaces their positions with respect to the positions in the absence of the interaction.

Remark

The most astounding property of the solitary wave is that two distinct solitary waves, i.e. solitary waves with distinct amplitudes and therefore with distinct velocities, interact according to the nonlinear KdV equation and yet emerge from interaction without any change of forms, a property possessed by the linear waves. For this behaviour, the name soliton has been given to these solitary waves.

The dimensional form of the solitary wave solution given by eqn (9.59) (note that eqn (10.65) describes the nondimensional form of the one-soliton solution) can be written as (see Mei (1983))

$$\tilde{\eta} = A \, \text{sech}^2 \sqrt{\frac{3A}{4h^3}} \, (X - CT),$$

and has been depicted in Fig. 10.1. Since $\tilde{\eta} > 0$ for all $X - CT$, the solution is a wave of elevation which is symmetric about $X - CT = 0$. It propagates in the medium without change of shape with velocity

$$C = C_0 \sqrt{1 + \frac{A}{h}} \simeq C_0 \left(1 + \frac{A}{2h} \right)$$

which is directly proportional to the amplitude A. The width of the wave, which is $2\pi\sqrt{(4h^3/3A)}$, is inversely proportional to $\sqrt{A}$.

In other words, the solitary wave propagates to the right with velocity C which is directly proportional to the amplitude, and has a width that is inversely proportional to the square root of the amplitude. Therefore taller solitons travel faster and are narrower than the shorter (or slower) ones.

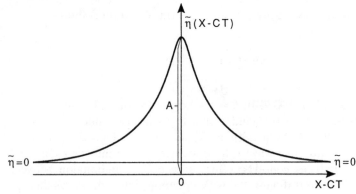

Fig. 10.1.   A solitary wave (soliton).

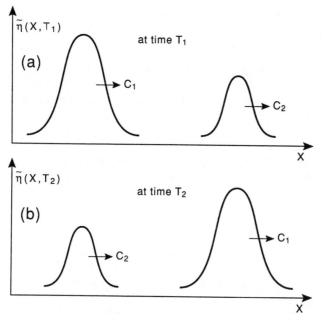

Fig. 10.2. Interactions of two solitons ($C_1 > C_2$ and $T_1 < T_2$).

They can overtake the shorter ones and surprisingly they emerge from the interaction without change of shape as shown in Fig. 10.2(a) and (b). Indeed, solitons resemble the behaviour of smooth and rigid particles during collisions.

10.9 Exercises

1. Show that the KdV equation (eqn (10.15)) satisfies the conservation law in the form $T_t + X_x = 0$ where (a) $T = v$, $X = -3v^2 + v_{xx}$, (b) $T = \frac{1}{2}v^2$, $X = -2v^3 + vv_x - \frac{1}{2}v_x^2$.

2. Show that the KdV equation $v_t + 6vv_x + v_{xxx} = 0$ satisfies the conservation law in the form $T_t + X_x = 0$ where (a) $T = v$, $X = 3v^2 + v_{xx}$, (b) $T = \frac{1}{2}v^2$, $X = 2v^3 + vv_{xx} - \frac{1}{2}v_x^2$, (c) $T = v^3 - \frac{1}{2}v_x^2$, $X = \frac{9}{2}v^4 + 3v^2v_{xx} + \frac{1}{2}v_{xx}^2 + v_xv_t$.

3. Show that the conservation laws for the associated KdV equation (eqn (10.22)) are (a) $(w)_t + (\frac{1}{3}w^3 + w_{xx})_x = 0$, (b) $(\frac{1}{2}w^2)_t + (\frac{1}{4}w^4 + ww_{xx} - \frac{1}{2}w_x^2)_x = 0$.

4. Show that the conservation laws for the equation $v_t - vv_x - v_{xxt} = 0$ are (a) $v_t - (\frac{1}{2}v^2 + v_{xt})_x = 0$, (b) $\frac{1}{2}(v^2 + v_x^2)_t - (\frac{1}{3}v^3 + vv_{xt})_x = 0$.

5. Show that the linear Schrödinger system $i(\psi)_t + \psi_{xx} = 0$, $-\infty < x < \infty$, $t > 0$, $\psi \to 0$ as $|x| \to \infty$, $\psi(x,0) = \psi(x)$ with $\int_{-\infty}^{\infty} |\psi|^2 \, dx = 1$, has the conservation law $(i|\psi|^2)_t + (\psi^*\psi_x - \psi\psi_x^*)_x = 0$, and the energy integral $\int_{-\infty}^{\infty} |\psi|^2 \, dx = 1$.

6. The hypergeometric function plays a very important role in the soliton problem. Determine a series of solution of the hypergeometric differential equation.

References

Ablowitz, M. J. and Segur, H. (1981). *Solitons and Inverse Scattering Transform*, SIAM, Philadelphia.

Abramowitz, M. and Stegun, I. A. (1965). *Handbook of Mathematical Functions*, Applied Mathematics Series 55, National Bureau of Standards, US Department of Commerce, Washington, DC.

Bhatnagar, P. L. (1979). *Nonlinear Waves in One Dimensional Dispersive Systems*, Clarendon Press, Oxford.

Bullough, R. K. and Caudrey, P. J. (1980). *Solitons*, Springer-Verlag, Berlin, New York.

Gardner, C. S., Greene, J. M., Kruskal, M. D. and Miura, R. M. (1967). Method for solving the Korteweg-de Vries equation. *Phys. Rev. Lett.*, 19, 1095–1096.

Gardner, C. S., Greene, J. M., Kruskal, M. D. and Miura, R. M. (1974). Korteweg-de Vries equation and generalizations. VI Methods for exact solution. *Commun. Pure Appl. Math.*, 27, 97–133.

Gel'fand, I. M. and Levitan, B. M. (1955). On the determination of a differential equation from its spectral function. *Am. Math. Soc. Transl. Series 2*, 1, 253–304.

Karpman, V. I. (1975). *Nonlinear Waves in Dispersive Media.*, Pergamon Press, Oxford.

Kay, I and Moses, H. E. (1956). Reflectionless transmission through dielectrics and scattering potentials. *J. Appl. Phys.*, 27, 1503–1508.

Lamb, Horace (1945). *Hydrodynamics*, 6th edn, Cambridge University Press, New York.

Landau, L. and Lifschitz, M. (1958). *Quantum Mechanics Non-relativistic Theory*, Pergamon Press, New York.

Mei, C. C. (1983). *The Applied Dynamics of Ocean surface Waves*, John Wiley, New York.

Miura, R. M. (1976). The Korteweg-de Vries equation: a survey of results. *SIAM Rev.*, 18, 412–459.

Miura, R. M., Gardner, C. S. and Kruskal, M. D. (1968). Korteweg-de Vries equation and generalizations. II. Existence of conservation laws and constants of the motion. *J. Math. Phys.*, 9, 1204–1209.

Scott, A. C., Chu, F. Y. E. and McLaughlin, D. W. (1973). The soliton: A new concept in applied science. *Proc. I. E. E*, 61, 1443–1483.

Scott-Russell, J. (1844). Report on waves. *Proc. R. Soc.*, Edinburgh, 319–320.

Whitham, G. B. (1967a). Nonlinear dispersion of water waves. *J. Fluid Mech.*, 27, 399–412.

Whitham, G. B. (1967b). Variational methods and applications to water waves. *Proc. R. Soc.*, A299, 6–25.

Whitham, G. B. (1974). *Linear and Nonlinear Waves*, Wiley, New York.

Zabusky, N. J. and Kruskal, M. D. (1965). Interactions of solitons in a collisonless plasma and the recurrence of initial states. *Phys. Rev. Lett.*, 15, 240–243.

Author index

Subject index